KU-796-464

Electric Power Applications of Fuzzy Systems

This book and other books may be purchased at a discount
from the publisher when ordered in bulk quantities. Contact:

IEEE Press Marketing
Attn: Special Sales
Piscataway, NJ 08855-1331
Fax: (732) 981-9334

For more information about IEEE PRESS products,
visit the IEEE Home Page: http://www.ieee.org/

© 1998 by the Institute of Electrical and Electronics Engineers, Inc.
345 East 47th Street, New York, NY 10017-2394

*All rights reserved. No part of this book may be reproduced in any form,
nor may it be stored in a retrieval system or transmitted in any form,
without written permission from the publisher.*

Printed in the United States of America

10   9   8   7   6   5   4   3   2   1

ISBN 0-7803-1197-3
IEEE Order Number: PC5666

Library of Congress Cataloging-in-Publication Data
Electric power applications of fuzzy systems / edited by Mohamed E. El
  -Hawary.
     p.   cm. – (IEEE Press power engineering series)
  Includes bibliographical references and index.
  ISBN 0-7803-1197-3 (alk. paper)
  1. Electric power systems—Control.   2. Fuzzy systems.   I. El
-Hawary, M. E.   II. Series.
TK1007.E43   1998
621.31—dc21   98–9885
                                                      CIP

D
621.317
ELE

Electric Power Applications of Fuzzy Systems

Edited by

Mohamed E. El-Hawary
Dalhousie University

**IEEE
PRESS**

IEEE Power Engineering Society, *Sponsor*

IEEE Press Power Systems Engineering Series
P. M. Anderson, *Series Editor*

The Institute of Electrical and Electronics Engineers, Inc., New York

IEEE Press
445 Hoes Lane, P.O. Box 1331
Piscataway, NJ 08855-1331

IEEE Press Editorial Board
Roger F. Hoyt, *Editor in Chief*

J. B. Anderson	A. H. Haddad	M. Padgett
P. M. Anderson	R. Herrick	W. D. Reeve
M. Eden	S. Kartalopoulos	G. Zobrist
M. E. El-Hawary	D. Kirk	
S. Furui	P. Laplante	

Kenneth Moore, *Director of Book Publishing*
John Griffin, *Senior Acquisitions Editor*
Marilyn Giannakouros, *Assistant Editor*

IEEE Power Engineering Society, *Sponsor*
PE-S Liaison to IEEE Press, Leo L. Grigsby

Cover design: William T. Donnelly, *WT Design*

IEEE PRESS POWER SYSTEMS ENGINEERING SERIES
Dr. Paul M. Anderson, *Series Editor*
Power Math Associates, Inc.

Series Editorial Advisory Committee

Dr. Roy Billinton
University of Saskatchewan

Dr. Atif S. Debs
Georgia Institute of Technology

Dr. M. El-Hawary
Dalhousie University

Mr. Richard G. Farmer
Arizona Public Service Company

Dr. Charles A. Gross
Auburn University

Dr. G. T. Heydt
Purdue University

Dr. George Karady
Arizona State University

Dr. Donald W. Novotny
University of Wisconsin

Dr. A. G. Phadke
Virginia Polytechnic and State University

Dr. Chanan Singh
Texas A & M University

Dr. E. Keith Stanek
University of Missouri–Rolla

Dr. J. E. Van Ness
Northwestern University

BOOKS IN THE IEEE POWER ENGINEERING SERIES

ANALYSIS OF ELECTRIC MACHINERY
Paul C. Krause and Oleg Wasynczuk, *Purdue University*
Scott D. Sudhoff, *University of Missouri at Rolla*
1994   Hardcover   584 pp   IEEE Order No. PC4556   ISBN 0-7803-1101-9

ANALYSIS OF FAULTED POWER SYSTEMS, Revised Printing
Paul M. Anderson, *Power Math Associates, Inc.*
1995   Hardcover   536 pp   IEEE Order No. PC5616   ISBN 0-7803-1145-0

ELECTRIC POWER APPLICATIONS OF FUZZY SYSTEMS
Mohamed E. El-Hawary, *Dalhousie University*
1998   Hardcover   500 pp   IEEE Order No. PC5666   ISBN 0-7803-1197-3

ELECTRIC POWER SYSTEMS: Design and Analysis, Revised Printing
Mohamed E. El-Hawary, *Technical University of Nova Scotia*
1995   Hardcover   808 pp   IEEE Order No. PC5606   ISBN 0-7803-1140-X

POWER SYSTEM CONTROL AND STABILITY, Revised Printing
Paul M. Anderson, *Power Math Associates, Inc.*
A. A. Fouad, *Iowa State University*
1993   Hardcover   480 pp   IEEE Order No. PC3798   ISBN 0-7803-1029-2

POWER SYSTEM STABILITY, VOLUMES I, II, III
An IEEE Press Classic Reissue Set
Edward Wilson Kimbark
1995   Softcover   1008 pp   IEEE Order No. PP5600   ISBN 0-7803-1135-3

PROTECTIVE RELAYING FOR POWER SYSTEMS, VOLUMES I & II
Stanley H. Horowitz, editor, *American Electric Power Service Corporation*
1992   Softcover   1184 pp   IEEE Order No. PP3228   ISBN 0-7803-0426-8

SUBSYNCHRONOUS RESONANCE IN POWER SYSTEMS
Paul M. Anderson, *Power Math Associates, Inc.*
B. L. Agrawal, *Arizona Public Service Company*
J. E. Van Ness, *Northwestern University*
1990   Hardcover   282 pp   IEEE Order No. PC2477   ISBN 0-87942-258-0

To Elizabeth Sarah,
a daughter, athlete, and team player;
this one is for you.

Contents

Chapter 5   **Fuzzy Logic Controller as a Power System Stabilizer 112**
             O. P. Malik, The University of Calgary
             K. A. M. El-Metwally, Cairo University

Contents

List of Contributors

Chapter 1   Introduction

M. E. El-Hawary
Dalhousie University
Halifax, Nova Scotia
B3J 2X4 Canada

Chapter 2   Fuzzy Systems: An Engineering Point of View

M. E. El-Hawary
Dalhousie University
Halifax, Nova Scotia
B3J 2X4 Canada

Chapter 3   Fuzzy Information Approaches to Equipment Condition Monitoring
and Diagnosis

K. Tomsovic and B. Baer
School of Electrical Engineering and Computer Science
Washington State University
Pullman, WA 99164-2752

M. A. El-Sharkawi, R. J. Marks II, and R. J. Streifel
Department of Electrical Engineering
University of Washington
Seattle, WA 98195

I. Kerszenbaum
Southern California Edison Company
Research Center
Irwindale, CA 91702

O. P. Malik
Department of Electrical and Computer Engineering
The University of Calgary
Calgary, Alberta
T2N 1N4 Canada

K. A. M. El-Metwally
Electric Machines and Power Department
Cairo University
Cairo, Egypt

T. Hiyama
Department of Electrical Engineering and Computer Science
Kumamoto University
Kumamoto
860 Japan

T. Hiyama
Department of Electrical Engineering and Computer Science
Kumamoto University
Kumamoto
860 Japan

C. N. Lu and R. C. Leou
Department of Electrical Engineering
National Sun Yat-Sen University
Kaohsiung, Taiwan
80424 Republic of China

V. Miranda
Instituto de Engenharia de Sistemas e Computadores
Universidade Porto
Porto
4000 Portugal

H. Matsumoto
Hitachi Research Laboratories
832-2 Horiguchi, Hitachinaka-shi
Ibaraki-ken
312 Japan

S. M. Shahidehpour and R. W. Ferrero
Department of Electrical and Computer Engineering
Illinois Institute of Technology
Chicago, IL 60616

Preface

Electric Power Applications of Fuzzy Systems is intended to offer an introduction to certain applications of fuzzy system theory to selected areas of electric power engineering. The idea is to introduce theoretical background material from a practical point of view and then proceed to explore a number of applications of fuzzy systems.

Fuzzy sets and logic were introduced in the mid-sixties in order to mathematically formalize the treatment of imprecise notions and concepts found in almost every decision-making situation. There has been a phenomenal increase in research activities aimed at implementing fuzzy concepts in many engineering applications. This has been facilitated by the publication of theoretically based books that explain the mathematical foundations of fuzzy sets and logic and, therefore, promote their usage in real-world applications such as in electric power systems.

Most recently, there has been a tremendous surge in research and application articles on applying fuzzy systems in electric power engineering. However, there are currently no books that put together a practical guide to the fundamentals and application aspects. This book is intended to fill this gap in the area of applied computational intelligence.

It is hoped that the book will attract not only the power engineer but also experts in the computational intelligence community, who would benefit from the exposure to the practical aspects of the problems discussed. We have assembled a distinguished panel of experts to write original chapters introducing and summarizing areas of applications. Our panel includes American, Canadian, Chinese, European, and Japanese contributors. All are well respected for their contributions to recent advances in power systems engineering.

The idea of this book is to put together under one cover original contributions by authors who have pioneered in the application of fuzzy system theory to the electric power engineering field. As such, each chapter represents both an introduction to and a state-of-the-art review of each application area.

Following the introduction of Chapter 1, we offer a summary of fundamentals of fuzzy systems in Chapter 2. The remainder of the book is devoted to individual applications. Chapter 3, on fuzzy information approaches to equipment condition monitoring and diagnosis, is written by Tomsovic and Baer. Equipment monitoring and maintenance aim at anticipating failures or accelerated aging in power equipment. Many of the indicators of equipment condition are imprecise and/or unreliable. The chapter emphasizes uncertainty modeling of diagnostic problems. Fuzzy information methods are employed to represent quantitatively the diagnostic capability of a system. Further, several methods are discussed for extracting information from test data and evaluating system performance.

In Chapter 4, El-Sharkawi, Marks II, Streifel, and Kerszenbaum discuss detection and localization of shorted turns in the DC field winding of turbine-generator rotors using novelty detection and fuzzified neural networks. Use of neural networks with fuzzy logic outputs and traveling-wave techniques is shown to be an accurate locator of shorted turns in turbo-generator rotors. The technique is extended to operational rotors by the use of a novelty filter.

Low-frequency oscillations are a common problem in large power systems. A power system stabilizer (PSS) can provide a supplementary control signal to the excitation system and/or the speed governor system of the electric-generating unit to damp these oscillations and to improve its dynamic performance. Chapters 5 and 6 are devoted to this topic.

In Chapter 5, on fuzzy logic controllers as power system stabilizers, Malik and El-Metwally describe the structure and design of a fuzzy logic controller and an algorithm to tune its parameters to achieve the desired performance. Two rule generation methods to automatically generate the fuzzy rule set are proposed.

In Chapter 6, Hiyama presents a study on fuzzy logic control schemes using polar information for stability enhancement. Here, advanced fuzzy logic control rules are introduced based on three-dimensional information of generator acceleration, speed, and phase angle. Stabilizing signals are revised at every sampling instant and fed back to the excitation control loop of the generator. The results have been implemented on a prototype tested on hydro units in the Japanese KEPCO system.

The second contribution of Hiyama is on fuzzy logic switching control of FACTS devices, presented in Chapter 7. Here a fuzzy logic switching control scheme is proposed for devices such as the thyristor-controlled series capacitor modules, static var compensators, and braking resistors.

In Chapter 8 the effects of uncertain load in real time and off-line power network modeling are discussed. Here, Lu and Leou describe different approaches for representing uncertain load in power systems.

Miranda treats power system reliability in Chapter 9, a fuzzy perspective of power system reliability. Here the author concentrates mainly on reliability assessment for mid- to long-term purposes. Three types of models are discussed. In Type I, only a fuzzy description of the system load is available. For Type II fuzzy reliability assessment, component reliability indices are fuzzy. Extending the concept, we can also deal with a Type III model.

In Chapter 10, Matsumoto discusses an operation support expert system for startup schedule optimization in fossil power plants. The chapter describes expectations of fuzzy theory in fossil power plant operation and two types of expert systems applied to operation support systems for a conventional one through type super-critical pressure fossil power plant and a gas and steam turbine combined-cycle power plant. Fuzzy models of expertise are introduced to modify the plant startup schedule. The most remarkable feature of this approach is that the quantitative optimum schedule is obtained through iterative modification of the schedule from a combination of qualitative knowledge and plant dynamic simulations.

In Chapter 11, Shahidehpour and Ferrero introduce a fuzzy system approach to short-term power purchases considering uncertain prices. Here, the authors present a methodology to evaluate power purchases in an uncertain environment. Power generation, line flows, and prices are considered as triangular fuzzy numbers. The proposed methodology computes the range of control variables that satisfies the set of constraints as well as a certain reduction in the operation cost. The range of control variables is correlated with the desired cost reduction (goal). The lower the desired cost reduction, the narrower the range of control variables that satisfies the set of constraints. The authors point out that the proposed method can be extended to include extra constraints related with operational practices in utilities.

The intended audience includes power system professionals as well as researchers and developers of software/hardware tools of fuzzy systems in particular and the computational intelligence community. It is expected that the reader is a graduate of an engineering/computer science degree program with a modest mathematical background.

Mohamed E. El-Hawary
DalTech of Dalhousie University

Acknowledgments

Our thanks to Paul M. Anderson, Editor of the Power Engineering Series of IEEE Press, and our Senior Acquisitions Editor, John Griffin, for their unprecedented help. While developing this project, Lisa Dayne and Marilyn Giannakouros were always there to prod and smooth out many rough edges. The contributions of our anonymous reviewers have improved the quality of the book considerably.

Finally, the Editor of this book wishes to express his love and gratitude to his journey-of-life partner (wife) Ferial El-Hawary for support and encouragement throughout the phases of this book.

Mohamed E. El-Hawary
Dalhousie University

Chapter 1

M. E. El-Hawary
Dalhousie University
Halifax, Nova Scotia
B3J 2X4 Canada

Introduction

Zadeh is credited with introducing the concept of fuzzy sets in 1965 as a mathematical means of describing vagueness in linguistics. The idea may be considered as a generalization of classical set theory. In the decade since Zadeh's pioneering paper on fuzzy sets [1], many theoretical developments in fuzzy logic took place in the United States, Europe, and Japan. From the mid-1970s to the present, however, Japanese researchers have done an excellent job of advancing the practical implementation of the theory; they have been a primary force in commercializing this technology. Much of the success of the new products associated with the fuzzy technology is due to fuzzy logic, and some is also due to the advanced sensors used in these products.

1.1. FUZZY SETS

The basic idea of fuzzy sets is quite simple. In a conventional (nonfuzzy, hard, or crisp) set, an element of the universe either belongs to or does not belong to the set. That is, the membership of an element is crisp—it is either yes (in the set) or no (not in the set).

A fuzzy set is a generalization of an ordinary set in that it allows the degree of membership for each element to range over the unit interval [0, 1]. Thus, the membership function of a fuzzy set maps each element of the universe of discourse to its range space, which, in most cases, is assumed to be the unit interval.

1

One major difference between crisp and fuzzy sets is that crisp sets always have unique membership functions, whereas every fuzzy set has an infinite number of membership functions that may represent it. This enables fuzzy systems to be adjusted for maximum utility in a given situation.

1.2. FUZZY SYSTEMS, COMPLEXITY, AND AMBIGUITY

Zadeh's principle of incompatibility was given in 1973 to explain why there is a need for a fuzzy system theory. The principle states, in essence, that as the complexity of a system increases, our ability to make precise and yet significant statements about its behavior diminishes until a threshold is reached beyond which precision and significance (or relevance) become almost mutually exclusive characteristics. This suggests that complexity and ambiguity (imprecision) are correlated: "The closer one looks at a real-world problem, the fuzzier becomes its solution" [2].

It is a characteristic of the way humans think to treat problems involving complexity and ambiguity in a subjective manner. Complexity generally stems from uncertainty in the form of ambiguity; these are features of most social, technical, and economic situations experienced on a daily basis. In considering a complex system, humans reason approximately about its behavior (a capability that computers do not have) and thus maintain only a generic understanding of the problem. This generality and ambiguity are adequate for a human to perceive and understand complex systems.

As one learns more and more about a system, its complexity decreases and understanding increases. As complexity decreases, the precision afforded by computational methods becomes more useful in modeling the system. For less complex systems, thus involving little uncertainty, closed-form mathematical expressions offer precise descriptions of the systems' behavior. For systems that are slightly more complex but for which significant data exist, model-free methods, such as computational neural networks, provide powerful and effective means to reduce some uncertainty through learning based on patterns in the available data.

There are virtually no problems for which we can say that the information content is known absolutely, that is, with no ignorance, no vagueness, no imprecision, no element of chance. Uncertain information can take on many different forms. There is uncertainty that arises because of complexity, for example, the complexity in the reliability evaluation of an electric distribution system. There is uncertainty that arises from ignorance, from chance, from various classes of randomness, from imprecision, from the inability to perform adequate measurements, from lack of knowledge, or from vagueness, like the fuzziness inherent in our natural language.

The nature of uncertainty in a system is an important consideration that the analyst should study prior to selecting an appropriate method to express the uncertainty. For most complex systems where few numerical data exist and where only ambiguous or imprecise information may be available, fuzzy reasoning offers a way to understand system behavior by allowing one to interpolate approximately between observed input and output situations. The imprecision in fuzzy models is generally quite high.

Fuzzy logic is based on the way the brain deals with inexact information. Fuzzy systems combine fuzzy sets with fuzzy rules to produce overall complex nonlinear behavior. Fuzzy systems are structured numerical estimators. They start from highly formalized insights about the structure of categories found in the real world and then express fuzzy if-then rules as some expert knowledge. Being numerical model-free estimators and dynamical systems, fuzzy systems are able to improve the intelligence of systems working in an uncertain, imprecise, and noisy environment.

Some of the information available in developing models of physical processes might be judgmental, perhaps an instinctive reaction on the part of the modeler, rather than hard quantitative information. Fuzzy reasoning allows us to incorporate intuition into a problem.

One prevalent way to convey information is our own means of communication: natural language. By its very nature, natural language is vague and imprecise; yet it is the most powerful form of communication and information exchange among humans. Despite the vagueness in natural language, humans have little trouble understanding one another's concepts and ideas; this understanding is not possible in communications with a computer, which requires extreme precision in its instructions.

To illustrate, consider the interpretation of the term *short-person*. To individual X a short person might be anybody below 5 ft 1 in. To individual Y, a short person is someone who is 5 ft 8 in. or shorter. What sort of meaning does the linguistic descriptor *short* convey to either of these individuals? It is surprising that, despite the potential for misunderstanding, the term short communicates sufficiently similar information to the two individuals, even if they are of considerably different heights themselves, and that understanding and correct communication are possible between them. Individuals X and Y, regardless of their own heights, do not require identical definitions of the term short to communicate effectively. A computer would require a specific height to compare with a preassigned value for "short." The underlying power of fuzzy set theory is that it uses linguistic variables, rather than quantitative variables, to represent imprecise concepts.

Fuzzy systems have been shown to be capable of modeling complex nonlinear processes to arbitrary degrees of accuracy. They have attracted growing interest of researchers in various scientific and engineering areas. The number and variety of applications of fuzzy systems have been increasing, ranging from consumer products and industrial process control to medical instrumentation, information systems, and decision analysis.

1.3. FUZZINESS AND PROBABILITY

Fuzziness is often confused with probability. The newcomer to the field often claims that fuzzy set theory is just another form of probability theory in disguise. Fuzzy set theory provides a means for representing uncertainties. Probability theory has been the primary tool for representing uncertainty in mathematical models. As a result, all uncertainty was assumed to follow the characteristics of random uncertainty.

Basic statistical analysis is founded on probability theory or stationary random processes, whereas most experimental results contain both random (typically noise) and nonrandom processes. One class of random processes or stationary random processes exhibits the following three characteristics:

1. The sample space on which the processes are defined cannot change from one experiment to another, that is, the outcome space cannot change.
2. The frequency of occurrence, or probability, of an event within that sample space is constant and cannot change from trial to trial or experiment to experiment.
3. The outcomes must be repeatable from experiment to experiment. The outcome of one trial does not influence the outcome of a previous or future trial.

However, fuzzy sets are not governed by these characteristics.

The outcomes of any particular realization of a random process are strictly a matter of chance; a prediction of a sequence of events is not possible. For a random process it is only possible given a precise description of its long-run averages.

As can be appreciated, not all uncertainty is random. Some forms of uncertainty are nonrandom and hence not suited to treatment or modeling by probability theory. In fact, one can argue that the predominant amount of uncertainty associated with complex systems is nonrandom in nature. Fuzzy set theory is an excellent tool for modeling the kind of uncertainty associated with vagueness, with imprecision, and/or with a lack of information regarding a particular element of the problem at hand.

The fundamental difference between fuzziness and probability is that fuzziness deals with deterministic plausibility, while probability concerns the likelihood of nondeterministic, stochastic events. Fuzziness is one aspect of uncertainty. It is the ambiguity (vagueness) found in the definition of a concept or the meaning of a term such as *comfortable temperature* or *well cooked*. However, the uncertainty of probability generally relates to the occurrence of phenomena, as symbolized by the concept of randomness. In other words, a statement is probabilistic if it expresses some kind of likelihood or degree of certainty or if it is the outcome of clearly defined but randomly occurring events. For example, the statements "There is a 50-50 chance that he will be there," "It will be sunny tomorrow," and "Roll the dice and get a six" demonstrate the uncertainty of randomness.

Hence, fuzziness and randomness differ in nature; that is, they are different aspects of uncertainty. The former conveys "subjective" human thinking, feelings, or language and the latter indicates an "objective" statistic in the natural sciences.

From the modeling point of view, fuzzy models and statistical models also possess philosophically different kinds of information: Fuzzy memberships represent similarities of objects to imprecisely defined properties, while probabilities convey information about relative frequencies. The quest for a method to quantify nonrandom uncertainty (imprecision, vagueness, fuzziness) in physical processes is the basic premise of fuzzy system theory, for to understand uncertainty in a system is to understand the system itself. As understanding improves, the fidelity in modeling improves.

1.4. WHEN IS A FUZZY FORMULATION APPROPRIATE?

Whenever precision is evident, for example, fuzzy systems are less efficient than more precise algorithms in offering a better understanding of the problem. Requiring precision in engineering models and products translates to requiring high cost and long lead times in production and development. For other than simple systems, expense is proportional to precision: More precision entails higher cost. When considering the use of fuzzy logic for a given problem, an engineer or scientist should ponder the need for exploiting the tolerance for imprecision. Not only does high precision dictate high costs but it also entails low tractability in a problem.

On the other hand, fuzzy systems can focus on modeling problems characterized by imprecise or ambiguous information. The following are situations where it is appropriate to formulate system problems within a fuzzy system framework:

1. In processes involving human interaction (e.g., human descriptive or intuitive thinking)
2. When an expert is available who can specify the rules underlying the system behavior and the fuzzy sets that represent the characteristics of each variable
3. When a mathematical model of the process does not exist, or exists but is too difficult to encode, or is too complex to be evaluated fast enough for real-time operation, or involves too much memory on the designated chip architecture
4. In processes concerned with continuous phenomena (e.g., one or more of the control variables are continuous) that are not easily broken down into discrete segments
5. When high ambient noise levels must be dealt with or it is important to use inexpensive sensors and/or low-precision microcontrollers.

The ability to use fuzzy system tools will allow one to address the vast majority of problems that have the preceding characteristics. Fuzzy formulations can help to achieve tractability, robustness, and lower solution cost.

Any field can be fuzzified and hence generalized by replacing the concept of a crisp set in the target field by the concept of a fuzzy set. Therefore, we can fuzzify some basic fields such as graph theory, arithmetic, and probability theory to develop fuzzy graph theory, fuzzy arithmetic, and fuzzy probability theory, respectively. Moreover, we can also fuzzify some applied fields such as neural networks, pattern recognition, and mathematical programming to obtain fuzzy neural networks, fuzzy pattern recognition, and fuzzy mathematical programming, respectively. The advantages of fuzzification include greater generality, higher expressive power, an enhanced ability to model real-world problems, and a methodology for exploiting the tolerance for imprecision.

1.5. APPLICATIONS OF FUZZY SYSTEMS

Fuzzy systems have superseded conventional technologies in many scientific applications and engineering systems. Fuzzy system techniques are applicable in areas such as control (the most widely applied area), pattern recognition (e.g., image, audio, signal processing), quantitative analysis (e.g., operations research, management), inference (e.g., expert systems for diagnosis, planning, and prediction; natural language processing; intelligent interface; intelligent robots; software engineering), and information retrieval (e.g., databases).

There has been rapid growth in the use of fuzzy logic in a wide variety of consumer products and industrial systems. Notable examples include the following:

1. In electric appliances, for instance, Matsushita builds a fuzzy washing machine that combines smart sensors with fuzzy logic. The sensors detect the color and kind of clothes present and the quantity of grit, and a fuzzy microprocessor selects the most appropriate combination from 600 available combinations of water temperature, detergent amount, and wash and spin cycle times.
2. Fisher, Sanyo, and others make fuzzy logic camcorders, which offer fuzzy focusing and image stabilization.
3. Mitsubishi manufactures an air conditioner that employs fuzzy systems to control temperature changes according to human comfort indices.

The number of fuzzy consumer products and fuzzy applications involving new patents is increasing so rapidly that it is not possible to offer a limited list of applications. It is this wealth of deployed, successful applications of fuzzy technology that is responsible for the current interest in fuzzy systems.

1.6. BEYOND FUZZY SYSTEMS: SOFT COMPUTING AND COMPUTATIONAL INTELLIGENCE

Viewed from a broad perspective, fuzzy logic is a constituent of an emerging research area, called soft computing, a term coined by Zadeh [2]. It is believed that the most important factor that underlies the marked increase in machine intelligence nowadays is the use of soft computing to imitate the human mind's ability to effectively employ approximate rather than exact modes of reasoning.

Unlike traditional hard computing whose prime goals are precision, certainty, and rigor, soft computing tolerates imprecision, uncertainty, and partial truths. The primary objective of soft computing is to take advantage of such tolerance to achieve tractability, robustness, a high level of machine intelligence, and lower costs. In addition to fuzzy logic, other principal constituents of soft computing are neural networks, probabilistic reasoning, which includes genetic algorithms, evolutionary programming, belief networks, chaotic systems, and parts of learning theory.

Among these, genetic algorithms and evolutionary programming are similar to neural networks in that they are based on low-level microscopic biological models.

They evolve toward finding better solutions to problems, just as species evolve toward better adaptation to their environments. It is also worth noting that fuzzy logic, neural networks, genetic algorithms, and evolutionary programming are also considered the building blocks of computational intelligence as conceived by James Bezdek [3]. Computational intelligence is low-level cognition in the style of the human mind and is in contrast to conventional (symbolic) artificial intelligence. In the partnership of fuzzy logic, neural networks, and probabilistic reasoning, fuzzy logic is concerned mainly with imprecision and approximate reasoning, neural networks with learning, and probabilistic reasoning with uncertainty. Since fuzzy logic, neural networks, and probabilistic reasoning are complementary rather than competitive, it is frequently advantageous to employ them in combination rather than exclusively. A number of excellent texts exist on fuzzy system theory [4–9].

1.7. FUZZY THEORY IN ELECTRIC POWER SYSTEMS

With the remarkable and successful penetration of fuzzy systems into manufacturing, appliances, and computer products, their applications in power systems are beginning to mature and receive wider acceptance in the electric power community. The application of fuzzy set theory to power systems is a relatively new area of research.

Concepts of fuzzy set theory were first introduced in solving power system long-range decision-making problems in the late 1970s. However, substantial interest in its applications to power areas is fairly recent. While conventional analytical solution methods exist for many problems in power system operation, planning, and control, their formulation of real-world problems suffers from restrictive assumptions. Even with these assumptions, solving large-scale power system problems is not trivial. Moreover, many uncertainties exist in a significant number of problems because power systems are large, complex, widely spread geographically, and influenced by unexpected events. These factors make it difficult to deal effectively with many power system problems through strictly conventional approaches alone. Therefore, areas of computational (artificial) intelligence emerged in recent years in power systems to complement conventional mathematical approaches and proved to be effective when properly coupled together.

In conceptualizing power system problems, the expert's empirical knowledge is generally expressed by language containing ambiguous or fuzzy descriptions. As a result, classical Boolean logic may not be a valid tool to represent such expertise. Fuzzy logic, on the other hand, is a natural choice for this purpose.

The growing number of publications on applications of fuzzy-set-based approaches to power systems indicates its potential role in solving power system problems. Results obtained so far are promising, but fuzzy set theory is not widely accepted. The reasons for its lack of acceptance include the following:

- Misunderstanding of the concept
- Excessive claims of some researchers
- Lack of implemented and available systems
- Its status as a new theory

Expert systems and other areas of artificial intelligence were introduced to solve power system operation planning and control problems. Expert systems are typically based on utilizing domain expert's knowledge. Some problems, such as diagnosis (especially transformer/generator malfunction diagnosis), alarm processing, and others, can be solved independently by expert system approaches. It is frequently difficult to make expert systems work efficiently because crisp representation of human empirical knowledge (usually expressed in natural languages and contain inherently uncertain representations) is very difficult and lacks flexibility.

Unexpected events and their uncertainties are traditionally represented by probability. However, it has recently been made clear that some of the uncertain factors are intrinsically of a fuzzy nature and are difficult to manage properly using probabilistic approaches.

There are problems in power systems that contain conflicting objectives. In power system operation, economy and security, maximum load supply, and minimum generating cost are conflicting objectives. The combination of these objectives by weighting coefficients is the traditional approach to such problems. Fuzzy set theory offers a better compromise and obtains solutions that cannot be easily found by weighting methods. The benefits of fuzzy set theory over traditional methods are as follows:

- Provides alternatives for the many attributes of objectives selected
- Resolves conflicting objectives by designing weights appropriate to a selected objective
- Provides the capability for handling ambiguity expressed in diagnostic processes, which involves symptoms and causes

Power system components have physical and operational limits that are usually described as hard inequality constraints in mathematical formulations. Enforcing minor violations of some constraints (practically acceptable) increases the computational burden and decreases the efficiency and may even prevent finding a feasible solution. In practice, certain slight violations of the inequality constraints are permissible. This means that there is not a clear constraint boundary and the constraints can be made soft. Traditionally, this problem has been managed by modifying either the objective function or the underlying iterative process. The fuzzy set approach inherently incorporates soft constraints and thus simplifies implementation of such considerations.

In the area of power system control, optimal control theory is often applied to design controllers to enhance system stability. Since power systems are large and nonlinear, simplifying assumptions are necessary in the design of such controllers. Due to model dependence, the controller's adaptability and robustness are problematic. Recently developed fuzzy-logic-based controllers show promise for robust performance and adaptive schemes.

With advances in fuzzy set theory and achievements made in applications to other areas, there is a need to provide a summary of advances in the applications of fuzzy theory to power system problems.

Momoh, Ma, and Tomsovic [10], suggest applying the following steps when fuzzy set theory is used to solve power system problems:

- *Description of original problem*: The problem to be solved should first be stated mathematically and linguistically.
- *Defining thresholds for variables*: For a given variable, there is a specific value with the greatest degree of satisfaction evaluated from empirical knowledge, and a certain deviation is acceptable with decreasing degree of satisfaction until there is a value that is completely unacceptable. The two values corresponding to the greatest and least degree of satisfaction are termed thresholds.
- *Fuzzy quantization*: Based on the threshold values already determined, proper forms of membership functions are constructed. The functions should reflect the change in degree of satisfaction with the change in variables evaluated by experts.
- *Selection of the fuzzy operations*: In terms of the practical decision-making process by human experts, a proper fuzzy operation is selected so that the results obtained are like those obtained by experts. The interpretation of results using fuzzy systems is based on domain experts' reasoning. Therefore, at this level a hybrid fuzzy set–expert system scheme is desirable. It helps to remove any ambiguity that may occur in problem solving.

1.8. SOME AREAS OF FUZZY APPLICATIONS IN POWER SYSTEMS

Fuzzy applications in electric power systems can be classified as dealing with the following:

- *Planning*: includes system expansion planning and long-midterm scheduling.
- *Operations*: includes security assessment, forecasting, controllers, and diagnosis.

1.8.1. Expansion Planning

There are a number of judgments based on experience and expert opinion that are crucial in decision making for power system expansion planning. Usually it is awkward to capture experience within the constraint formulations of conventional optimization models. This is because many factors, such as load demand levels, new stations locations, environmental effects, and so on, have a decisive effect on the decision making and yet are difficult to represent deterministically. Further, the objectives and constraints are uncertain or competing. In fact, the decision-making process in power system expansion planning is to a large extent qualitative and can be described more flexibly and intuitively by fuzzy set concepts.

1.8.2. Long- and Midterm Scheduling

Power system long- and midterm scheduling problems, such as annual maintenance scheduling, seasonal fuel scheduling, and midterm operation mode studies, are solved by various optimal and heuristic methods. The problems are characterized by many complications and uncertainties. Accurate formulations are often difficult and conventional optimization methods are not efficient. Moreover, it is more reasonable to represent constraints as soft (the fuzzy degree of satisfaction expression permits an engineering acceptable violation of constraints) than as hard. The combination of conventional methods with fuzzy sets may constitute an effective approach to solve these problems.

Fuzzy systems can play an important role in power system operation and planning optimization. Several publications have used fuzzy sets to manage conflicting objectives and soft constraints. This scheme not only makes the problem formulation more flexible, but if applied correctly, it can also improve the computational efficiency.

1.8.3. Dynamic Security Assessment

Dynamic security assessment (DSA) is a major application in power system operation and many techniques have been proposed for its solution. Since system security level varies according to engineering-economic considerations, the reliability of protective relays, the quality of the models used, and the risk of various faults, it may not be reasonable to say that the system is definitely secure or not using conventional binary logic. Present practice in DSA is to conduct off-line studies for a wide range of likely system conditions and network configurations. In on-line applications the results of off-line studies are not directly available but operators are allowed to rely on their own judgment and knowledge acquired in off-line studies and experience gathered over a long period of time. Fuzzy set theory may be useful in building an expert system for DSA based on the operator's empirical knowledge.

1.8.4. Load Forecasting

Power system load is influenced by many factors, such as weather, economic and social activities, and different load components (residential, industrial commercial, etc.). By analysis of only historical load data, it is difficult to obtain accurate load forecasts. The relation between load and independent variables is complex, and it is not always possible to fit the load curve using statistical models. Expert system approaches have shown advantages over conventional methods. The numerical aspects and uncertainties of this problem appear suitable for fuzzy methodologies.

1.8.5. Fuzzy Logic Controllers

Based on the number of publications on the subject, power system control problems are the most popular areas for fuzzy-set-based approaches. In fact, many achievements in the field of fuzzy control have been seen in other industries.

Fuzzy logic controllers are mainly used for power system excitation and converter controls. In traditional controller design, a system model needs to be constructed and control laws are derived based on analysis of the model. Because of nonlinearity, it is nearly always necessary to linearize the system model, and then the linear controllers are used to control the nonlinear system. One advantage of fuzzy logic over other forms of knowledge-based controllers lies in the interpolative nature of fuzzy control rules. The overlapping fuzzy antecedents to the control rules provide transitions between the control actions of different rules. Because of this interpolative quality, fuzzy logic controllers usually require far fewer rules than other knowledge-based controllers. Fuzzy logic controllers have received much attention in recent years, since they are more model independent, show high robustness, and can adapt.

1.8.6. Diagnosis

Human experts play central roles in troubleshooting or fault analysis. In power systems, it is required to diagnose equipment malfunctions as well as disturbances. The information available to perform equipment malfunction diagnosis is most of the time incomplete. In addition, the conditions that induce faults may change with time. Subjective conjectures based on experience are necessary. Accordingly, the expert systems approach has proved to be useful. As stated previously, fuzzy theory can lend itself to the representation of knowledge and the building of an expert system.

1.9. OUTLINE OF THE BOOK

Our discussion begins with Chapter 2 where we offer a summary of fundamentals of fuzzy systems. In Chapter 3, it is pointed out that utilities invest a good deal of time and resources in equipment monitoring and maintenance to anticipate failures or accelerated aging in power equipment. Such monitoring includes regular insulation condition tests for switching devices, reactors, power transformers, generator windings, and so on. In general, many of the indicators of equipment condition are imprecise and/or unreliable. Engineers must have considerable experience with a particular test before that test becomes useful. Several utilities have developed expert systems to codify this experience and improve knowledge of the breakdown process. The authors begin by a review of elements of fuzzy logic as a prelude to discussing possibility theory. This is followed by an example representing a simplified version of a transformer diagnostic and monitoring system. A proposed implementation for processing diagnostic information is given and evaluated. The chapter emphasizes uncertainty modeling of diagnostic problems. Fuzzy information methods are employed to represent quantitatively the diagnostic capability of a system. Further, several methods are discussed for extracting information from test data and evaluating system performance.

In Chapter 4, El-Sharkawi, Marks II, Streifel, and Kerszenbaum discuss detection and localization of shorted turns in the DC field winding of turbine-generator rotors using novelty detection and fuzzified neural networks. Use of neural networks with fuzzy logic outputs and traveling-wave techniques is shown to be an accurate locator of shorted turns in turbo-generator rotors. The technique also applies to transformers and other devices containing symmetrical windings. The technique is extended to operational rotors by the use of a novelty filter. The forms of shorted-turn detection show great promise in the monitoring of high-speed turbo-generators. Since introduction of shorts into on-line rotors is not possible, the explicit verification of any fault detection technique is not possible without the availability of a machine with a suspected shorted turn. Signature signals could be collected before the suspect rotor is dismantled for maintenance. After the rotor is repaired and brought back on line, healthy signature signals can be collected. The healthy regions and thresholds can then be established using the proposed techniques. The signature signals recorded before correction of the fault can then be processed by the detection algorithm.

Low-frequency oscillations are a common problem in large power systems. A power system stabilizer (PSS) can provide supplementary control signals to the excitation system and/or the speed governor system of the electric-generating unit to damp these oscillations and to improve its dynamic performance. Chapters 5 and 6 are devoted to this important topic. In Chapter 5, Malik and El-Metwally describe the structure and design of a fuzzy logic controller, and an algorithm to tune its parameters to achieve the desired performance is described. Two rule generation methods to automatically generate the fuzzy rule set are proposed. The application of the fuzzy logic controller as a power system stabilizer is investigated by simulation studies on a single-machine infinite-bus system and on a multimachine power system. Implementation of the fuzzy-logic-based power system stabilizer on a microcontroller and results of experimental studies on a physical model of a power system illustrate the effectiveness of the fuzzy-logic-based controller. Results of simulation and experimental studies look promising.

Continuing with power system stabilization, in Chapter 6, Hiyama presents a study on fuzzy logic power system stabilizers using polar information. Here, advanced fuzzy logic control rules are introduced based on three-dimensional information of generator acceleration, speed, and phase angle. Stabilizing signals are revised at every sampling instant and fed back to the excitation control loop of the generator. Simulation and laboratory studies are reported to demonstrate the technique's effectiveness. The results have been implemented on a prototype tested on hydro units in the Japanese KEPCO system.

The second contribution of Hiyama is on fuzzy logic switching control for FACTS devices, presented as Chapter 7. Here a fuzzy logic switching control scheme is proposed for devices such as the thyristor-controlled series capacitor modules, static var compensators (SVCs), and braking resistors (BRs). The stable region of operation is highly enlarged by the fuzzy logic switching of flexible AC transmission system (FACTS) devices. The coordination with power system stabilizers is shown to be effective to enlarge the stable region. The coordination between the BR and the SVC is also effective when the rating of the BR is small. Through the simulations, the robustness of the proposed control scheme is verified.

In Chapter 8 the effects of uncertain load in real-time and off-line power network modeling are discussed. Lu and Leou describe different approaches for representing uncertain load in power systems. In practice, load data can only be known within some finite precision. This being more the case as the study represents conditions that are more distant into the future. In order to avoid a large amount of repeated calculations and to take into account load uncertainties, it seems beneficial to represent the uncertain load using two approaches discussed in this chapter.

In building network models for power system analysis, if knowledge of network parameters is not complete but operation involves repetition of events with fixed laws governing it, one could represent the information by probability distributions. However, if the knowledge available is not sufficient to build the density function of the distribution but one still has qualitative information on the data to be represented, a possibility representation or use of fuzzy numbers could be useful. Probability theory and possibility theory are two complementary views of uncertainty. Uncertainties in loads or generations that do not have definite probability distribution can be incorporated in power system models by using the possibility approach to give a better representation of system behavior. As a consequence of dealing with fuzzy events, a whole set of load scenarios is analyzed at one time, avoiding the need for expensive simulation studies.

Miranda treats power system reliability in Chapter 9. Here the author concentrates mainly on reliability assessment for mid- to long-term purposes. Three types of models are discussed. In type I, only a fuzzy description of the system load is available. For type II fuzzy reliability assessment, component reliability indices are fuzzy. Extending the concept, we can also deal with a type III model.

In Chapter 10, Matsumoto discusses an operation support expert system for startup schedule optimization in fossil power plants. The chapter describes expectations of fuzzy theory in fossil power plant operation and two types of expert systems applied to operation support systems for a conventional one-through-type supercritical pressure fossil power plant and a gas and steam turbine combined-cycle power plant. Fuzzy models of expertise are introduced to modify the plant startup schedule. The most remarkable feature of this approach is that the quantitative optimum schedule is obtained through iterative modification of the schedule from a combination of qualitative knowledge and plant dynamic simulations. Simulation results with these two operation support expert systems demonstrate that plants are started quickly and accurately through the optimum startup schedule. Operator work is reduced in monitoring and executing startup schedules with functions for on-line assessment and off-line learning of plant dynamics. Convergence of the schedule optimization is fast compared with conventional operations research techniques. The expert systems are expected to contribute to reducing operators' work burden and to harmonize machine operation not only for economical aspects but also for environmental advantages.

In Chapter 11, Shahidehpour and Ferrero present a methodology to evaluate power purchases in an uncertain environment. Power generation, line flows, and prices are considered as triangular fuzzy numbers. Local generation and power purchases are control variables in the optimization procedure. The proposed methodology computes the range of control variables that satisfy the set of constraints as well as a certain reduction in the operation cost. The range of control variables is correlated with the desired cost reduction (goal). The lower the desired cost reduction, the narrower the range of control variables that satisfy the set of constraints. The utility decision maker reduces the goal iteratively until no feasible solution is found, obtaining the lowest operation cost while satisfying the operational constraints. The degree of acceptance of variables in the problem can be measured with the left and right spreads of the fuzzy numbers. When more uncertain variables are introduced in the problem, the degree of uncertainty of the obtained solution grows. The authors point out that the proposed method can be extended to include extra constraints related with operational practices in utilities.

References

[1] L. A. Zadeh, "Fuzzy Sets," *Information and Control*, Vol. 8, 1965, pp. 338–353.

[2] L. A. Zadeh, "Outline of a New Approach to the Analysis of Complex Systems and Decision Processes," *IEEE Transactions on Systems, Man, and Cybernetics*, Vol. SMC-1, 1973, pp. 28–44.

[3] J. Bezdek, *Pattern Recognition with Fuzzy Objective Function Algorithms*, Plenum, New York, 1981.

[4] D. Dubois and H. Prade, *Fuzzy Sets and Systems: Theory and Applications*, Academic, San Diego, 1980.

[5] G. Klir and T. A. Folger, *Fuzzy Sets, Uncertainty, and Information*, Prentice-Hall, Englewood Cliffs, NJ, 1988.

[6] C-T. Lin and C. S. G. Lee, *Neural Fuzzy Systems*, Prentice-Hall, Englewood Cliffs, NJ, 1996.

[7] B. Kosko, *Neural Networks and Fuzzy Systems*, Prentice-Hall, Englewood Cliffs, NJ, 1992.

[8] T. J. Ross, *Fuzzy Logic with Engineering Applications*, McGraw-Hill, New York, 1995.

[9] H. Zimmermann, *Fuzzy Set Theory and Its Applications*, 3rd ed., Kluwer Academic, Dordrecht, 1996.

[10] J. A. Momoh, X. W. Ma, and K. Tomsovic, "Overview and Literature Survey of Fuzzy Set Theory in Power Systems," *IEEE Transactions on Power Systems*, Vol. 10, No. 3, 1995, pp. 1676–1690.

M. E. El-Hawary
Dalhousie University
Halifax, Nova Scotia
B3J 2X4 Canada

Chapter 2

Fuzzy Systems: An Engineering Point of View

This chapter is devoted to a thorough discussion of fundamentals of fuzzy set and logic from the point of view of engineering systems. The coverage is restricted to elements of the theory that are common to all applications. Note that each subsequent chapter introduces additional elements of the theory as needed to make each chapter as self-contained as possible.

2.1. CHAPTER OVERVIEW

In Section 2.2, we introduce principal concepts and mathematical notions of fuzzy set theory—a theory of classes of objects with nonsharp boundaries. We first view fuzzy sets as a generalization of classical crisp sets by extending the range of the membership function (or characteristic function) from [0, 1] to all real numbers in the interval [0, 1]. A number of notions of fuzzy sets such as representation, support, α-cuts, convexity, and fuzzy numbers are then introduced. The resolution principle, which can be used to expand a fuzzy set in terms of its α-cuts, is discussed.

In Section 2.3, we develop basic definitions for set-theoretic and algebraic operations on fuzzy sets. It is shown that the only basic axioms not common to both crisp and fuzzy sets are the two excluded middle laws. All other operations are common to both crisp and fuzzy sets. For many situations in reasoning, the excluded middle laws do present something of a constraint. Aside from the difference of set membership being an infinite-valued idea as opposed to a binary-valued quantity, fuzzy sets are handled and treated in the same mathematical form as are crisp sets. The minimum operator is used to represent the intersection of fuzzy sets, interpreted as the logical "and." Moreover, the maximum operator is used to represent the union, interpreted as the logical "or."

Alternative operators have also been proposed. The proposals vary with respect to the generality or adaptability of the operators and the degree to which and how they are justified. In Section 2.4, we discuss two basic classes of operators referred to as triangular norms and conorms and the class of averaging operators. Each class contains parameterized as well as nonparameterized operators.

Section 2.5 is devoted to the celebrated extension principle intended to generalize crisp mathematical concepts to a fuzzy framework The principle extends point-to-point mappings to mappings for fuzzy sets and offers a powerful tool for fuzzy set theory. Any mathematical relationship between crisp elements can be extended to cover fuzzy entities. We demonstrate the application for fuzzy numbers. We introduce the concept of shape function and the *LR* representation of fuzzy numbers to facilitate doing extended arithmetic operations. The section is concluded with a discussion of interval arithmetic and its generalization to fuzzy numbers.

The chapter concludes with coverage of fuzzy relations and their types in Section 2.6. The concept of fuzzy composition is introduced to allow defining relation equations and the corresponding inverse relations and their solution.

2.2. FUZZY SETS AND MEMBERSHIP

Fuzzy logic can be viewed as a superset of conventional (Boolean) logic that has been extended to handle the concept of partial truth-truth values between "completely true" and "completely false." In 1965, Zadeh [1] suggested that set membership is the key to decision making when dealing with uncertainty. Before we begin our discussion, it is instructive to refer to Zadeh's statement (p. 338):

> *The notion of a fuzzy set provides a convenient point of departure for the construction of a conceptual framework which parallels in many respects the framework used in the case of ordinary sets, but is more general than the latter and, potentially, may prove to have a much wider scope of applicability. Essentially, such a framework provides a natural way of dealing with problems in which the source of imprecision is the absence of sharply defined criteria of class membership rather than the presence of random variables.*

2.2.1. Membership Functions

A classical (crisp, or hard) set is a collection of distinct objects, defined in such a manner as to separate the elements of a given universe of discourse into two groups: those that belong (members) and those that do not belong (nonmembers). The transition of an element between membership and nonmembership in a given set in the universe is abrupt and well defined. The crisp set can be defined by the so-called characteristic function. Let U be a universe of discourse; the characteristic function $\mu_A(x)$ of a crisp set A in U takes its values in $[0, 1]$ and is defined such that $\mu_A(x) = 1$ if x is a member of A (i.e., $x \in A$) and 0 otherwise; that is,

$$\mu_A(x) = \begin{cases} 1 & \text{if and only if } x \in A \\ 0 & \text{if and only if } x \notin A \end{cases}$$

Note that the boundary of the set A is rigid and sharp and performs a two-class dichotomization (i.e., $x \in A$ or $x \notin A$) and that the universe of discourse U is a crisp set.

Zadeh introduced fuzzy sets as an extension and generalization of the basic concepts of crisp sets. He extended the notion of binary membership to accommodate various "degrees of membership" on the real continuous interval $[0, 1]$, where the endpoints 0 and 1 conform to no membership and full membership, respectively, just as the characteristic function does for crisp sets. Here the infinite number of values in between the endpoints can represent various degrees of membership for an element x in some set on the universe. Seen in this light, then, a fuzzy set introduces vagueness by eliminating the sharp boundary dividing members from nonmembers in the group. Thus, the transition between full membership and nonmembership is gradual rather than abrupt. This transition among various degrees of membership accommodates the fact that the boundaries of fuzzy sets are vague and ambiguous.

A fuzzy set $\tilde{A}$ in the universe of discourse U is defined as a set of ordered pairs

$$\tilde{A} = \{(x, \mu_{\tilde{A}}(x)) | x \in U\}$$

where $\mu_{\tilde{A}}(x)$ is called the membership function (or characteristic function) of $\tilde{A}$ and $\mu_{\tilde{A}}(x)$ is the grade (or degree) of membership of x in $\tilde{A}$, which indicates the degree that x belongs to $\tilde{A}$.

EXAMPLE 2.1

Let U be the real line and let the crisp set A represent the test scores achieving an A grade, where the characteristic function is

$$\mu_A(x) = \begin{cases} 1 & x \geq 75 \\ 0 & x < 75 \end{cases}$$

which is shown in Figure 2.1. Now let the fuzzy set $\tilde{A}$ represent the test scores achieving an excellent grade. Then we have

$$\tilde{A} = \{(x, \mu_{\tilde{A}}(x)) | x \in U\}$$

where the membership function can be expressed as

$$\mu_{\tilde{A}}(x) = \frac{1}{1 + e^{\alpha |x - x_0|^p}}$$

which is shown in Figure 2.2. We comment here that α is a scaling factor and that increasing p changes the definition of the term to become less fuzzy, or sharper. At x_0, $\mu(x_0) = 0.5$. Note that apart from using different parameters p and α, we can use other expressions to represent the membership function for fuzzy set $\tilde{A}$ in this example. For example, we may have

$$\mu_{\tilde{A}} = \begin{cases} 0 & x \leq 75 \\ [1 + (x - 75)^{-2}]^{-1} & x > 75 \end{cases}$$

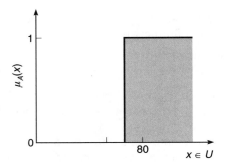

Figure 2.1 Characteristic function of crisp set A in Example 2.1.

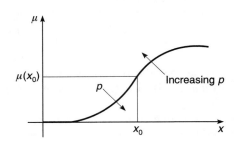

Figure 2.2 Characteristic function of fuzzy set $\tilde{A}$ in Example 2.1.

Assignment of the membership function of a fuzzy set is subjective in nature; however, it cannot be assigned arbitrarily. A qualitative estimation reflecting a given ordering of the elements in $\tilde{A}$ may be sufficient. In addition, estimating a membership function is not straightforward.

If there is no confusion between a fuzzy set $\tilde{A}$ and a crisp set A, the tilde above the fuzzy set A will be eliminated to simplify the notation.

Before continuing the discussion, the following definitions are needed:

1. *Support of a fuzzy set*: The support of a fuzzy set A is the crisp set of all $x \in U$ such that $\mu(x) > 0$. That is,

$$\text{supp}(A) = \{x \in U | \mu_A > 0\}$$

   An empty fuzzy set has empty support; that is, the membership function assigns 0 to all elements of the universal set U.

2. *Singleton*: A fuzzy set A whose support is a single point in U with $\mu_A(x) = 1$ is referred to as a fuzzy singleton.

3. *Crossover point*: The element $x \in U$ at which $\mu_A(x) = 0.5$ is called the crossover, or breakeven, point.

4. *Kernel*: The kernel of a fuzzy set A consists of the elements x whose membership grade is 1; that is,

$$\text{ker}(A) = \{x | \mu_A(x) = 1\}$$

5. *Height of a fuzzy set*: The height of a fuzzy set A is the supremum of (x) over U; that is,

$$\text{Height of } A = \text{Height}(A) = \sup_x \mu_A(x)$$

6. *Normalized fuzzy set*: A fuzzy set is normalized when the height of the fuzzy set is unity [i.e., $\text{Height}(A) = 1$]; otherwise it is subnormal. A nonempty fuzzy set A can always be normalized by dividing $\mu(x)$ by the height of A.

2.2.2. Representations of Membership Functions

The representation of a fuzzy set can be expressed in terms of the support of the fuzzy set. For a discrete universe of discourse $U = \{x_1, x_2, \ldots, x_n\}$, a fuzzy set A can be represented using the ordered-pairs concept and is written as

$$A = \{(x_1, \mu_A(x_1)), (x_2, \mu_A(x_2)), \ldots, (x_n, \mu_A(x_n))\}$$

Using the support of a fuzzy set A, we can simplify the representation of a fuzzy set A as

$$A = \frac{\mu_1}{x_1} + \frac{\mu_2}{x_2} + \frac{\mu_3}{x_3} + \cdots + \frac{\mu_n}{x_n} = \sum \frac{\mu_i}{x_i}$$

where $+$ indicates the union of the elements and μ_i is the grade of membership of x_i, that is, $\mu_i = \mu_A(x_i) > 0$.

Using the support of a fuzzy set to represent a fuzzy set, we consider only those elements in the universe of discourse that have a nonzero degree of membership grade in the fuzzy set. If U is not discrete but is an interval of real numbers, we can use the notation

$$A = \int_U \frac{\mu_A(x)}{x}$$

where $\int$ indicates the union of the elements in A.

In both notations, the horizontal bar is not a quotient but rather a delimiter. The numerator in each term is the membership value in set A associated with the element of the universe indicated in the denominator. In the first notation, the summation symbol is not for algebraic summation but rather denotes the collection or aggregation of each element; hence the plus sign in the first notation is not the algebraic add but a function-theoretic union. In the second notation the integral sign is not an algebraic integral but a continuous function-theoretic union notation for continuous variables.

2.2.3. α-Cuts

Another important concept and property of fuzzy sets is the resolution principle, which requires us to understand α-cuts (or α-level sets.) An α-cut (or α-level set) of a fuzzy set $\tilde{A}$ is a crisp set A that contains all the elements of the universe U that have a membership grade in $\tilde{A}$ greater than or equal to α. That is,

$$A_a = \{x \in U | \mu_A(x) \geq \alpha, \alpha \in (0, 1]\}$$

If $A_\alpha = \{x \in U | \mu_A(x) > \alpha\}$, then A_α is called a strong α-cut. Furthermore, the set of all levels $\alpha \in (0, 1]$ that represents distinct α-cuts of a given fuzzy set A is called a level set of A. That is

$$\Pi_A = \{\alpha | \mu_A(x) = \alpha \text{ for some } x \in U\}$$

It is clear that if $\alpha \leq \beta$, then $A_\beta \subseteq A_\alpha$.

With this understanding of α-cuts, we shall introduce an important idea in fuzzy set theory, called the resolution principle, which indicates that a fuzzy set A can be expanded in terms of its α-cuts.

2.2.4. Resolution Principle

Let A be a fuzzy set in the universe of discourse U. Then the membership function of A can be expressed in terms of the characteristic functions of its α-cuts according to

$$\mu_A(x) = \sup_{\alpha \in (0,1]} [\alpha \wedge \mu_{A_\alpha}(x)] \; \forall x \in U$$

where $\wedge$ denotes the minimum operation and $\mu_{A_\alpha}(x)$ is the characteristic function of the crisp set A_α:

$$\mu_{A_\alpha} = \begin{cases} 1 & \text{if } x \in A_\alpha \\ 0 & \text{otherwise} \end{cases}$$

This leads to the following representation of a fuzzy set A using the resolution principle. Let A be a fuzzy set in the universe of discourse U. Let αA_α denote a fuzzy set with the membership function

$$\mu_\alpha A_\alpha(x) = [\alpha \wedge \mu_{A_\alpha}(x)] \, \forall x \in U$$

Then the resolution principle states that the fuzzy set A can be expressed in the form

$$A = \bigcup_{\alpha \in \Pi_A} \alpha A_\alpha$$

or

$$A = \int_0^1 \alpha A_\alpha$$

The resolution principle indicates that a fuzzy set A can be decomposed into $\alpha A_\alpha, \alpha \in (0, 1]$. On the other hand, a fuzzy set A can be retrieved as a union of its αA_α, which is called the representation theorem. In other words, a fuzzy set can be expressed in terms of its α-cuts without resorting to the membership function. This concept is illustrated in Figure 2.3.

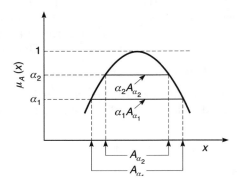

Figure 2.3 Decomposition of a fuzzy set.

2.2.5. Convexity

Convexity of fuzzy sets plays an important role in fuzzy set theory, and it can be defined in terms of α-cuts or membership functions. A fuzzy set is convex if and only if each of its α-cuts is a convex set. Or, equivalently, a fuzzy set A is convex if and only if

$$\mu_A[\lambda x_1 + (1 - \lambda)x_2] \geq \min(\mu_A(x_1), \mu_A(x_2))$$

where $x_1, x_2 \in U, \lambda \in (0, 1]$. This can be interpreted as follows: Take two elements x_1 and x_2 in a fuzzy set A and draw a connecting straight line between them; then the membership grade of all the points on the line must be greater than or equal to the minimum of $\mu_A(x_1)$ and $\mu_A(x_2)$. For example, the fuzzy set A in Figure 2.4(a) is convex, but it is not normalized. The fuzzy set B in Figure 2.4(b) is not convex, but it is normalized. Note that the convexity definition does not imply that the membership function of a convex fuzzy set is a convex function [see Fig. 2.4(c)].

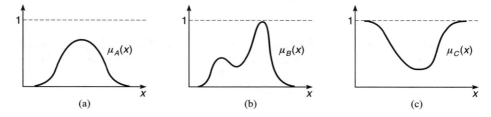

(a)   (b)   (c)

Figure 2.4 Convex and nonconvex fuzzy sets.

2.2.6. Fuzzy Numbers

A convex, normalized fuzzy set defined on the real line R whose membership function is piecewise continuous or, equivalently, each α-cut is a closed interval, is called a fuzzy number. Two typical fuzzy numbers are the S function and the Π function, which are, respectively, defined by

$$S(x: a, b) = \begin{cases} 0 & \text{for } x < a \\ 2\left(\dfrac{x - a}{b - a}\right)^2 & \text{for } a \leq x < \frac{1}{2}(a + b) \\ 1 - 2\left(\dfrac{x - b}{b - a}\right)^2 & \text{for } \frac{1}{2}(a + b) \leq x < b \\ 1 & \text{for } x \geq b \end{cases}$$

$$\Pi(x: a, b) = \begin{cases} S(x; b - a, b) & \text{for } x < b \\ 1 - S(x; b, b + a) & \text{for } x > b \end{cases}$$

The S and Π functions are shown in Figures 2.5(a) and (b), respectively. In $S(x; a, b)$, the crossover point is $\frac{1}{2}(a + b)$. In $\Pi(x; a, b)$, b is the point at which Π is unity, while

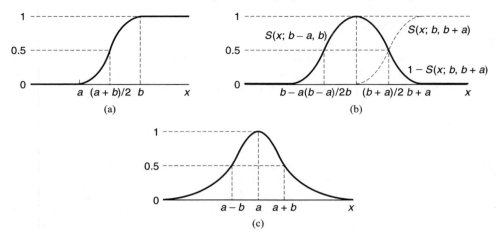

Figure 2.5 Typical fuzzy number shapes.

the two crossover points are $b - \frac{1}{2}a$ and $b + \frac{1}{2}a$. The separation between these two crossover points, a, is the bandwidth. Sometimes the Π function is simply defined as

$$\Pi'(x; a, b) = \frac{1}{1 + [x - a/b]^2}$$

which is shown in Figure 2.5(c), where $2b$ is the bandwidth. The two functions make it convenient to express the membership function of a fuzzy subset of the real line in terms of these "standard" functions whose parameters can be adjusted accordingly to fit a specified membership function approximately.

2.2.7. Cardinality

Similar to the cardinality of a crisp set, which is defined as the number of elements in the crisp set, the cardinality (or scalar cardinality) of a fuzzy set A is the summation of the membership grades of all the elements of x in A. That is,

$$|A| = \sum_{x \in U} \mu_A(x)$$

The relative cardinality of A is given as

$$|A|_{\text{rel}} = \frac{|A|}{|U|}$$

where $|U|$ is finite. The relative cardinality evaluates the proportion of elements of U having the property A when U is finite.

2.3. SET-THEORETIC AND ALGEBRAIC OPERATIONS FOR FUZZY SETS

In this section we introduce some basic set-theoretic definitions and operations for fuzzy sets. Throughout the discussion we let A and B be fuzzy sets in the universe of discourse U. We begin by defining the intersection and union of fuzzy sets.

2.3.1. Intersection

Let the fuzzy set C be the intersection of the two fuzzy sets A and B:

$$C = A \cap B$$

Then the membership function of the intersection $\mu_C(x)$ is defined by

$$\mu_C(x) = \min\{\mu_A(x), \mu_B(x)\}$$

or

$$\mu_{A \cap B}(x) \equiv \mu_A(x) \wedge \mu_B(x)$$

where $\wedge$ indicates the minimum operation. It is clear that

$$A \cap B \subseteq A \text{ and } A \cap B \subseteq B$$

The shaded area in Figure 2.6 shows the membership function of the intersection of two sets A and B.

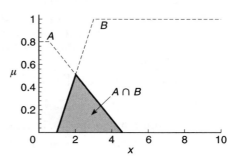

Figure 2.6 The intersection of fuzzy sets A and B.

2.3.2. Union

Let the fuzzy set C be the union of the two fuzzy sets A and B:

$$C = A \cup B$$

Then the membership function of the union $\mu_C(x)$ is defined by

$$\mu_C(x) = \max\{\mu_A(x), \mu_B(x)\}$$

or

$$\mu_{A \cup B}(x) \equiv \mu_A(x) \vee \mu_B(x)$$

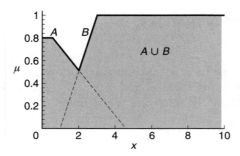

Figure 2.7 The union of fuzzy sets A and B.

where $\vee$ indicates the maximum operation. It is clear that

$$A \subseteq A \cup B \quad \text{and} \quad B \subseteq A \cup B$$

The shaded area in Figure 2.7 shows the membership function of the union of the two sets A and B of Figure 2.6.

It is interesting to note that min and max are not the only operators that could have been selected to represent the intersection or union, respectively, of fuzzy sets. Bellman and Giertz [2] treated this issue from an axiomatic point of view. They interpreted the intersection as "logical and" and the union as "logical or" and the fuzzy set A as the statement "the element x belongs to set A," which can be accepted as more or less true. Their line of reasoning begins by considering two statements S and T for which the truth values are μ_S and μ_T, respectively, with $\mu_S, \mu_T \in [0, 1]$. The truth value of the "and" and "or" combination of these statements $\mu(S \text{ and } T)$ and $\mu(S \text{ or } T)$, both from $[0, 1]$, are interpreted as the values of the membership functions of the intersection and union respectively, of S and T. As a result, we are dealing with two real-valued functions f and g such that $\mu_{S \text{ and } T} = f(\mu_S, \mu_T)$ and $\mu_{S \text{ or } T} = g(\mu_S, \mu_T)$.

Bellman and Giertz [2] proposed the following constraints on f and g:

1. Both f and g are nondecreasing and continuous in μ_S and μ_T.
2. Both f and g are symmetric:

$$f(\mu_S, \mu_T) = f(\mu_T, \mu_S) \qquad g(\mu_S, \mu_T) = g(\mu_T, \mu_S)$$

3. The terms $f(\mu_S, \mu_T)$ and $g(\mu_T, \mu_S)$ are strictly increasing in μ_S.
4. The meaning of the statements $f(\mu_S, \mu_T) \leq \min(\mu_T, \mu_S)$ and $g(\mu_S, \mu_T) \geq \max(\mu_T, \mu_S)$ is that accepting the truth of the statement "S and T" requires more, and accepting the truth of the statement "S or T" less, than accepting S or T alone as true.
5. Moreover, $f(1, 1) = 1$ and $g(0, 0) = 0$.
6. Logically equivalent statements must have equal truth values, and fuzzy sets with the same contents must have the same membership functions, that is, S_1 and (S_2 and S_3) is equivalent to (S_1 and S_2) or (S_1 and S_3) and therefore must be equally true.

From a formal point of view, the preceding constraints resolve to the following relations:

1. Commutativity:

$$\mu_S \wedge \mu_T = \mu_T \wedge \mu_S \qquad \mu_S \vee \mu_T = \mu_T \vee \mu_S$$

2. Associativity:

$$(\mu_S \wedge \mu_T) \wedge \mu_U = \mu_T \wedge (\mu_S \wedge \mu_U)$$

$$(\mu_S \vee \mu_T) \vee \mu_U = \mu_T \vee (\mu_S \vee \mu_U)$$

3. Distributivity:

$$\mu_S \wedge (\mu_T \vee \mu_U) = (\mu_S \wedge \mu_T) \vee (\mu_S \wedge \mu_U)$$

$$\mu_S \vee (\mu_T \wedge \mu_U) = (\mu_S \vee \mu_T)(\mu_S \vee \mu_U)$$

4. The relations $\mu_S \wedge \mu_T$ and $\mu_S \vee \mu_T$ are continuous and nondecreasing in each component.
5. The relations $\mu_S \wedge \mu_S$ and $\mu_S \vee \mu_S$ are strictly increasing in μ_S.
6. $\mu_S \wedge \mu_T \leq \min(\mu_T, \mu_S)$
   $\mu_S \vee \mu_T \geq \max(\mu_T, \mu_S)$
7. Idempotence:

$$1 \wedge 1 = 1 \qquad 0 \vee 0 = 0$$

Bellman and Giertz prove that

$$\mu_{S \wedge T} = \min(\mu_S, \mu_T) \quad \text{and} \quad \mu_{S \vee T} = \max(\mu_S, \mu_T)$$

2.3.3. Complement

Given a normalized fuzzy set A, its complement is denoted by $\bar{A}$, whose membership function is given by

$$\mu_{\bar{A}}(x) = 1 - \mu_A(x)$$

We return to Bellman and Giertz's axioms [2]. It seems reasonable to assume that if statement S is true, its complement non-S is false, or if $\mu_S = 1$, then $\mu_{\text{non } S} = 0$, and vice versa. The function h (as complement in analogy to f and g for intersection and union) should also be continuous and monotonically decreasing. We also wish to have the complement of the complement to be the original statement (to conform with conventional logic and set theory). These requirements, however, are not enough to determine uniquely the mathematical form of the complement. Bellman and Giertz require in addition that $\mu_S(0.5) = 0.5$. Other assumptions are possible.

2.3.4. Equality

Two fuzzy sets A and B are equal if and only if their membership functions are equal:

$$\mu_A(x) = \mu_B(x)$$

Therefore, if $\mu_A(x) \neq \mu_B(x)$, for some $x \in U$, then $A \neq B$.

2.3.5. Subset

The set A is a subset of B; that is, $A \subset B$ if and only if

$$\mu_A(x) \leq \mu_B(x)$$

If $A \subseteq B$ and $A \neq B$, then A is a proper subset of B; that is, $A \subset B$. The definition of subset is crisp. To check the degree that A is a subset of B, we can use the subsethood measure:

$$S(A, B) \equiv \deg(A \subseteq B) \equiv \frac{|A \cap B|}{|A|}$$

As in the case of crisp sets, we have the double-negation law (involution) and DeMorgan's laws for fuzzy sets:

- Double-negation law (involution):

$$\bar{\bar{A}} = A$$

- DeMorgan's laws:

$$\overline{A \cup B} = \bar{A} \cup \bar{B} \qquad \overline{A \cap B} = \bar{A} \cup \bar{B}$$

However, the laws of the excluded middle (i.e., $E \cup \bar{E} = U$) and the law of contradiction (i.e., $E \cap \bar{E} = \Phi$) of the crisp set E are no longer true and valid in fuzzy sets. That is, for fuzzy set A,

$$A \cup \bar{A} \neq U \qquad A \cap \bar{A} \neq \Phi$$

This means that because of the lack of precise boundaries, complementary sets are overlapping and cannot cover the universal set U perfectly. On the contrary, these two laws are the necessary characteristics of crisp sets, which bring together a crisp set E and its complement to provide the whole set; that is, nothing exists between E and $\bar{E}$.

The other properties of fuzzy set operations that are common to crisp set operations are given by the following:

Commutativity	$A \cup B = B \cup A, A \cap B = B \cap A$
Associativity	$(A \cup B) \cup C = A \cup (B \cup C)$
	$(A \cap B) \cap C = A \cap (B \cap C)$
Distributivity	$A \cap (B \cup C) = (A \cap B) \cup (A \cap C)$
	$A \cup (B \cap C) = (A \cup B) \cap (A \cup C)$
Absorption	$A \cup (A \cap B) = A$
	$A \cap (A \cup B) = A$
Idempotence	$A \cup A = A, A \cap A = A$

However, it is noted that the above discussion about the properties of fuzzy set operations is based on the definitions of complement, intersection, and union operations given earlier. There are other definitions of these operations, and hence some properties mentioned above may fail to hold while others may become true for various definitions. When we choose union and intersection to combine fuzzy sets, we have to give up either the excluded-middle laws or distributivity and idempotency.

2.3.6. Algebraic Operations on Fuzzy Sets

We now introduce some popular algebraic operations on fuzzy sets.

Algebraic Sum of Two Fuzzy Sets (probabilistic sum). The algebraic sum of two fuzzy sets A and B (denoted by $A + B$) is defined as

$$\mu_{A+B}(x) \equiv \mu_A(x) + \mu_B(x) - \mu_A(x) \cdot \mu_B(x)$$

Algebraic Product of Two Fuzzy Sets. The algebraic product of two fuzzy sets, $A \cdot B$, is defined by

$$\mu_{A \cdot B}(x) = \mu_A(x) \cdot \mu_B(x)$$

Bounded Sum of Two Fuzzy Sets. The bounded sum of two fuzzy sets, $A \oplus B$, is defined by

$$\mu_{A \oplus B}(x) \equiv \min\{1, \mu_A(x) + \mu_B(x)\}$$

The sum $\mu_A(x) + \mu_B(x)$ may exceed unity, in which case we pick 1 as the value of the bounded sum.

Bounded Difference of Two Fuzzy Sets. The bounded difference of two fuzzy sets, $A \ominus B$, is defined by

$$\mu_{A \ominus B}(x) \equiv \max\{0, \mu_A(x) - \mu_B(x)\}$$

Again, since the difference $\mu_A(x) - \mu_B(x)$ can have a negative value, we take zero as the value of the membership function in this case.

2.3.7. Fuzzy Power Set

The collection of all fuzzy sets and fuzzy subsets on the universe of discourse U is denoted as the fuzzy power set $P(U)$. Since all fuzzy sets can overlap, then it is clear that the cardinality $|P(U)|$ of the fuzzy power set is infinite.

Cartesian Product of Two Fuzzy Sets. The Cartesian product of fuzzy sets is defined by considering fuzzy sets $A_1, A_2, A_3, \ldots, A_n$ in the corresponding universes of discourse $U_1, U_2, U_3, \ldots, U_n$. It is a fuzzy set in the product space $U_1 \times U_2 \times \cdots \times U_n$ with a membership function given by

$$\mu_{A_1 \times A_2 \times \cdots \times A_n}(x_1, x_2, \ldots, x_n) \equiv \min\{\mu_{A_1}(x_1), \mu_{A_2}(x_2), \ldots, \mu_{A_n}(x_n)\}$$

The* mth *Power of a Fuzzy Set. The mth power of a fuzzy set A is a fuzzy set with the membership function

$$\mu_{A^m}(x) = [\mu_A(x)]^m$$

Note that since the value of the membership function is less than unity, taking the mth powers results in lower values of the membership function.

2.4. AGGREGATION

In Section 2.3, the min operator was used to represent the intersection of fuzzy sets, interpreted as the logical "and." Moreover, the maximum operator was used to represent the union, interpreted as the logical "or." Alternative operators have also been proposed. These proposals vary with respect to the generality or adaptability of the operators as well as to the degree to which and how they are justified. The way in which justification was handled ranges from intuitive arguments to empirical or axiomatic justification. Adaptability ranges from uniquely defined concepts via parameterized "families" of operators to general classes of operators that satisfy certain properties.

In this section, we discuss the two basic classes of operators: operators for the intersection and union of fuzzy sets, referred to as triangular norms (t-norms) and triangular conorms (t-conorms), and the class of averaging operators, which model connectives for fuzzy sets between t-norms and t-conorms. Each class contains parameterized as well as nonparameterized operators.

2.4.1. *t*-Norms

To represent the intersection of fuzzy sets, Zadeh [1] suggested the min operator and the algebraic product $A \cdot B$. The "bold intersection" was modeled by the "bounded sum" defined earlier. The minimum product and bounded-sum operators are part of the triangular, or t-norm, operators. We define t-norms as single-valued, two-argument functions of the form

$$t: [0,1] \times [0,1] \rightarrow [0,1]$$

such that

$$\mu_{A \cap B}(x) = t[\mu_A(x), \mu_B(x)]$$

where the function $t(\cdot)$ satisfies the following conditions:

1. Boundary conditions:

$$t(0,0) = 0, t(\mu_A(x), 1) = t(1, \mu_A(x)) = \mu_A(x)$$

2. Monotonicity: If

$$\mu_A(x) \leq \mu_C(x) \quad \text{and} \quad \mu_B(x) \leq \mu_D(x)$$

then

$$t(\mu_A(x), \mu_B(x)) \leq t(\mu_C(x), \mu_D(x))$$

3. Commutativity:

$$t(\mu_A(x), \mu_B(x)) = t(\mu_B(x), \mu_A(x))$$

4. Associativity:

$$t(\mu_A(x), t(\mu_B(x), \mu_C(x))) = t(t(\mu_A(x), \mu_B(x)), \mu_C(x))$$

The functions t define a general class of intersection operators for fuzzy sets. Since the operators in this class of t-norms are, in particular, associative, then it is possible to compute the membership values for the intersection of more than two fuzzy sets by recursively applying a t-norm operator.

2.4.2. *t*-Conorms (or *s*-Norms)

For the union of fuzzy sets, the maximum operator, the algebraic sum, and the "bold union" modeled by the bounded sum have been suggested. Corresponding to the class of intersection operators, a general class of aggregation operators (or the union of fuzzy sets called triangular conorms or t-conorms (sometimes referred to as s-norms) is defined in an analogous manner. The max operator, algebraic sum, and bounded-sum t-conorms are maps of the form

$$s: [0, 1] \times [0, 1] \rightarrow [0, 1]$$

such that

$$\mu_{A \cup B}(x) = s[\mu_A(x), \mu_B(x)]$$

where the function $s(\cdot)$ satisfies the following conditions:

1. Boundary conditions:

$$s(1, 1) = 1, s(\mu_A(x), 0) = s(0, \mu_A(x)) = \mu_A(x)$$

2. Monotonicity: If

$$\mu_A(x) \le \mu_C(x) \quad \text{and} \quad \mu_B(x) \le \mu_D(x)$$

then

$$s(\mu_A(x), \mu_B(x)) \le s(\mu_C(x), \mu_D(x))$$

3. Commutativity:

$$s(\mu_A(x), \mu_B(x)) = s(\mu_B(x), \mu_A(x))$$

4. Associativity:

$$s(\mu_A(x), s(\mu_B(x), \mu_C(x))) = s(s(\mu_A(x), \mu_B(x)), \mu_C(x))$$

Here, t-norms and t-conorms are related in a sense of logical duality. A t-conorm can be defined as a two-argument function mapping from $[0, 1] \times [0, 1]$ to $[0, 1]$ such that the function t, defined as

$$t(\mu_A(x), \mu_B(x)) = 1 - s(1 - \mu_A(x), 1 - \mu_B(x))$$

TABLE 2.1   SOME NONPARAMETRIC t-NORMS (FUZZY INTERSECTIONS)
                AND t-CONORMS (FUZZY ANIONS)

	t-Norm		t-Conorm
minimum	$t_3 = \min\{a, b\}$	maximum	$s_3 = \max\{a, b\}$
Hamacher product	$t_{2.5} = \dfrac{ab}{a + b - ab}$	Hamacher sum	$s_{2.5} = \dfrac{a + b - 2ab}{1 - ab}$
Algebraic product	$t_2 = ab$	Algebraic sum	$s_2 = a + b - ab$
Einstein product	$t_{1.5} = \dfrac{ab}{2 - [a + b - ab]}$	Einstein sum	$s_{1.5} = \dfrac{a + b}{1 + ab}$
Bounded difference	$t_1 = \max\{0, a + b - 1\}$	Bounded sum	$s_1 = \min\{1, a + b\}$
Drastic product	$t_w = \begin{cases} \min\{a, b\} & \text{if } \max\{a, b\} = 1 \\ 0 & \text{otherwise} \end{cases}$	Drastic sum	$s_w = \begin{cases} \max\{a, b\} & \text{if } \min\{a, b\} = 0 \\ 1 & \text{otherwise} \end{cases}$

is a t-norm. So, any t-conorm can be generated from a t-norm t through this transformation. More generally, for suitable negation operators like the complement operator for fuzzy sets, defined as $n(\mu_A(x)) = 1 - \mu_A(x)$, pairs of t-norms t and t-conorms s satisfy the following generalization of DeMorgan's law:

$$s(\mu_A(x), \mu_B(x)) = n(t(n(\mu_A(x)), n(\mu_B(x))))$$

$$t(\mu_A(x), \mu_B(x)) = n(s(n(\mu_A(x)), n(\mu_B(x))))$$

The t-norms and t-conorms are also used to define other operations.

Typical dual pairs of nonparametric t-norms and t-conorms are as displayed in Table 2.1. Note that in order to simplify the notation we use $a \equiv \mu_A(x)$ and $b \equiv \mu_B(x)$.

The relations among various t-norms (fuzzy intersections) and t-conorms (fuzzy union) are governed by the following system of inequalities:

$$t_w \leq t_1 \leq t_{1.5} \leq t_2 \leq t_{2.5} \leq t_3 \qquad s_w \geq s_1 \geq s_{1.5} \geq s_2 \geq s_{2.5} \geq s_3$$

Note that this order implies that for any fuzzy sets A and B in the universe of discourse with membership values between 0 and 1, any intersection operator that is a t-norm is bounded by the min operator and the drastic product operator. The t-conorm is bounded by the max operator and the drastic sum operator, respectively.

2.4.3. Parametric Families of t-Norms and Conorms

In many practical situations, it may be desirable to extend the range of the preceding operators with a view of adapting them to their application context. A great deal of work has been done by many authors who created parameterized families of t-norms and t-conorms, often maintaining the associative property. For illustration purposes, we review some interesting parameterized operators. Some of these operators and their equivalence to the logical "and" and "or," respectively, have been justified axiomatically.

2.4.4. Hamacher's Family of Aggregation Operators

We begin by sketching the axioms on which the Hamacher operator is based to be able to compare the axiomatic system of Bellman and Giertz (min/max) [2] with that of the Hamacher operator (which is essentially a family of product operators).

Hamacher Intersection. The intersection of two fuzzy sets A and B is defined in terms of the positive parameter γ as

$$t_H(a, b, \gamma) = \mu_{A \cap B}(x) = \frac{ab}{\gamma + (1 - \gamma)(a + b - ab)}$$

where $a = \mu_A(x)$ and $b = \mu_B(x)$. Hamacher's basic axioms to derive a mathematical model for the "and" operator are as follows:

1. The operator $\wedge$ is associative, that is,

$$A \wedge (B \wedge C) = (A \wedge B) \wedge C$$

2. The operator $\wedge$ is continuous.
3. The essential difference between the Hamacher operator and the Bellman-Giertz axioms is that the operator $\wedge$ is invective in each argument, that is,

$$(A \wedge B) = (A \wedge C) \Rightarrow B = C$$
$$(A \wedge B) = (C \wedge B) \Rightarrow A = C$$

4. Here, $\mu_A(x) = 1 \Rightarrow \mu_{A \wedge A}(x) = 1$.

For $\gamma \to 1$, this reduces to the algebraic product.

Hamacher Union. The union of two fuzzy sets A and B is defined in terms of the parameter γ' as

$$s_H(a, b, \gamma') = \mu_{A \cup B}(x) = \frac{a + b + (\gamma' - 1)ab}{1 + \gamma'ab}$$

where $\gamma' \geq -1$ and $\gamma' = \gamma - 1$. The Hamacher union operator reduces to the algebraic sum.

2.4.5. Yager's Family of Aggregation Operators

Yager defined another triangular family of operators in terms of the parameter p.

Yager Intersection. The intersection of fuzzy sets A and B is defined as

$$t_Y(a,b,p) = \mu_{A \cap B}(x) = 1 - \min\{1, [(1-a)^p + (1-b)^p]^{1/p}\} \qquad p \geq 1$$

This intersection operator converges to the min operator for $p \rightarrow \infty$. For $p = 1$ this intersection operator reduces to the bold intersection or bounded product given by $\mu = \max\{0, (A + B - 1)\}$. The Yager intersection has a number of interesting properties, including

$$t_Y(a,0,p) = 0 \qquad t_Y(a,1,p) = a$$

Yager Union. This is a typical parametric t-conorm defined by the function

$$s_Y(a,b,p) = \mu_{A \cup B}(x) = \min\{1, [a^p + b^p]^{1/p}\} \qquad p \geq 1$$

This union operator converges to the maximum operator for $p \rightarrow \infty$, and for $p = 1$, the Yager operator reduces to the bold union; that is, the bounded sum $\mu = \min\{1, a + b\}$. Additional properties of Yager union include

$$s_Y(a,1,p) = 1 \qquad s_Y(a,0,p) = a$$

Both Yager operators satisfy the DeMorgan laws and are commutative, associative for all p, and monotonically nondecreasing in $\mu(x)$; they also include the classical cases of binary logic. They are, however, not distributive.

2.4.6. Dubois and Prade's Family of Aggregation Operators

Dubois and Prade also proposed a commutative and associative parameterized family of aggregation operators in terms of a parameter α.

Dubois and Prade Intersection. Here we have

$$t_{\text{DP}}(a,b,\alpha) = \mu_{A \cap B}(x) = \frac{ab}{\max(a,b,\alpha)}$$

An equivalent representation is given by

$$t_{\text{DP}}(a,b,\alpha) = \begin{cases} \mu_{A \cap B}(x) = \dfrac{ab}{\alpha} & 0 \leq a, b \leq \alpha \\[2mm] \min\{a,b\} & \alpha \leq a, b \leq 1 \end{cases}$$

From the above equation, it can be seen that the parameter α acts as a threshold. This intersection operator decreases with respect to α and lies between $\min\{a,b\}$ (which is the resulting operation for $\alpha = 1$) and the algebraic product ab (for $\alpha = 1$).

Dubois and Prade Union. Dubois and Prade suggested the operation

$$s_{\text{DP}}(a,b,p) = \frac{a + b - ab - \min\{a, b, (1-\alpha)\}}{\max\{(1-a), (1-b), \alpha\}}$$

2.4.7. Schweizer and Sklar Family of Aggregation Operators

Schweizer and Sklar proposed the two operators

$$t_{SS}(a, b, r) = \max\{0, a^{-r} + b^{-r} - 1\}^{-1/r}$$

$$s_{SS}(a, b, r) = 1 - \max\{0, (1-a)^{-r} + (1-b)^{-r} - 1\}^{-1/r}$$

Here $r \in (-\infty, \infty)$

2.4.8. Dombi Family of Aggregation Operators

Dombi proposed the two operators

$$t_D(a, b, \lambda) = \frac{1}{1 + [(1/a - 1)^\lambda + (1/b - 1)^\lambda]^{1/\lambda}}$$

$$s_D(a, b, \lambda) = \frac{1}{1 + [(1/a - 1)^{-\lambda} + (1/b - 1)^{-\lambda}]^{-1/\lambda}}$$

For this system $\lambda \in (0, \infty)$.

2.4.9. Frank Family of Aggregation Operators

Frank suggested the aggregation operators

$$t_F(a, b, \vartheta) = \log_\vartheta \left[1 - \frac{(\vartheta^a - 1)(\vartheta^b - 1)}{\vartheta - 1}\right]$$

$$s_F(a, b, \vartheta) = 1 - \log_\vartheta \left[1 + \frac{(\vartheta^{1-a} - 1)(\vartheta^{1-b} - 1)}{\vartheta - 1}\right]$$

Here $\vartheta \in (0, \infty)$.

We note that in conventional set theory and binary logic the definitions of intersection ("and") and union ("or") are uniquely defined. But in fuzzy set theory and fuzzy logic, there are many proposed definitions. The reason for this is simple: Many operators behave in exactly the same manner if the degrees of membership are restricted to the values 0 or 1. If this restriction is relaxed, the operators lead to different results.

If we consider that "and" and "or" are only limiting special cases, then we realize that there are many ways to combine fuzzy sets with fuzzy statements. Many authors have proposed general connectives that are important in many application areas such as fuzzy decision analysis. The general operators do not distinguish between the intersection and union of fuzzy sets and are called averaging operators.

2.4.10. Averaging Operators

All t-norms are smaller than the min operator intersection function and all t-conorms are greater than the max operator union function. Naturally, there is a class of aggregation operators lying between the min operator and the max operator. A straightforward approach for aggregating fuzzy sets (e.g., in the context of decision making) would be to use the aggregating procedures based on the idea of trade-offs between conflicting goals when compensation is allowed, and the resulting trade-offs lie between the most optimistic lower bound and the most pessimistic upper bound; that is, they map between the minimum and the maximum degree of membership of the aggregated sets. Therefore they are called averaging operators. Operators such as the weighted and conventional arithmetic or geometric mean are examples of non-parametric averaging operators. In reality, they are adequate models for human aggregation procedures in decision environments and have empirically performed quite well. Hence, averaging operators are aggregation operators for which

$$h: [0, 1]^n \rightarrow [0, 1]$$

such that

$$\mu_A(x) = h(\mu_{A1}(x), \mu_{A2}(x), \ldots, \mu_{An}(x))$$

The generalized means operator is defined as

$$h_\alpha(a_1, a_2, \ldots, a_n) = \left[\frac{a_1^\alpha + a_2^\alpha + \cdots + a_n^\alpha}{n} \right]^{1/\alpha}$$

where $\alpha > 0$. It can be verified that when $\alpha \rightarrow -\infty$, h_α becomes $\min\{a_1, \ldots, a_n\}$, and when $\alpha \rightarrow \infty$, it becomes $\max\{a_1, \ldots, a_n\}$. Hence, the generalized means covers the entire interval between the min and the max operators. An important extension of the generalized means is the weighted generalized means, defined by

$$h_\alpha(a_i, w_i) = \left[\sum w_i a_i^\alpha \right]^{1/\alpha}$$

The w_i are weights expressing the relative importance of the aggregated sets; the sum of weights is restricted to 1. The weighted generalized means are useful in decision-making problems where different criteria differ in importance (weights).

2.4.11. "Fuzzy And" and "Fuzzy Or" Operators

The fuzzy aggregation operators "fuzzy and" and "fuzzy or" suggested by Werners [3] combine the minimum and maximum operators, respectively, with the arithmetic mean. The combination of these operators leads to very good results with respect to empirical data and allows compensation between the membership values of the aggregated sets. The "fuzzy and" operator is defined as

$$t_W(a, b, \beta) = \beta \min\{a, b\} + \tfrac{1}{2}(1 - \beta)(a + b)$$

with $\beta \in [0, 1]$.

The "fuzzy or" operator is defined as

$$s_W(a, b, \beta) = \beta \max\{a, b\} + \tfrac{1}{2}(1 - \beta)(a + b)$$

The parameter β indicates how close the operator is to the strict logical meaning of "and" and "or," respectively. For $\beta = 1$, the "fuzzy and" becomes the minimum operator, and the "fuzzy or" reduces to the maximum operator. Here, $\beta = 0$ yields the arithmetic mean for both "fuzzy and" and "fuzzy or."

Additional averaging aggregation procedures are symmetric summation operators, which, like the arithmetic or geometric mean operators, indicate some degree of compensation, but in contrast to the latter, these procedures are not associative. Examples of symmetric summation operators are the operators M_1, M_2 and N_1, N_2, known as symmetric summations and symmetric differences, respectively. Here the aggregation of two fuzzy sets A and B is pointwise defined as follows:

$$M_1 = \frac{a + b - ab}{1 + a + b - 2ab} \qquad M_2 = \frac{ab}{1 + a - b + 2ab}$$

$$N_1 = \frac{\max\{a, b\}}{1 + |a - b|} \qquad N_2 = \frac{\min\{a, b\}}{1 + |a - b|}$$

The above-mentioned averaging operators correspond to fixed compensation between the logical "and" and the logical "or."

2.4.12. "Compensatory And" Operator

In order to describe a variety of phenomena in decision situations, several operators with different compensations are necessary. Zimmermann and Zysno [4] proposed an operator referred to as "compensatory and" that is more general in the sense that the compensation between intersection and union is expressed by a parameter γ. The "compensatory and" operator is defined as

$$\mu_{ZY} = (ab)^{1-\gamma}[1 - (1 - a)(1 - b)]^{\gamma}$$

This γ operator is clearly a combination of the algebraic product (modeling the logical "and") and the algebraic sum (modeling the "or"). It is pointwise invective (except at zero and one), continuous, monotonous, and commutative. It also satisfies the DeMorgan laws and follows the truth tables of binary logic. The parameter indicates where the actual operator is located between the logical "and" and "or."

Other operators following the idea of parameterized compensation are defined by taking linear convex combinations of noncompensatory operators modeling the logical "and" and "or." The aggregation of two fuzzy sets A and B by the convex combination between the min and max operators is defined as

$$\mu_1(a, b, \gamma) = \gamma \min\{a, b\} + (1 - \gamma)\max\{a, b\}$$

Combining the algebraic product and algebraic sum, we obtain the operation

$$\mu_2(a, b, \gamma) = \gamma \cdot \{a \cdot b\} + (1 - \gamma) \cdot \{a + b - a \cdot b\}$$

This class of operators is again in accordance with binary logic truth tables. But Zimmermann and Zysno showed that the "compensatory and" operator is more adequate in human decision making than are these operators.

2.4.13. Guidelines for Choosing Appropriate Aggregation Operators

Presently, we discuss guidelines for deciding on which of the multitude of aggregation operators to use. The following eight criteria for classifying operators may be helpful in selecting the appropriate connective for a specific task.

1. *Axiomatic strength*: We have the axioms of Bellman-Giertz and Hamacher, respectively, who wanted their operators to satisfy. A specific operator is deemed better than others if it satisfies less limiting axioms.

2. *Empirical fit*: If fuzzy set theory is used as a modeling language for real situations or systems, the operators must be appropriate models of real-system behavior, and this can normally be proven only by empirical testing.

3. *Adaptability*: If one wants to use a very small number of operators to model many situations, then these operators have to be adaptable to the specific context. This can, for instance, be achieved by parameterization. Thus min and max operators cannot be adapted at all. They are acceptable only in situations in which they fit. (Of course, they have other advantages, such as numerical efficiency.) In contrast, Yager's operators or the γ operator can be adapted to certain contexts by setting the p's or γ's appropriately.

4. *Numerical efficiency*: If one compares the min operator with, for instance, Yager's intersection operator, it becomes quite obvious that the latter requires more computational effort than the former. In practice, this might be quite important, in particular when large problems have to be solved.

5. *Compensation*: The logical "and" does not allow for compensation at all; that is, an element of the intersection of two sets cannot compensate for a low degree of belonging to one of the intersected sets by a higher degree of belonging to another of them. In binary logic, one cannot compensate by the higher truth of one statement for the lower truth of another statement when combining them by "and."

    Compensation, in the context of aggregation operators for fuzzy sets, means the following: Given that the degree of membership to the aggregated fuzzy set is

$$\mu_{AGG}(x_k) = f(a(x_k), b(x_k)) = k$$

    f is compensatory if $\mu_{AGG}(x_k) = f(a(x_k), b(x_k)) = k$ is obtainable for a different a by a change in b. Thus the min operator is not compensatory, while, for example, the product operator and the γ operator are.

6. *Range of compensation*: If one would use a convex combination of min and max operators, a compensation could obviously occur in the range between min and max. The product operator allows compensation in the open interval $(0, 1)$. In general, the larger the range of compensation, the better the compensatory operator.

7. *Aggregating behavior*: If we consider normal or subnormal fuzzy sets, the degree of membership in the aggregated set depends very frequently on the

number of sets combined. If we combine fuzzy sets by the product operator, for instance, each additional fuzzy set "added" will normally decrease the resulting aggregate degrees of membership. This might be a desirable feature. It might, however, also be inadequate, since for formal reasons the resulting degree of membership may be required to be nonincreasing.

8. *Required scale level of membership functions*: The scale level (nominal, interval, ratio, or absolute) on which membership information can be obtained depends on a number of factors. Different operators may require different scale levels of membership information to be admissible. In general, the operator that requires the lowest scale level is the most preferable from the point of view of information gathering.

2.5. EXTENSION PRINCIPLE AND ITS APPLICATIONS

The extension principle is intended to generalize crisp mathematical concepts to the fuzzy set framework and extends point-to-point mappings to mappings for fuzzy sets. The principle introduced by Zadeh in 1978 [5] is an important tool of fuzzy set theory. It offers a facility for any function f that maps an n-tuple $(x_1, x_2, \ldots, x_n)$ in the crisp set U to a point in the crisp set V to be generalized to mapping n fuzzy subsets in U to a fuzzy subset in V. Therefore, any mathematical relationship between nonfuzzy elements can be extended to cover fuzzy entities. The extension principle is also helpful in treating set-theoretic operations for higher order fuzzy sets. We first state the extension principle and then demonstrate some of its applications.

Given a function $f: U \to V$ and a fuzzy set A in U, where $A = \{\mu_1/x_1, \mu_2/x_2, \ldots, \mu_n/x_n\}$, the extension principle states that

$$f(A) = f\left\{\frac{\mu_1}{x_1}, \frac{\mu_2}{x_2}, \ldots, \frac{\mu_n}{x_n}\right\}$$

$$= \left\{\frac{\mu_1}{f(x_1)}, \frac{\mu_2}{f(x_2)}, \ldots, \frac{\mu_n}{f(x_n)}\right\}$$

If f maps more than one element of U to the same element y in V, then the maximum among their membership grades is taken. That is,

$$\mu_{f(A)}(y) = \max_{\substack{x_i \in U \\ f(x_i) = y}} [\mu_i(x_i)]$$

where x are the elements that are mapped to the same y.

In many instances, the function f of interest maps n-tuples in U to a point in V. Let U be a Cartesian product of universes $U = U_1 \times U_2 \times \cdots \times U_n$ and $A_1, \ldots, A_n$ be n fuzzy sets in $U_1, \ldots, U_n$, respectively. The function f maps an n-tuple $(x_1, \ldots, x_n)$ in the crisp set U to a point y in the crisp set V; that is, $f(x_1, \ldots, x_n)$. The extension principle allows the function $f(x_1, \ldots, x_n)$ to be extended to act on the n fuzzy subsets of $U, A_1, \ldots, A_n$, such that

$$B = f(A)$$

where B is the fuzzy image (fuzzy set) of $A_1, \ldots, A_n$ through $f(\cdot)$. The fuzzy set B is defined by

$$B = \{(y, \mu_B(y)) | y = f(x_1, x_2, \ldots, x_n), (x_1, x_2, \ldots, x_n) \in U\}$$

with

$$\mu_B(y) = \sup_{\substack{(x_1, x_2, \ldots, x_n) \in U \\ y = f(x_1, x_2, \ldots, x_n)}} \min[\mu_{A_1}(x_1), \mu_{A_2}(x_2), \ldots, \mu_{A_n}(x_n)]$$

with an additional condition that $\mu_B(y) = 0$ if there exists no $(x_1, x_2, \ldots, x_n) \in U$ such that $y = f(x_1, x_2, \ldots, x_n)$.

2.5.1. Fuzzy Numbers

A fuzzy number $\tilde{M}$ is defined as a convex normalized fuzzy set with a piecewise continuous membership function with exactly one point x_0 with $\mu_M(x_0) = 1$. In practical applications, we often relax this last requirement and allow trapezoidal membership functions to be used to represent a fuzzy number.

Positive (Negative) Fuzzy Numbers. A positive (negative) fuzzy number is one whose membership function is zero for all negative (positive) values of the independent variable x. This is expressed symbolically as $\mu_M(x) = 0, \forall x < 0, (\forall x > 0)$.

Increasing (Decreasing) Two-Argument Operations. An operation $*$ operating on two variables is referred to as increasing (decreasing) if

$$x_1 > y_1 \quad \text{and} \quad x_2 > y_2 \Rightarrow x_1 * x_2 > y_1 * y_2 (x_1 * x_2 < y_1 * y_2)$$

Addition and multiplication are increasing operations.

2.5.2. Algebraic Operations for Fuzzy Numbers

The extension principle can be employed to define algebraic operations for fuzzy numbers. The extended-addition and extended-multiplication operations are denoted by $\oplus$ and $\otimes$. If M and N are fuzzy numbers with continuous membership functions from the real line to $[0, 1]$ and $*$ is a continuous and increasing operation on two variables, then the membership function of the fuzzy number $M * N$ is continuous and is given by

$$\mu(z) = \sup_{z = x*y} \min[\mu_M(x), \mu_N(y)]$$

If the operation $*$ is commutative and/or distributive, then the extended operation Y is commutative and/or distributive.

In order to find some important fuzzy number operations, we recall that for a fuzzy number the extension principle reduces to

$$\mu_{f(M)}(z) = \sup_{x \in f^{-1}(z)} \mu_M(x)$$

As a result, we have the following important results:

- To obtain the opposite of a fuzzy number M denoted by $-M$, we let $f(x) = -x$; then we get

$$-M = \{(x, \mu_M(-x)) | x \in X\}$$

- To obtain the inverse of a fuzzy number M denoted by M^{-1}, we let $f(x) = 1/x$; then we get

$$M^{-1} = \left\{ \left(x, \mu_M \left[\frac{1}{x} \right] \right) | x \in X \right\}$$

- To obtain the scalar multiple of a fuzzy number M denoted by λM, we let $f(x) = \lambda x$; then we get

$$\lambda M = \{(x, \mu_M(\lambda \cdot x)) | x \in X\}$$

Properties of Extended Addition. Addition is an increasing operation, and thus the extended sum of fuzzy numbers M and N is itself a fuzzy number. The operation is associative and distributive. Moreover, we have

$$\ominus(M \oplus N) = (\ominus M) \oplus (\ominus N)$$

$$M \oplus (\ominus M) \neq 0$$

The last result means that for extended addition there is no inverse element.

Properties of the Extended Product. Multiplication is an increasing operation for positive real numbers (decreasing for negative real numbers), and thus the extended product of positive fuzzy numbers M and N is itself a positive fuzzy number. Moreover, the extended product of two negative fuzzy numbers is a positive fuzzy number. The product of a positive fuzzy number and a negative fuzzy number is a negative fuzzy number. The extended-product operation is associative and distributive. Moreover, we have

$$\ominus(M) \otimes N = \ominus(M \otimes N)$$

$$M \otimes M^{-1} \neq 0$$

The last result means that for the extended-product operation there is no inverse element. In addition, if M is either a positive or negative fuzzy number and N and P are both either positive or negative fuzzy numbers, then we have

$$M \otimes (N \oplus P) = (M \otimes N) \oplus (M \otimes P)$$

Extended Subtraction. While subtraction is neither an increasing nor a decreasing operation, the extended-subtraction operation can be expressed as $M \ominus N = M \oplus (\ominus N)$. Using the extension principle, we can show that

$$\mu_{M \ominus N}(z) = \sup_{z = x+y} \min(\mu_M(x), \mu_{-N}(y))$$

Whenever M and N are fuzzy numbers, so will be $M \ominus N$.

Extended Division. While division is neither an increasing nor a decreasing operation, the extended-division operation can be expressed as an extended product involving the inverse of N. Using the extension principle, we can show that

$$\mu_{M:N}(z) = \sup_{z=x\cdot y} \min\left\{\mu_M(x), \mu_N\left(\frac{1}{y}\right)\right\}$$

Whenever M and N are strictly positive fuzzy numbers, so will the outcome of their extended division.

2.5.3. *LR* Representation of Fuzzy Sets

If we do not place further restrictions on the nature and type of membership functions of fuzzy numbers considered, evaluating the outcome of extended operations with fuzzy numbers will involve rather extensive computational burden. Dubois and Prade [6] suggested a special type of representation for fuzzy numbers based on functions called shape functions, defined in the following.

Shape (Reference) Functions. Shape functions are denoted as either L (for left) or R (for right) functions and satisfy the following requirements:

- The L function is decreasing with $L(0) = 1$ and $L(x) < 1, \forall x > 0$.
- Either $L(x) > 0$ for $\forall x < 1, L(1) = 0$, or $L(x) > 0$ for $\forall x$ and $\lim_{x\to\infty} L(x) = 0$.

Example L functions are as follows:

1. $L(x) = \begin{cases} 1 & x \in [-1, 1], \\ 0 & \text{elsewhere.} \end{cases}$
2. $L(x) = \max(1, 1 - |x|^p), p \geq 0$.
3. $L(x) = e^{-|x|^p}, p \geq 0$.
4. $L(x) = 1/(1 + |x|^p), p \geq 0$.

2.5.4. *LR*-Type Fuzzy Numbers

A fuzzy number is defined to be of the LR type if there are reference functions L and R and positive scalars α (left spread), β (right spread), and m (mean) such that

$$\mu_M(x) = \begin{cases} L\left(\dfrac{m-x}{\alpha}\right) & \text{for } x \leq m \\ R\left(\dfrac{x-m}{\beta}\right) & \text{for } x \geq m \end{cases}$$

As the spread increases, M becomes fuzzier and fuzzier. Symbolically we write

$$M = (m, \alpha, \beta)_{LR}$$

2.5.5. Extended Arithmetic Operations via *LR*-Type Numbers

Given two fuzzy numbers defined in terms of LR functions, extended operations such as addition and multiplication can be performed in a rather easy manner. Assume that M_1 and M_2 are fuzzy numbers of the LR type:

$$M_1 = (m_1, \alpha_1, \beta_1)_{LR} \qquad M_2 = (m_2, \alpha_2, \beta_2)_{LR}$$

Addition.

$$(m_1, \alpha_1, \beta_1)_{LR} + (m_2, \alpha_2, \beta_2)_{LR} = (m_s, \alpha_s, \beta_s)_{LR}$$

$$m_s = m_1 + m_2$$

$$\alpha_s = \alpha_1 + \alpha_2$$

$$\beta_s = \beta_1 + \beta_2$$

The mean of the sum is equal to the sum of the means, and each of the spreads of the sum are the sum of the respective spreads.

Opposite of a Fuzzy Number. The formula for the opposite of a fuzzy number is

$$-M_1 = -(m_1, \alpha_1, \beta_1)_{LR} = (-m_1, \beta_1, \alpha_1)_{RL}$$

Note that the references are exchanged.

Subtraction of Fuzzy Numbers. The formula for the subtraction of two fuzzy numbers is

$$(m_1, \alpha_1, \beta_1)_{LR} \ominus (m_2, \alpha_2, \beta_2)_{RL} = (m_d, \alpha_d, \beta_d)_{LR}$$

$$m_d = m_1 - m_2$$

$$\alpha_d = \alpha_1 + \beta_2$$

$$\beta_d = \beta_1 + \alpha_2$$

Multiplication of Fuzzy Numbers. The product of two fuzzy numbers of the LR-type is not, in general, an LR-type number. Approximate formulas can be derived, however, if we assume that α_1 and α_2 are small with respect to the means m_1 and m_2. The formulas depend on whether the fuzzy numbers are positive or negative. We have three cases.

Case 1—$M_1 > 0$ and $M_2 > 0$: In this case, we have

$$(m_1, \alpha_1, \beta_1)_{LR} \otimes (m_2, \alpha_2, \beta_2)_{LR} = (m_p, \alpha_p, \beta_p)_{LR}$$

$$m_p = m_1 m_2$$

$$\alpha_p = m_2 \alpha_1 + m_1 \alpha_2$$

$$\beta_p = m_2 \beta_1 + m_1 \beta_2$$

The mean of the product is the product of the means. The spreads of the product are sums of the original spreads weighted by the means of the original fuzzy numbers.

Case 2—$M_1 < 0$ and $M_2 < 0$: In this case, we have

$$(m_1, \alpha_1, \beta_1)_{LR} \otimes (m_2, \alpha_2, \beta_2)_{LR} \approx (m_p, \alpha_p, \beta_p)_{RL}$$

$$m_p = m_1 m_2$$

$$\alpha_p = -m_2 \beta_1 - m_1 \beta_2$$

$$\beta_p = -m_2 \alpha_1 - m_1 \alpha_2$$

The mean of the product is the product of the means. Note that the references are exchanged.

Case 3—$M_1 < 0$ and $M_2 > 0$: In this case, we have

$$(m_1, \alpha_1, \beta_1)_{RL} \otimes (m_2, \alpha_2, \beta_2)_{LR} \approx (m_p, \alpha_p, \beta_p)_{RL}$$

$$m_p = m_1 m_2$$

$$\alpha_p = m_2 \alpha_1 - m_1 \beta_2$$

$$\beta_p = m_2 \beta_1 - m_1 \alpha_2$$

The mean of the product is the product of the means. Note that the references are not as expected.

When the spreads are not small compared with mean values, alternative approximation formulas can be used. One such a formula for positive M_1 and M_2 is given by

$$(m_1, \alpha_1, \beta_1)_{LR} \otimes (m_2, \alpha_2, \beta_2)_{LR} = (m_p, \alpha_p, \beta_p)_{LR}$$

$$m_p = m_1 m_2$$

$$\alpha_p = m_2 \alpha_1 + m_1 \alpha_2 - \alpha_1 \alpha_2$$

$$\beta_p = m_2 \beta_1 + m_1 \beta_2 + \beta_1 \beta_2$$

This function gives correct values of the product for at least three points: $(m_1 m_2, 1)$, $[(m_1 - \alpha_1)(m_2 - \alpha_2), L(1)]$, and $[(m_1 + \beta_1)(m_2 + \beta_2), R(1)]$.

Inverse of a Fuzzy Number. The exact inverse of an *LR*-type fuzzy number is neither an *LR* nor an *RL* type. But if we consider only a neighborhood of $1/m$, then we have

$$\frac{1 - mx}{\alpha x} \approx \frac{(1/m - x)}{\alpha/m^2}$$

In this case, M^{-1} is approximately of the RL type, and we have

$$[m, \alpha, \beta]_{LR}^{-1} \approx [m^{-1}, \beta m^{-2}, \alpha m^{-2}]_{RL}$$

Division of Fuzzy Numbers. For positive M_1 and M_2 we have the approximate formula

$$\frac{(m_1, \alpha_1, \beta_1)_{LR}}{(m_2, \alpha_2, \beta_2)_{RL}} \approx \left(\frac{m_1}{m_2}, \frac{m_1 \beta_2 + m_2 \alpha_1}{m_2^2}, \frac{m_1 \alpha_2 + m_2 \beta_1}{m_2^2}\right)_{LR}$$

Similar formulas can be given when M_1 or M_2 or both are negative.

Flat Fuzzy Numbers (LR-Type Fuzzy Intervals). A flat fuzzy number (fuzzy interval) is defined to be of the LR type if there are reference functions L and R and positive scalars α (left spread), β (right spread), and parameters $\overline{m}$ and $\underline{m}$ such that

$$\mu_M(x) = \begin{cases} L\left(\dfrac{\underline{m} - x}{\alpha}\right) & \text{for} \quad x \le \underline{m} \\ 1 & \text{for } \underline{m} \le x \le \overline{m} \\ R\left(\dfrac{x - \overline{m}}{\beta}\right) & \text{for} \quad x \ge \overline{m} \end{cases}$$

Symbolically we write

$$M = (\underline{m}, \overline{m}, \alpha, \beta)_{LR}$$

A trapezoidal fuzzy number is obtained by letting

$$L(x) = R(x) = \max(0, 1 - x)$$

Formulas for extended arithmetic operations using flat fuzzy numbers can be obtained. For example, for the case of addition we have

$$(\underline{m}_1, \overline{m}_1, \alpha_1, \beta_1)_{LR} \oplus (\underline{m}_2, \overline{m}_2, \alpha_2, \beta_2)_{LR} = (\underline{m}_s, \overline{m}_s, \alpha_s, \beta_s)_{LR}$$

$$\underline{m}_s = \underline{m}_1 + \underline{m}_2$$

$$\overline{m}_s = \overline{m}_1 + \overline{m}_2$$

$$\alpha_s = \alpha_1 + \alpha_2$$

$$\beta_s = \beta_1 + \beta_2$$

We also have an approximate multiplication formula:

$$(\underline{m}_1, \overline{m}_1, \alpha_1, \beta_1)_{LR} \otimes (\underline{m}_2, \overline{m}_2, \alpha_2, \beta_2)_{LR} = (\underline{m}_P, \overline{m}_P, \alpha_P, \beta_P)_{LR}$$

$$\underline{m}_P = \underline{m}_1 \underline{m}_2$$

$$\overline{m}_P = \overline{m}_1 \overline{m}_2$$

$$\alpha_P = \underline{m}_1 \alpha_2 + \underline{m}_2 \alpha_1$$

$$\beta_P = \overline{m}_1 \beta_2 + \overline{m}_2 \beta_1$$

The preceding equations are valid for $M_1 > 0$ and $M_2 > 0$.

2.5.6. Interval Arithmetic

Many areas of science and engineering involve uncertain values or inaccurate data from measuring instruments that are usually specified in terms of intervals. Mathematical operations are then performed on these intervals to get a reliable estimate of the measurements (again in the form of intervals). Interval analysis (or interval arithmetic), commonly applied in physics, is the branch of mathematics dealing with these computations.

Because the concept of fuzzy numbers includes the interval as a special case, fuzzy arithmetic is an extension of interval arithmetic. Interval analysis only serves to delegate the inaccuracies of measuring instruments, in the form of intervals, back to the magnitudes estimated by the measurements.

The idea of a fuzzy number can be perceived as an extension of the concept of intervals. As a result of the ramping of membership in a fuzzy set, fuzzy arithmetic has more expressive power than interval arithmetic. Instead of considering intervals at only one unique level, fuzzy numbers consider them at several levels or, more generally, at all levels from 0 to 1.

We discuss the representation of uncertain data by intervals and their basic mathematical operations. Then we extend the concept of intervals to fuzzy numbers and their computations.

Assume that values or data obtained from a scientific instrument are uncertain and position this value inside a closed interval on the real line R; that is, this uncertain value is inside an interval of confidence of R, $x \in [a_1, a_2]$, where $a_1 \leq a_2$. This indicates that we are certain that the value x is greater than or equal to a_1 and smaller than or equal to a_2. Symbolically, we represent an interval by $A = [a_1, a_2]$, and if an uncertain value x is inside this closed interval, we can also use set notation to express it as

$$A = [a_1, a_2] = \{x | a_1 \leq x \leq a_2\}$$

In general, the numbers a_1 and a_2 are finite. However, in some cases, $a_1 = -\infty$ and/or $a_2 = \infty$. If a value x is certain and is a singleton in R, it can be expressed in the form of intervals as $x = [x, x]$. For instance, $0 = [0, 0]$. In other words, an ordinary number k can be written in the form of an interval as $k = [k, k]$.

When considering intervals, we have four types:

$[a_1, a_2] = \{x | a_1 \leq x \leq a_2\}$: closed interval
$[a_1, a_2) = \{x | a_1 \leq x < a_2\}$: closed at left endpoint and open at right endpoint
$(a_1, a_2] = \{x | a_1 < x \leq a_2\}$: open at left endpoint and closed at right endpoint
$(a_1, a_2) = \{x | a_1 < x < a_2\}$: open interval

Here brackets and parentheses are used to denote closed and open endpoints, respectively.

We presently discuss some mathematical operations on intervals of confidence such as addition $(+)$, subtraction $(-)$, multiplication $(\cdot)$, division $(:)$, max operation $(\vee)$, and min operation $(\wedge)$.

Addition and Subtraction. Let $A = [a_1, a_2]$ and $B = [b_1, b_2]$ be two intervals of confidence in R. If $x \in [a_1, a_2]$ and $y \in [b_1, b_2]$ then $x + y \in [a_1 + b_1, a_2 + b_2]$. Symbolically, we write

$$A(+)B = [a_1, a_2](+)[b_1, b_2] = [a_1 + b_1, a_2 + b_2]$$

For the same two intervals of confidence, their subtraction is

$$A(-)B = [a_1, a_2](-)[b_1, b_2] = [a_1 - b_2, a_2 - b_1]$$

that is, we subtract the larger value in $[b_1, b_2]$ from a_1 and the smaller value in $[b_1, b_2]$ from a_2.

Opposite (Image) of A. If $x \in [a_1, a_2]$, then its opposite (image) $-x \in [-a_2, -a_2]$. That is, if $A = [a_1, a_2]$, then its opposite (image) is $\bar{A} = [-a_2, -a_1]$. Note that

$$A(+)\bar{A} = [a_1, a_2](+)[-a_2, -a_1] = [a_1 - a_2, a_2 - a_1] \neq 0$$

Using the concept of image, subtraction becomes addition of an image. Thus we have

$$A(-)B = A(+)\bar{B} = [a_1, a_2](+)[-b_2, -b_1] = [a_1 - b_2, a_2 - b_1]$$

Multiplication and Division. Consider two intervals of confidence $A = [a_1, a_2]$, and $B = [b_1, b_2]$ in R^+, where R^+ is the nonnegative real line. The product of these two intervals is

$$A(\cdot)B = [a_1, a_2](\cdot)[b_1, b_2] = [a_1 b_1, a_2 b_2]$$

If $A, B \subset R$, then $A(\cdot)B$ is more complicated and has nine possible combinations. If we multiply an interval by a nonnegative number k, then for all $A \subset R^+$, we have

$$k \cdot A = [k, k](\cdot)[a_1, a_2] = [ka_1, ka_2]$$

For the division of two intervals of confidence in R^+ we have

$$A(:)B = [a_1, a_2](:)[b_1, b_2] = \left[\frac{a_1}{b_2}, \frac{a_2}{b_1} \right]$$

If $b_1 = 0$, then the upper bound increases to ∞. If $b_1 = b_2 = 0$ then the interval of confidence is extended to ∞.

Inverse A^{-1}. If $x \in [a_1, a_2] \subset R_0^+$ where R_0^+ is the positive real line, then its inverse is $(1/x) \in [1/a_2, 1/a_1]$ and

$$A^{-1} = [a_1, a_2]^{-1} = \left[\frac{1}{a_2}, \frac{1}{a_1} \right]$$

Using the concept of inverse, division becomes multiplication of an inverse. That is,

$$A(:)B = A(\cdot)B^{-1} = [a_1, a_2](\cdot)\left[\frac{1}{b_2}, \frac{1}{b_1} \right] = \left[\frac{a_1}{b_2}, \frac{a_2}{b_1} \right]$$

For division by a nonnegative number $k > 0$, it is equivalent to $(1/k)(\cdot)A$. That is,

$$A(:)k = A(\cdot)\left[\frac{1}{k}, \frac{1}{k}\right] = \left[\frac{a_1}{k}, \frac{a_2}{k}\right]$$

Interval arithmetic can be interpreted in the following way. When we add or multiply two crisp numbers, the result is a crisp singleton. When we add or multiply two intervals, these operations are carried out on an infinite number of combinations of pairs of crisp singletons from each of the two intervals. As a result, in this sense, an interval is expected as the result. In the simplest case, when we multiply two intervals containing only positive real numbers, the solution interval is found by taking the product of the two lowest values from each of the intervals to form the solution's lower bound and by taking the product of the two highest values from each of the intervals to form the solution's upper bound. Although we can see that an infinite number of combinations of products between these two intervals exist, we need only the endpoints of the intervals to find the endpoints of the solution.

Maximum and Minimum Operations. Consider two intervals of confidence $A, B \subset R$. Their max and min operations are respectively defined as

$$A(\vee)B = [a_1, a_2](\vee)[b_1, b_2] = [a_1 \vee b_1, a_2 \vee b_2]$$

$$A(\wedge)B = [a_1, a_2](\wedge)[b_1, b_2] = [a_1 \wedge b_1, a_2 \wedge b_2]$$

It is obvious that subtraction and division are neither commutative nor associative.

Generalizing Interval Arithmetic to Fuzzy Numbers. A fuzzy number is a normal, convex fuzzy subset on the real line whose membership function is piecewise continuous. The fuzzy subset can be thought of as a crisp set with moving boundaries. In this sense, a convex membership function defining a fuzzy subset can be described by the intervals associated with different levels of α-cuts. That is, every α-cut, $\alpha \in [0, 1]$, of a fuzzy number A is a closed interval of R, and the highest value of the membership function is equal to unity. As a consequence, given two fuzzy numbers A and B in R, for a specific $\alpha_1 \in [0, 1]$, we will obtain two closed intervals $A_{\alpha_1} \equiv [a_1^{(\alpha_1)}, a_2^{(\alpha_1)}]$ from fuzzy number A and $B_{\alpha_1} \equiv [b_1^{(\alpha_1)}, b_2^{(\alpha_1)}]$ from fuzzy number B. Interval arithmetic can be applied to these two closed intervals. Therefore, fuzzy number representation can be considered an extension of the concept of intervals.

Instead of considering intervals at only one unique level, in fuzzy numbers they are considered at several levels, with each of these levels corresponding to each α-cut of the fuzzy numbers. To indicate that we are considering or dealing with arithmetic operations on all levels of the closed interval of fuzzy numbers, we use the notation $A_\alpha \equiv [\alpha_1^{(\alpha_1)}, \alpha_2^{(\alpha_1)}]$ to represent a closed interval of a fuzzy number A at an α-level.

Let us extend the discussion of interval arithmetic for closed intervals to fuzzy numbers. Let $(\Diamond)$ denote an arithmetic operation on fuzzy numbers such as addition $(+)$, subtraction $(-)$, multiplication $(\times)$, or division $(:)$. Using the extension principle, the result $A(\Diamond)B$, where A and B are two fuzzy numbers, can be obtained:

$$\mu_{A(\Diamond)B}(z) = \sup_{z = x \Diamond y} \{\mu_A(x), \mu_B(y)\}$$

where $x, y \in R$. For the min operation ($\wedge$) and the max operation ($\vee$), we have

$$\mu_{A(\Diamond)B}(z) = \sup_{z=x\Diamond y} \{\mu_A(x)\Diamond\mu_B(y)\}$$

Using the α-cuts, we have

$$(A(\Diamond)B)_\alpha = A_\alpha(\Diamond)B_\alpha \quad \text{for all } \alpha \in [0,1]$$

where $A_\alpha \equiv [a_1^{(\alpha)}, a_2^{(\alpha)}]$ and $B_\alpha \equiv [b_1^{(\alpha)}, b_2^{(\alpha)}]$. Note that for $\alpha_1, \alpha_2 \in [0,1]$ if $\alpha_1 > \alpha_2$, then $A_{\alpha_1} \subset A_{\alpha_2}$. The right-hand side of the preceding equation denotes the arithmetic operation on the α-cuts of A and B (or closed intervals of R).

Extending the addition and subtraction operations on intervals to two fuzzy numbers A and B in R, we have

$$A_\alpha(+)B_\alpha = [a_1^{(\alpha)} + b_1^{(\alpha)}, a_2^{(\alpha)} + b_2^{(\alpha)}]$$

$$A_\alpha(-)B_\alpha = [a_1^{(\alpha)} - b_2^{(\alpha)}, a_2^{(\alpha)} - b_1^{(\alpha)}]$$

Extending the multiplication and division operations on intervals to two fuzzy numbers, we have

$$A_\alpha(\cdot)B_\alpha = [a_1^{(\alpha)}b_1^{(\alpha)}, a_2^{(\alpha)}b_2^{(\alpha)}]$$

$$A_\alpha(:)B_\alpha = \left[\frac{a_1^{(\alpha)}}{b_2^{(\alpha)}}, \frac{a_2^{(\alpha)}}{b_1^{(\alpha)}}\right] \qquad b_2^{(\alpha)} > 0$$

Operations on fuzzy numbers have the following properties:

1. If A and B are fuzzy numbers on the real axis, then $A(+)B$ and $A(-)B$ are also fuzzy numbers.
2. If A and B are fuzzy numbers in R^+, then $A(\cdot)B$ and $A(:)B$ are also fuzzy numbers.
3. There are no image and inverse fuzzy numbers A and A^{-1}, respectively, such that

$$A(+)\overline{A} = 0 \quad \text{and} \quad A(\cdot)A^{-1} = 1$$

4. The following inequalities are true:

$$[A(-)B](+)B \neq A \quad \text{and} \quad A(:)B(\cdot)B \neq A$$

5. Suppose that A, B, and C are fuzzy numbers in R^+; then the following distributive property applies:

$$[A(+)B](\cdot)C = [A(\cdot)C](+)[B(\cdot)C]$$

Note, however, that

$$[A(\cdot)B](+)C \neq [A(+)C](\cdot)[B(+)C]$$

Equations with fuzzy numbers cannot be solved in the usual manner because of the lack of additive and multiplicative inverses.

Min and Max. We extend the min and max operations of intervals to fuzzy numbers. Two fuzzy numbers A and B in R are comparable if

$$a_1^{(\alpha)} \le b_1^{(\alpha)} \quad \text{and} \quad a_2^{(\alpha)} \le b_2^{(\alpha)} \quad \text{for all } \alpha \in [0, 1]$$

and we can write $A \le B$. The fuzzy minimum and fuzzy maximum of two fuzzy numbers A and B in R are denoted as $A \wedge B$ and $A \vee B$, respectively, and are defined respectively by

$$A_\alpha(\wedge)B_\alpha \equiv [\min(a_1^{(\alpha)}, b_1^{(\alpha)}), \min(a_2^{(\alpha)}, b_2^{(\alpha)})] = [a_1^{(\alpha)} \wedge b_1^{(\alpha)}, a_2^{(\alpha)} \wedge b_2^{(\alpha)}]$$

$$A_\alpha(\vee)B_\alpha \equiv [\max(a_1^{(\alpha)}, b_1^{(\alpha)}), \max(a_2^{(\alpha)}, b_2^{(\alpha)})] = [a_1^{(\alpha)} \vee b_1^{(\alpha)}, a_2^{(\alpha)} \vee b_2^{(\alpha)}]$$

2.6. FUZZY RELATIONS

A crisp binary relation indicates the presence or absence of association, or interaction, between elements of two sets. Fuzzy binary relations generalize crisp binary relations to represent various degrees of association between elements. Degrees of association can be represented by membership grades in a fuzzy binary relation much in the same manner that degrees of set membership are represented in the fuzzy set. As a result, fuzzy binary relations are indeed fuzzy sets. This section introduces basic concepts and operations on fuzzy relations, and both the relation-relation composition and the set-relation composition are discussed based on max-min and min-max compositions. The section concludes with a discussion of fuzzy relation equations that play important roles in many application areas such as fuzzy system analysis, design of fuzzy controllers, and fuzzy pattern recognition.

2.6.1. Basics of Fuzzy Relations

The idea of relations involves essentially the discovery of relations between observations and variables. The traditional crisp relation is based on the notion that everything is either related or unrelated. Thus, a crisp relation represents the presence or absence of interactions between the elements of two or more sets. By generalizing this concept to allow for various degrees of interactions between elements, we obtain fuzzy relations. As a result, a fuzzy relation is based on realizing that everything is related to some extent or is unrelated. To be specific, crisp and fuzzy relations are defined in terms of subsets.

A crisp relation among crisp sets $X_1, X_2, \ldots, X_n$ is a crisp subset on the Cartesian product $X_1 \times X_2 \times \cdots \times X_n$. This relation is denoted by $R(X_1, X_2, \ldots, X_n)$. Hence

$$R(X_1, X_2, \ldots, X_n) \subset X_1 \times X_2 \times \cdots \times X_n$$

where

$$X_1 \times X_2 \times \cdots \times X_n = \{(x_1, x_2, \ldots, x_n) | x_i \in X_i \quad \text{for all } i \in \{1, 2, \ldots, n\}\}$$

This implies that the relation R exists among $\{X_1, X_2, \ldots, X_n\}$ if the tuple $(x_1, x_2, \ldots, x_n)$ is in the set $R(X_1, X_2, \ldots, X_n)$; otherwise the relation R does not exist among $\{X_1, X_2, \ldots, X_n\}$. In the simplest case, consider two crisp sets X_1 and X_2. Then

$$R(X_1, X_2) = \{((x_1, x_2), \mu_R(x_1, x_2)) | (x_1, x_2) \in X_1 \times X_2\}$$

is a fuzzy relation on $X_1 \times X_2$. A fuzzy relation is a fuzzy set defined on the Cartesian product of crisp sets $\{X_1, X_2, \ldots, X_n\}$, where tuples $(x_1, x_2, \ldots, x_n)$ may have varying degrees of membership $\mu_R(x_1, x_2, \ldots, x_n)$ within the relation. That is,

$$R(X_1, X_2, \ldots, X_n) = \int_{X_1 \times X_2 \times \cdots \times X_N} \mu_R(x_1, x_2, \ldots, x_n)/(x_1, x_2, \ldots, x_n) \qquad x_i \in X_i$$

It is clear that a fuzzy relation is essentially a fuzzy set.

Binary (Fuzzy) Relation. A special fuzzy relation called a binary (fuzzy) relation plays an important role in fuzzy set theory. This is a fuzzy relation between two sets X and Y denoted by $R(X, Y)$. When $X \neq Y$, binary relations $R(X, Y)$ are referred to as bipartite graphs. When $X = Y$, R is a binary fuzzy relation on a single set X, and the relations are referred to as directed graphs, or digraphs, and are denoted by $R(X, X)$ or $R(X)$. There are convenient forms of representation of binary fuzzy relations $R(X, Y)$ in addition to the membership function. Let $X = \{x_1, x_2, \ldots, x_n\}$ and $Y = \{y_1, y_2, \ldots, y_n\}$. First, the fuzzy relation $R(X, Y)$ can be expressed by an $n \times m$ matrix as

$$R(X, Y) = \begin{bmatrix} \mu_R(x_1, y_1) & \mu_R(x_1, y_2) & \cdots & \mu_R(x_1, y_m) \\ \mu_R(x_2, y_1) & \mu_R(x_2, y_2) & \cdots & \mu_R(x_2, y_m) \\ \vdots & \vdots & \ddots & \vdots \\ \mu_R(x_n, y_1) & \mu_R(x_n, y_2) & \cdots & \mu_R(x_n, y_m) \end{bmatrix}$$

Such a matrix is called a fuzzy matrix.

A binary fuzzy relation can also be expressed by a graph called a fuzzy graph. Each element in X and Y (denoted by x_i and y_i, respectively) corresponds to a node in the fuzzy graph. Elements with nonzero membership grades in $R(X, Y)$ are represented in the graph by links (arcs) connecting the respective nodes. These links are labeled with the value of the membership grade $\mu_R(x_i, y_i)$. When $X \neq Y$, each of the sets X and Y can be represented by a set of nodes such that nodes corresponding to one set are clearly distinguished from nodes representing the other set. In this case, the link connecting two nodes is an undirected link and the fuzzy graph is an undirected binary graph, which is called a bipartite graph in graph theory. On the other hand, when $X = Y$, the fuzzy graph can be reduced to a simpler graph in which only one set of nodes corresponding to set X is used. In this case, directed links are used and a node is possibly connected to itself. Such graphs are called directed graphs.

The domain of a binary fuzzy relation $R(X, Y)$ is the fuzzy set dom $R(X, Y)$ with the membership function

$$\mu_{\text{dom}R}(x) = \max_{y \in Y} \mu_R(x, y) \qquad \forall x \in X$$

The range of a binary fuzzy relation $R(X, Y)$ is the fuzzy set ran $R(X, Y)$ with the membership function

$$\mu_{\text{ran}R}(y) = \max_{x \in X} \mu_R(x, y) \qquad \forall y \in Y$$

Analogous to the height of a fuzzy set, the height of a fuzzy relation R is a number $H(R)$ defined by

$$H(R) = \sup_{y \in Y} \sup_{x \in X} \mu_R(x, y)$$

If $H(R) = 1$, then R is a normal fuzzy relation; otherwise it is a subnormal fuzzy relation.

Similar to the resolution principle for fuzzy sets, every binary fuzzy relation $R(X, Y)$ can be represented in its resolution form:

$$R = \bigcup_{\alpha \in \Lambda_R} \alpha R_\alpha$$

where R_α is the level set of R. Hence, a fuzzy relation can be represented as a series of crisp relations comprising its α-cuts, each scaled by the value α.

2.6.2. Operations on Fuzzy Relations

Since the fuzzy relation from X to Y is a fuzzy set in $X \times Y$, operations on fuzzy sets can be extended to fuzzy relations. We introduce some operations that are specific to fuzzy relations. Given a fuzzy relation $R(X, Y)$, let $[R \downarrow Y]$ denote the projection of R onto Y. Then $[R \downarrow Y]$ is a fuzzy set (relation) in Y whose membership function is defined by

$$\mu_{[R \downarrow Y]}(y) = \max_x \mu_R(x, y)$$

Since the standard max operation is a t-conorm, this projection can be generalized by replacing the max operator with any t-conorm. This projection concept can be extended to an n-ary relation $R(X_1, X_2, \ldots, X_n)$.

Given a fuzzy relation $R(X)$ or a fuzzy set R on X, let $[R \uparrow Y]$ denote the cylindric extension of R into Y. Then $[R \uparrow Y]$ is a fuzzy relation in $X \times Y$ with the membership function defined by

$$\mu_{[R \uparrow Y]}(x, y) = \mu_R(x) \quad \text{for every } x \in X, y \in Y$$

The cylindric extension produces the "largest" fuzzy relation (in the sense of membership grades of elements of the extended Cartesian product) that is compatible with the given projection. The cylindric extension thus derives its most nonspecific n-dimensional relation from one of its r-dimensional projections, where $n > r$.

For a fuzzy relation $R(X, Y)$, the inverse fuzzy relation $R^{-1}(X, Y)$ is defined by

$$\mu_{R^{-1}}(y, x) \equiv \mu_R(x, y) \quad \text{for all } (x, y) \in X \times Y$$

Composition of Fuzzy Relations. Another important operation of fuzzy relations is the composition of fuzzy relations. Basically, there are two types of composition operators: max-min and min-max compositions. These compositions can be applied to both relation-relation compositions and set-relation compositions. We consider relation-relation compositions.

Let $P(X, Y)$ and $Q(Y, Z)$ be two fuzzy relations on $X \times Y$ and $Y \times Z$, respectively. The max-min composition of $P(X, Y)$ and $Q(Y, Z)$, denoted by $P(X, Y) \circ Q(Y, Z)$, is defined by

$$\mu_{P \circ Q}(x, z) \equiv \max_{y \in Y} \min[\mu_P(x, y), \mu_Q(y, z)] \quad \text{for all } x \in X, z \in Z$$

Dual to the max-min composition is the min-max composition. The min-max composition $P(X, Y)$ and $Q(Y, Z)$, denoted as $P(X, Y) \Diamond Q(Y, Z)$, is defined by

$$\mu_{P \Diamond Q}(x, z) \equiv \min_{y \in Y} \max[\mu_P(x, y), \mu_Q(y, z)] \quad \text{for all } x \in X, z \in Z$$

From the above definition, it is clear that the following relation holds:

$$\overline{P(X, Y) \Diamond Q(Y, Z)} = \overline{P(X, Y)} \circ \overline{Q(Y, Z)}$$

The max-min composition is the most commonly used. The max-min composition for fuzzy relations can be interpreted as indicating the strength of the existence of a relational chain between the elements of X and Z.

Generalized Compositions. The max-min composition can be generalized to other compositions by replacing the min operator with any *t*-norm operator. In particular, when the algebraic product is adopted, we have the max product composition denoted as $P(X, Y) \bullet Q(Y, Z)$ and defined by

$$P(X, Y) \bullet Q(Y, Z) \equiv \max_{y \in Y}[\mu_P(x, y) \cdot \mu_Q(y, z)]$$

Similarly, the min-max composition can be generalized to other compositions by replacing the max operator with any *t*-conorm operator. Since the min operator is a maximal *t*-norm and the max operator is the smallest *t*-conorm, we obtain the following inequalities characterizing the range of all possible results of compositions:

$$\max_{y \in Y} t[\mu_P(x, y), \mu_Q(y, z)] \leq \max_{y \in Y} \min[\mu_P(x, y), \mu_Q(y, z)]$$

$$\min_{y \in Y} s[\mu_P(x, y), \mu_Q(y, z)] \geq \min_{y \in Y} \max[\mu_P(x, y), \mu_Q(y, z)]$$

Similar to the properties of the composition of crisp binary relations, the max-min, max product, and min-max compositions have the following properties:

$$P \circ Q \neq Q \circ P$$

$$[P \circ Q]^{-1} = Q^{-1} \circ P^{-1}$$

$$(P \circ Q) \circ R = P \circ (Q \circ R)$$

Relational Joint. A similar operator on two binary fuzzy relations yields triples instead of pairs. Let $P(X, Y)$ and $Q(Y, Z)$ be two binary fuzzy relations. The relational joint of P and Q, denoted as $P * Q$ is defined by

$$\mu_{P*Q}(x, y, z) = \min[\mu_P(x, y), \mu_Q(y, z)] \quad \text{for each } x \in X,\, z \in Z$$

Like the max-min compositions, the min operator can be generalized by replacing it by any t-norm operator.

2.6.3. Types of Binary Fuzzy Relations

In this section a number of important types of binary fuzzy relations are identified.

Reflexive and Related Relations. A fuzzy relation $R(X, X)$ is reflexive if and only if

$$\mu_R(x, x) = 1 \quad \forall x \in X$$

If this is not the case, for some $x \in X$, then R is irreflexive. If the requirement is not met for all $x \in X$, then the relation is referred to as antireflexive.

A fuzzy relation $R(X, X)$ is ε reflexive if and only if

$$\mu_R(x, x) \geq \varepsilon \quad \forall x \in X$$

A fuzzy relation $R(X, X)$ is weakly reflexive if and only if

$$\mu_R(x, y) \leq \mu_R(x, x)$$
$$\mu_R(y, x) \leq \mu_R(x, x) \quad \forall x,\, y \in X$$

The following properties hold for max-min compositions:

- If P is reflexive and Q is an arbitrary binary fuzzy relation, then

$$P \circ Q \supseteq Q \quad Q \circ P \supseteq Q$$

- If P is reflexive, then

$$P \subseteq P \circ P$$

- If P and Q are reflexive relations, so is $P \circ Q$.

Symmetric and Related Relations. A fuzzy relation $R(X, X)$ is symmetric if and only if

$$\mu_R(x, y) = \mu_R(y, x)$$

Alternatively,

$$R(X, Y) = R(Y, X)$$

If this is not the case, for some $x \in X$, then R is asymmetric. If the requirement is not met for all members of the support of the relation, then the relation is referred to as antisymmetric. If it is not satisfied for all $x, y \in X$, then $R(X, X)$ is called strictly antisymmetric.

A fuzzy relation $R(X, X)$ is perfectly antisymmetric if and only if, for $x \neq y$,

$$\mu_R(x, y) > 0$$

Then

$$\mu_R(y, x) = 0 \qquad \forall x, y \in X$$

The following properties hold for max-min compositions:

- If P and Q are symmetric, then $P \circ Q$ is symmetric if

$$P \circ Q = Q \circ P$$

- If P is symmetric, so is each power of P.

Transitive Relations. A fuzzy relation $R(X, X)$ is max-min transitive if and only if

$$R \circ R \subset R$$

It is easy to see that $\mu_{R \circ R}(x, y) \leq \mu_R(x, y)$.

The following properties hold for max-min compositions:

- If P is symmetric and transitive, then

$$\mu_R(x, y) \leq \mu_R(x, x)$$

- If P and Q are transitive and $P \circ Q = Q \circ P$, then $P \circ Q$ is transitive.
- If P is reflexive and transitive, then $P \circ P = P$.

When a fuzzy relation is nontransitive or antitransitive, we may want to find a transitive fuzzy relation that contains $R(X, X)$ and is minimum in the sense of membership grade. This is the transitive closure of a fuzzy relation, denoted by $R_T(X, X)$; it is a fuzzy relation that is transitive and contains $R(X, X)$, and its elements have the smallest possible membership grades. When X has n elements, the transitive closure $R_T(X, X)$ can be obtained by

$$R_T(X, X) = R \cup R^2 \cup R^3 \cup \cdots \cup R^n$$

where

$$R^k = R^{k-1} \circ R \qquad k \geq 2$$

It is noted that we do not need to calculate all the $R^k, k = 2, 3, \ldots, n$, in finding the transitive closure of R. We can unify only the terms $R, R^2, \ldots, R^k$ when $R^k = R^{k-1} \circ R$.

Binary Crisp Equivalence Relations. A binary crisp relation $C(X, X)$ is called an equivalence relation if it satisfies the following conditions:

Reflexivity: $(x, x) \in C(X, X)$ for all $x \in X$.

Symmetry: If $(x, y) \in C(X, X)$, then $(y, x) \in C(X, X)$, $x, y \in X$.

Transivity: If $(x, y) \in C(X, X)$ and $(y, z) \in C(X, X)$, then $(x, z) \in C(X, X)$, $x, y, z \in X$.

Equivalence Class. For each element x in X, one can define a crisp set A_x that contains all the elements of X that are related to x by the equivalence relation. That is,

$$A_x = \{ y | (x, y) \in C(X, X) \}$$

Here, A is a subset of X. The element x is itself contained in A_x because of the reflexivity. Because $C(X, X)$ is transitive and symmetric, each member of A_x is related to all the other members of A_x. This set A_x is called an equivalence class of $C(X, X)$ with respect to x.

Partition. The family of all such equivalence classes defined by the relation forms a partition on X and is denoted by $\pi(C)$.

Similarity Relations. Similarity relations generalize equivalence relations in binary crisp relations to binary fuzzy relations. A binary fuzzy relation that is reflexive, symmetric, and transitive is known as a similarity relation. The concept of a similarity relation is extremely powerful in grouping elements into crisp sets whose members are "similar" to each other to some specified degree. Note that when this degree of similarity is 1, the grouping is then an equivalence class. This concept is clear when the similarity relation is represented in its resolution form, that is,

$$R = \bigcup_{\lambda \in \Lambda_R} \lambda R_\lambda$$

Then each λ-cut R_z is an equivalence relation representing the presence of similarity between the elements to the degree λ. For this equivalence relation R_λ there is a partition on X, $\pi(R_\lambda)$. Hence, each similarity relation is associated with the set

$$\pi(R) = \{ \pi(R_\lambda) | \lambda \in \Lambda_R \}$$

of partitions.

Similar to the equivalence class in an equivalence relation, we can also define a similarity class in a similarity relation. Let R be a similarity relation for $X = \{ x_1, x_2, \ldots, x_n \}$. Then the similarity class for X_i, A_{x_i}, is a fuzzy set in X with a membership function

$$\mu_{A_{x_i}}(x_j) = \mu_R(x_i, x_j)$$

Hence, the similarity class for an element x represents the degree to which all the other members of X are similar to x.

Resemblance Relations. A binary fuzzy relation that is reflexive and symmetric is called a resemblance relation. It is clear that similarity relations are a special case of resemblance relations. For example, the binary relation "look-alike" is reflexive and symmetric but not transitive, and therefore it is not a similarity relation but a resemblance relation.

Let us express a resemblance relation $R(X, X)$ in its resolution form $R = \bigcup_{\lambda \in \Lambda_R} \lambda R_\lambda$. Similar to the equivalence class in equivalence relations, a resemblance class corresponding to the λ-cut R_λ of the resemblance relation R can be defined as A_λ, a maximal subset of X such that $\mu_R(x, y) \geq \lambda$ for all $x, y \in A_\lambda$. The family consisting of all the resemblance classes A_λ is called a λ-cover of X with respect to R_λ. Hence, for each λ-cut there is an a λ-cover of X that corresponds to the partition $\pi(C)$ of an equivalence relation C. However, because of the lack of transitivity, the resemblance classes in an a λ-cover might not be disjoint. Furthermore, corresponding to the partition tree in similarity relations, there is a complete λ-cover tree consisting of all λ-covers, $\lambda \in \Lambda_R$.

Since the only difference between resemblance relations and similarity relations is transitivity, we can obtain a similarity relation from a resemblance relation by applying the transitive closure R_T to the resemblance relation R.

Fuzzy Partial Ordering. A binary fuzzy relation R on a set X is a fuzzy partial ordering if and only if it is reflexive, antisymmetric, and transitive. A relation like "slightly less than or equal to" is a fuzzy partial ordering. Every partial ordering R can be represented by a directed graph called a Hasse diagram. This diagram can be derived from the simple fuzzy graph representation of R by omitting the directed link connecting a node to itself and the link $x_i \rightarrow x_j$ if there exist other nodes $x_{k_1}, x_{k_2}, \ldots, x_{k_n}$, such that $x_i \rightarrow x_{k_1} \rightarrow x_{k_i} \rightarrow \cdots \rightarrow x_{k_n} \rightarrow x_j$, where n can be equal to 1. Hence in a Hasse diagram, if there is a link from x_i to x_j, then x_j is called an immediate successor of x_i and x_i is an immediate predecessor of x_j.

For a given partial ordering $R(X, X)$ two important fuzzy sets are associated with each element x in X. They are the dominating class of x and the dominated class of x. Given a fuzzy partial ordering $R(X, X)$ on X, the dominating class for an element x of X is denoted by $R_{\geq[x]}$ and is a fuzzy set in X defined by

$$R_{\geq[x]}(y) = \mu_R(x, y) \qquad y \in Y$$

Similarly, the dominated class for element x of X is denoted by $R_{\leq[x]}$ and is a fuzzy set in X defined by

$$R_{\leq[x]}(y) = \mu_R(y, x) \qquad y \in Y$$

Element x is called a maximal or undominated element if

$$\mu_R(x, y) = 0 \qquad x \neq y, y \in X$$

Element x is called a minimal or undominating element if

$$\mu_R(y, x) = 0 \qquad x \neq y, y \in X$$

Let R be a fuzzy partial ordering on X and let A be a crisp subset of X. The fuzzy upper bound of A is the fuzzy set $U(A)$ defined by

$$U(A) = \bigcap_{x \in A} R_{\geq[x]}$$

where $\bigcap$ is a fuzzy intersection (t-norm) operation. The least upper bound of the set A, if it exists, is the unique element in $U(A)$ such that

$$\mu_{U(A)}(x) > 0 \qquad \text{and} \qquad \mu_R(x, y) > 0$$

for all elements y in the support of $U(A)$. That is, the least upper bound of A is the smallest element of $U(A)$. In the same way, the fuzzy lower bound of A is the fuzzy set $L(A)$ defined by

$$L(A) = \bigcap_{x \in A} R_{\leq [x]}$$

The greatest lower bound of the set A, if it exists, is the unique element in $L(A)$ such that

$$\mu_{L(A)}(x) > 0 \quad \text{and} \quad \mu_R(y, x) > 0$$

for all elements y in the support of $L(A)$. In other words, the greatest lower bound of A is the largest element of $L(A)$.

2.6.4. Fuzzy Relation Equations

Fuzzy relation equations play an important role in areas such as fuzzy system analysis, design of fuzzy controllers, decision-making processes, and fuzzy pattern recognition. The concept is associated with the concept of composition of binary fuzzy relations, which includes both the set-relation composition and the relation-relation composition. We deal with the max-min composition because it has been studied extensively and has been utilized in numerous applications. Let A be a fuzzy set in X and $R(X, Y)$ be a binary fuzzy relation in $X \times Y$. The set-relation composition of A and R, $A \circ R$, results in a fuzzy set in Y. Let us denote the resulting fuzzy set as B. Then we have the so-called fuzzy relation equation

$$A \circ R = B$$

whose membership function is

$$\mu_B(y) = \mu_{A \circ R}(y) \equiv \max_{x \in X} \min[\mu_A(x), \mu_R(x, y)]$$

If we view R as a fuzzy system, A as a fuzzy input, and B as a fuzzy output, then we can consider the fuzzy relation equation as describing the characteristics of a fuzzy system via its fuzzy input-output relation. Hence, given a fuzzy input A to a fuzzy system R, a fuzzy output B can be decided by the fuzzy relation equation. The basic problem concerning the fuzzy relation equation is that given any two items of A, B, and R, one can find the third. When A and R are given, the fuzzy set B can be easily determined.

Inverse Fuzzy Relation Equations. Next, we study more difficult problems known as inverse relation equations. We have two types:

- *Problem 1*: Given A and B, determine R such that $A \circ R = B$.
- *Problem 2*: Given R and B, determine A such that $A \circ R = B$.

Unfortunately, a fuzzy relation equation is not linear, and the inverse cannot provide a unique solution, in general, or any solution in some situations. Since the solutions for the preceding problems may not exist, we first need to check the solvability of these equations or the existence of their solutions.

EXISTENCE OF SOLUTION TO PROBLEM 1. Problem 1 has solution(s) if and only if the height of the fuzzy set A is greater than or equal to the height of the fuzzy set B:

$$\max_{x \in X} \mu_A(x) \geq \mu_B(y) \quad \text{for all } y \in Y$$

THE α-OPERATION AND α-COMPOSITION. In order to solve problem 1 (and problem 2 later), we need to introduce the α-operation. For any $a, b \in [0, 1]$, the α-operator is defined as

$$a \alpha b = \begin{cases} 1 & \text{if } a \leq b \\ b & \text{if } a > b \end{cases}$$

For fuzzy sets A and B in X and Y; respectively, the α-composition of A and B forms a fuzzy relation

$$A \xleftrightarrow{\alpha} B$$

defined by

$$\mu_{A \xleftrightarrow{\alpha} B} \equiv \mu_A(x) \alpha \mu_B(y) = \begin{cases} 1 & \mu_A(x) \leq \mu_B(y) \\ \mu_B(y) & \mu_A(x) > \mu_B(y) \end{cases}$$

Furthermore, the α-composition of a fuzzy relation R and a fuzzy set B is denoted by

$$R \xleftrightarrow{\alpha} B$$

and is defined by

$$\mu_{R \xleftrightarrow{\alpha} B} \equiv \min_{y \in Y} [\mu_R(x, y) \alpha \mu_B(y)]$$

USEFUL PROPERTIES. With the above α-operator and α-composition, the following properties will be useful for determining the solutions of problem 1. Let R be a fuzzy relation on $X \times Y$. For any fuzzy sets A and B in X and Y, respectively, we have

$$R \subseteq A \xleftrightarrow{\alpha} (A \circ R) \qquad A \circ (A \xleftrightarrow{\alpha} B) \subseteq B$$

CHARACTERIZING THE SOLUTION TO PROBLEM 1. If the solution to problem 1 exists, then the largest R (in the sense of set-theoretic inclusion) that satisfies the fuzzy relation equation $A \circ R = B$ is

$$\overset{*}{R} = A \xleftrightarrow{\alpha} B$$

whose membership function is given by

$$\mu_{A \xleftrightarrow{\alpha} B} \equiv \mu_A(x) \alpha \mu_B(y) = \begin{cases} 1 & \mu_A(x) \leq \mu_B(y) \\ \mu_B(y) & \mu_A(x) > \mu_B(y) \end{cases}$$

Next, consider problem 2; that is, given R and B, determine A such that $A \circ R = B$.

NONEXISTENCE OF SOLUTION TO PROBLEM 2. Problem 2 has no solution if the following inequality holds:

$$\max_{x \in X} \mu_R(x, y) < \mu_B(y) \quad \text{for some } y \in Y$$

This allows us, in certain cases, to determine quickly that problem 2 has no solution. However, the converse is only a necessary and not a sufficient condition for the existence of a solution to problem 2.

MORE USEFUL RELATIONS. Let R be a fuzzy relation on X and Y. For any fuzzy sets A and B in X and Y, respectively, we have:

$$(R \xleftrightarrow{\alpha} B) \circ R \subseteq B \qquad A \subseteq R \xleftrightarrow{\alpha} B(A \circ R)$$

CHARACTERIZING THE SOLUTION TO PROBLEM 2. If the solution to problem 2 exists, then the largest fuzzy set A that satisfies the fuzzy relation equation $A \circ R = B$ is

$$\overset{*}{A} = R \xleftrightarrow{\alpha} B$$

whose membership function is given by

$$\mu_{R \xleftrightarrow{\alpha} B} \equiv \min_{y \in Y} [\mu_R(x, y) \alpha \mu_B(y)]$$

References

[1] L. A. Zadeh, "Fuzzy Sets," *Information and Control*, Vol. 8, 1965, pp. 338–353.
[2] R. Bellman and M. Giertz, "On the Analytic Formalism of the Theory of Fuzzy Sets," *Information Science,* Vol. 5, 1973, pp. 149–156.
[3] B. Werners, "Aggregation Models in Mathematical Programming," in *Mathematical Models for Decision Support*, G. Mitra (Ed.), Birkhäuser, Berlin, 1988, pp. 295–319.
[4] H-J. Zimmermann and P. Zysno, "Latent Connectives in Human Decision Making," *Fuzzy Sets and Systems*, Vol. 4, No. 1, 1980, pp. 37–51.
[5] L. A. Zadeh, "Fuzzy Sets as a Basis for a Theory of Possibility," *Fuzzy Sets and Systems*, Vol. 1, No. 1, 1978, pp. 3–28.
[6] D. Dubois and H. Prade, "Fuzzy Real Algebra, Some Results," *Fuzzy Sets and Systems*, Vol. 2, 1979, pp. 327–348.

K. Tomsovic
B. Baer
*School of Electrical Engineering
and Computer Science
Washington State University
Pullman, WA 99164-2752*

Chapter 3

Fuzzy Information Approaches to Equipment Condition Monitoring and Diagnosis

3.1. INTRODUCTION

Equipment condition monitoring plays a crucial role in the overall integrity of the power system. As a result, utilities invest significant time and finances into equipment monitoring and maintenance in order to anticipate failures or accelerated aging in power equipment. Such monitoring includes regular insulation condition tests for switching devices, reactors, power transformers, generator windings, and so on. In general, many of the indicators of equipment condition are imprecise and/or unreliable. Engineers must have considerable experience with a particular test before that test becomes useful. Several utilities have developed expert systems to codify this experience and improve knowledge of the breakdown process.

This work emphasizes the uncertainty modeling of diagnostic problems. Fuzzy information methods are employed to represent quantitatively the diagnostic capability of a system. Further, several methods are discussed for extracting information from test data and evaluating system performance. This research proposes that systematic representation of uncertainty can lead to significant improvements in diagnostic capability.

Power system security depends on properly functioning and maintained equipment. An understanding of the failure mechanisms and expected lifetimes of equipment is needed for both operations and planning. In recent years, utilities have begun to focus more attention on the costs and importance of diagnostic and maintenance practices as evidenced, for example, by the interest in reliability-centered maintenance (RCM). There have also been several attempts at developing software tools. Diagnosis and maintenance tend to be experience-based skills so that these efforts have focused on expert system developments [1, 2]. There have also been efforts aimed at model-based reasoning approaches [3]. While a number of these systems have been quite successful and are in regular use, a further understanding of knowledge representation and uncertainty is needed. In this study, a theoretical framework is explored that complements the expert system approach with analytical techniques based on fuzzy mathematics. The objective of this framework is both to simplify the software design and to improve performance of a diagnostic system operating under the uncertainty inherent in realistic data.

Imprecision is inherent to any complex diagnostic problem. That is, rarely is there a single observation or measurement that definitively indicates impending failure. Experience with a piece of equipment or diagnostic technique is necessary to overcome this imprecision and perform effective diagnosis. In the power system, this uncertainty is concerned with variations in aging mechanisms, incomplete understanding of different stresses (electrical, chemical, and thermal), incomplete data on the stresses, and limits in measurements of incipient failures. Thus, many of the diagnostic expert systems developed within power systems have had to model uncertainty in the reasoning process. Modeling uncertainty in expert systems focuses on representations that are meaningful to experts, allows propagation of uncertainties along extended chains of reasoning, and eases implementation of large knowledge bases.

Several techniques for representing uncertainty in expert systems have been proposed in the artificial intelligence (AI) literature, including Bayesian analysis and certainty measures [4]. For the most part, these techniques are ad hoc methods that emphasize simplifying coding of the uncertainty. This work begins from a fundamental model of uncertainty based on fuzzy mathematics and leads to a rule-based representation for expert system development. Techniques are developed that show the most effective method for extracting information from an observation and suggest actions to take that will lead to the most coherent conclusion. Fuzzy mathematics applications within power systems have been proposed in several areas [5]. In particular, there have been several applications to transformer diagnosis of fuzzy set methods. Elsewhere [6, 7], fuzzy logic is used to implement dissolved-gas analysis methods. An acoustic technique for finding partial discharges applied fuzzy logic to representation of uncertainties [8]. The techniques developed in this chapter have also been applied to transformer diagnostics and condition monitoring [9–11], and thus, examples in this chapter will focus on this problem.

This chapter is organized as follows. The diagnostic framework is discussed and requirements for a model of uncertainty are presented. An introduction to fuzzy mathematics with emphasis on the lesser known fuzzy information aspects is then given. Several detailed examples show the usefulness of the proposed technique. Implementation and representation issues are discussed. Learning methods and performance improvement of a diagnostic expert system are explored. A method for performance evaluation and improvement is proposed within the developed fuzzy set framework. Some directions for further research are discussed.

3.2. EXPERT SYSTEMS AND EQUIPMENT DIAGNOSTICS

The condition of power system equipment is fundamental to the secure operation of the power system. This hardware includes cables, generators, insulators, protection devices, switch gears, and transformers. The life of electrical equipment is primarily determined by the insulation [12]. As a result, utilities invest significant time and finances into assessing insulation condition and anticipating failures or accelerated aging. Assessment of insulation condition varies from the informal (e.g., visual inspection of transmission line insulators) to the sophisticated (e.g., acoustic measurements for detection of partial discharges in power transformers). Despite advances in understanding aging mechanisms, it can be said that there are many good indicators of aging or impending failure but few definitive tests. The focus of this section is to identify the features of an expert system that are needed for effective representation of equipment monitoring and diagnostics.

To begin, consider the widely used chemical test for power transformers of insulating oil called dissolved-gas analysis (DGA). (Note, the examples in this chapter will refer to transformer diagnostics in order to clarify the developed approach. This is merely convenience as the authors are most familiar with this type of analysis; the developed framework is general.) In DGA, relative concentrations of several hydrocarbons and other gases are measured. High concentrations of certain gases are indicative of fault conditions. The relative gas concentrations give an indication of the fault type [13]. Dissolved-gas analysis is typical of equipment diagnostic methods in several ways. First, it requires a broad assessment of several external influences. Specifically, complete analysis requires a history of the transformer loading, knowledge of fault currents experienced by the transformer, a trend analysis of previous DGA tests on this transformer, and an understanding of similarly manufactured transformers. Much of this information is approximate and some may not be available at all. Further, DGA results in several pieces of information that may or may not be consistent. For example, a high concentration of one gas may be ignored if other gas concentrations do not indicate a fault developing. Finally, this diagnostic test may be supplemented with other tests, for example, an analysis of insulation paper based on furfural levels.

It is important to note that most diagnostic tests have a particular focus. For example, acoustic tests are directed at detecting partial discharges while DGA is broader and can find indications of either thermal or electrical breakdowns. The key point in the above is that tests provide information in different forms, operate on different subsets of the universe, and have inherent uncertainties.

3.2.1. Representing Diagnostic Information

In this work, a standard rule base of if-then relations is implemented. Each relates the results of a specific diagnostic test and the conclusions of an expert based on that test. It is desired that for these relations, the following holds:

1. Any single missing piece of data or error in any single relation will not invalidate the analysis (although it could easily reduce the accuracy of the final result).

2. Every relation and uncertainty value can be found individually.
3. Uncertainty values can be propagated locally. That is, the uncertainty values can be updated based on sequential evaluation of the rules. This is not intended to restrict global algorithms for distributing evidence but to constrain the complexity of the calculations.

The first assumption can be viewed similar to error tolerance conditions for detecting bad measurements in state estimation. The assumption is of particular importance here because of the possibility of large errors in the measurements and/or relations. The last two assumptions ensure that inputting knowledge to the system can be done incrementally. Development and testing of large knowledge bases require that there are not strong interdependencies among rules so incremental improvements can be made. Notice, this disallows the use of conditional probabilities since each probability cannot be determined independently.

Each rule represents one or more tests and relates these tests to equipment condition:

IF measurement is *A THEN* equipment condition is *B*

It will be useful to specify the importance or necessity of this measurement as well so that the rule structure becomes:

IF measurement is *A THEN* equipment condition is *B*
AND measurement is necessary to degree *C* in order to reach conclusion

For example, in DGA,

IF the methane (CH_4) concentration in the transformer oil is high
THEN this is indicative of either low-energy discharge or local overheating
AND the presence of methane is somewhat important to conclude this

Condition *B* may also be some intermediary value that is propagated to other rules. Thus, the following information is represented in each rule:

1. Relevant measurement (e.g., gas concentration, degree of polymerization of insulation paper, oil moisture content) and indicated equipment condition
2. Acceptable range for the measured quantity, which includes any uncertainty associated with this measurement or acceptable range
3. Importance of the measurement in determining the condition of the equipment

Diagnostic knowledge may be represented by a large number of these rules so that the overall uncertainty must be calculated to reach a conclusion. As in any rule-based system, the rules are chained together by what is called the inference engine. In this work, the important consideration of the inference engine is the methods by which the uncertainties are propagated among rules in the reasoning process.

Finally, it should be noted that many tests require prefiltering or other numerical computations. These have not been overlooked but are represented here as part of the measurement. The following section presents the developed technique within the above desired constraints.

3.3. THE FUZZY INFORMATION APPROACH

As the basics of fuzzy sets are widely available, only a fairly brief review is given in the following. Less well-known are the methodologies associated with fuzzy measures and fuzzy information theory. These areas will be developed more fully to highlight the application of these techniques to diagnostic problems. Several examples are given in the following section to clarify the application of these techniques and design issues. More extensive treatment of fuzzy mathematics can be found elsewhere [14, 15].

3.3.1. Fundamentals of Fuzzy Logic

Each element of a fuzzy set is an ordered pair containing a set element and the degree of membership in the fuzzy set. A higher membership value can be said to indicate that an element more closely matches the characteristic feature of the set. For fuzzy set A,

$$A = \{(x, \mu_A(x)) | x \in X\} \tag{3.1}$$

where X is the universe, $\mu_A(x)$ represents the membership function, and $\mu: X \to [0, 1]$. For example, one could define a membership function for the set of numbers much less than 100 as follows:

$$\mu_{\ll 100}(x) = \frac{1}{1 + x^2/100}$$

Typically, the following definitions of fundamental logical operations (intersection, union, and complement) on sets are used:

$$\mu_{A \cap B}(x) = \min(\mu_A(x), \mu_B(x)) \tag{3.2}$$

$$\mu_{A \cup B}(x) = \max(\mu_A(x), \mu_B(x)) \tag{3.3}$$

$$\mu_{\bar{A}}(x) = 1 - \mu_A(x) \tag{3.4}$$

The above operators satisfy certain desired properties. For example, if set containment is defined as $A \subseteq B$ if $\forall x \in X$, $\mu_A(x) \leq \mu_B(x)$, then the following always holds: $A \subseteq A \cup B$ and $A \cap B \subseteq A$. Depending on the application, other operators from within the triangular norm and conorm classes may be more appropriate than the above minimum and maximum functions [16]. For example, the framework of Dombi [17] is used in this work. Specifically,

$$\mu_{A \cap B}(x) = \frac{1}{1 + [(1/\mu_A(x) - 1)^\lambda + (1/\mu_B(x) - 1)^\lambda]^{1/\lambda}} \tag{3.5}$$

with $\lambda \geq 1$. Increasing the parameter λ will increase the emphasis on the smaller membership value. One can define the union operation by allowing $\lambda \leq -1$. Notice as $|\lambda| \to \infty$, (3.5) approaches either (3.2) or (3.3), and in practice, values of λ larger than 2 or 3 render this form essentially equivalent to the minimum and maximum.

Finally where it is useful to generate a crisp (nonfuzzy) set from a fuzzy set, one can define an α-cut as

$$A_\alpha = \{x | \mu_A(x) \geq \alpha\} \tag{3.6}$$

so that A_α is a crisp set containing all elements of the fuzzy set A that have at least a membership degree of α.

3.3.2. Fundamentals of Fuzzy Measures

In assessment of a system state, uncertainty will arise either from the measurement or from incomplete knowledge of the system. This type of uncertainty is most often modeled as random noise and managed with probability methods. Fuzzy measures are introduced here as a generalization of probability measures such that the additivity restriction is removed. Specifically, a fuzzy measure G is defined over the power set of the universe X [designated as $\mathscr{P}(X)$]:

$$G: \mathscr{P}(X) \to [0, 1]$$

where:

- $G(\varnothing) = 0$ and $G(X) = 1$. (Boundary conditions.)
- $\forall A, B \in \mathscr{P}(X)$, if $A \subseteq B$, then $G(A) \leq G(B)$. (Monotonicity.)
- For any sequence $A_1 \subseteq A_2 \subseteq \cdots \subseteq A_n$, $\lim_{i \to \infty} G(A_i) = G(\lim_{i \to \infty} A_i)$. (Continuity.)

Here, $\varnothing$ is the empty set.

There are three particularly interesting cases with this definition of a fuzzy measure: probability, belief (a lower bound of the probability), and plausibility (an upper bound of the probability). If the following additivity condition is satisfied, then G is a probability measure, represented by P:

$$P\left(\bigcup_{i=1}^{n} A_i\right) = \sum_{i=1}^{n} P(A_i) - \sum_{i=1}^{n}\sum_{j=i+1}^{n} P(A_i \cap A_j) + \cdots + (-1)^n P(A_1 \cap A_2 \cap \cdots \cap A_n) \tag{3.7}$$

If this equality is replaced by (3.8) below, then G is called a belief measure and represented by Bel:

$$\text{Bel}\left(\bigcup_{i=1}^{n} A_i\right) \geq \sum_{i=1}^{n} \text{Bel}(A_i) - \sum_{i=1}^{n}\sum_{j=i+1}^{n} \text{Bel}(A_i \cap A_j) \\ + \cdots + (-1)^n \text{Bel}(A_1 \cap A_2 \cap \cdots \cap A_n) \tag{3.8}$$

Finally a plausibility measure results if the following holds instead of (3.7) or (3.8):

$$\text{Pl}\left(\bigcup_{i=1}^{n} A_i\right) \leq \sum_{i=1}^{n} \text{Pl}(A_i) - \sum_{i=1}^{n} \sum_{j=i+1}^{n} \text{Pl}(A_i \cup A_j) + \cdots + (-1)^n \text{Pl}(A_1 \cup A_2 \cup \cdots \cup A_n)$$

$$(3.9)$$

It can be shown that this leads to the following relation for plausibility and belief measures:

$$\text{Bel}(A) + \text{Pl}(\bar{A}) = 1 \qquad (3.10)$$

Finally it is useful to summarize these expressions in the following way, $\forall A \in X$:

- $\text{Bel}(A) + \text{Bel}(\bar{A}) \leq 1$.
- $\text{Pl}(A) + \text{Pl}(\bar{A}) \geq 1$.
- $P(A) + P(\bar{A}) = 1$.
- $\text{Pl}(A) \geq P(A) \geq \text{Bel}(A)$.

These expressions and consideration of the forms in (3.9) and (3.10) lead to the interpretation of belief representing supportive evidence and plausibility representing nonsupportive evidence. This is best illustrated by considering the state descriptions for each of the measures when nothing is known about the system (the state of *total ignorance*). A plausibility measure would be one for all nonempty sets; and belief would be zero for all sets excepting the universe X. Conversely, it would be typical in probability to assume a uniform distribution so that all states were equally likely. Thus, an important difference in the use of fuzzy measures is in terms of representing what is unknown. The use of the above structure in this work focuses on incrementally finding a solution to a problem by initially assuming all equipment states are possible (plausibility measure of 1) but no specific state can be assumed (belief measure of 0). That is, one begins from the state of ignorance. As evidence is gathered during diagnosis, supportive evidence will increase the belief values of certain events and nonsupportive evidence will decrease the plausibility of other events. Mathematically, of course, supportive evidence is equivalent to nonsupportive evidence on the complement and the distinction does not need to be made. Still, this description provides a natural way of representing tests that are geared either toward supporting or refuting specific hypotheses.

3.3.3. Bodies of Evidence and Information Measures

The fuzzy sets and measures framework defined above provides the fundamentals for representing uncertainty. To reach decisions and use the representative powers of fuzzy sets require further manipulative techniques to extract information and apply knowledge to the data. In this section, a generalized framework called a body of evidence is defined to provide a common representation for information. Evidence will be gathered and represented in terms of fuzzy relations (sets) and fuzzy measures and then translated to the body-of-evidence framework. Techniques will be employed to extract the most reliable information from the evidence. Let the body of evidence be represented as

$$m: \mathscr{P}(X) \rightarrow [0, 1]$$

with

- $m(\varnothing) = 0$. (Boundary condition.)
- $\sum_{A \in \mathscr{P}(X)} m(A) = 1$. (Additivity.)

It is important to emphasize that $m(A)$ is not a measure but rather can be used to generate a measure or, conversely, to be generated from a measure. A specific basic assignment over $\mathscr{P}(X)$ is often referred to as a body of evidence. Based on the above axioms, it can be shown [14] that

$$\text{Bel}(A) = \sum_{B \subseteq A} m(B) \tag{3.11}$$

$$\text{Pl}(A) = \sum_{B \cap A \neq \varnothing} m(B) \tag{3.12}$$

and conversely that

$$m(A) = \sum_{B \subseteq A} (-1)^{|A-B|} \text{Bel}(A) \tag{3.13}$$

where $|\cdot|$ is set cardinality. These equations show us another view of belief and plausibility. Belief measures the evidence that can completely (from the set containment) explain a hypothesis. Plausibility measures the evidence that can at least partially (from the nonempty intersection) explain a hypothesis. In many cases, one wants to combine information from independent sources. Evidence can be "weighted" by the degree of certainty among bodies of evidence. Such an approach leads to the Dempster rule of combination, where, given two independent bodies of evidence m_1 and m_2 and a set $A \neq \varnothing$,

$$m_{1,2}(A) = \frac{\sum_{B \cap C = A} m_1(B) \cdot m_2(C)}{1 - K} \tag{3.14}$$

where

$$K = \sum_{B \cap C = \varnothing} m_1(B) \cdot m_2(C) \tag{3.15}$$

The factor K ensures that the resulting body of evidence is normalized in case there exists evidence that is unreconcilable (evidence on mutually exclusive sets).

It is reasonable to assume that certain bodies of evidences provide greater clarity of information than others. In order to assess the quality of information in a body of evidence several entropylike calculations are used. These assessments will be used to characterize the degree of conflicting evidence as well as the specificity of evidence. For example, they can be used to determine the amount of information gain obtained from an observation. Define the following:

- Confusion:

$$C(m) = - \sum_{m(A) \neq 0} m(A) \log \mathrm{Bel}(A) \qquad (3.16)$$

- Dissonance:

$$E(m) = - \sum_{m(A) \neq 0} m(A) \log \mathrm{Pl}(A) \qquad (3.17)$$

- Vagueness:

$$V(m) = \sum_{m(A) \neq 0} m(A) \log |A| \qquad (3.18)$$

These measures provide an assessment of the quality of information in the basic assignment. Furthermore, they are useful in informing the user of the quality of the conclusion obtained by analysis.

3.4. AN EXTENDED EXAMPLE

It is often difficult to understand the relationship between the fuzzy mathematics and the implementation of a useful expert system. In this section, several points are highlighted through the use of an extended example. This example represents a simplified version of the transformer diagnostic and monitoring system implemented elsewhere [9–11]. Some caution is in order in that the examples have been simplified to the degree they no longer fully represent the actual physical situation. Further, the emphasis in this chapter is on the manipulation of fuzzy membership values without providing justification for the values. Justification of the fuzzy values is taken up more carefully in the next section. The reader interested in further discussion on the transformer diagnostic technique should refer to the literature [2, 3, 6–11].

EXAMPLE 3.1: Computing Belief and Plausibility
Let the possible conclusions (set elements) that the expert system can reach for the transformer condition be as follows:

X_1: The transformer has an electrical fault.
X_2: The transformer has a thermal fault.
X_3: The transformer paper insulation has significantly aged.
X_4: The transformer is operating normally.
$X = \{X_1, X_2, X_3, X_4\}$

Assume an engineer wishes to determine the problem with the transformer by entering data from several tests as evidence. Evidence values are given for three tests in Table 3.1.
Using the data in Table 3.1, the corresponding belief and plausibility values can be computed from (3.11) and (3.12), as follows:

TABLE 3.1   BASIC ASSIGNMENT FOR EXAMPLE 3.1

Test 1: DGA	Test 2: Frequency Response	Test 3: Visual Inspection
$m_1(\{X_1, X_2\}) = 0.4$	$m_2(\{X_1\}) = 0.3$	$m_3(\{X_1, X_2, X_3\}) = 0.4$
$m_1(\{X_2, X_3\}) = 0.3$	$m_2(\{X_2, X_3\}) = 0.2$	$m_3(X_1, X_3) = 0.2$
$m_1(\{X_4\}) = 0.1$	$m_2(\{X_1, X_2, X_4\}) = 0.2$	$m_3(\{X_4\}) = 0.2$
$m_1(X) = 0.2$	$m_2(\{X_3\}) = 0.15$	$m_3(X) = 0.2$
	$m_2(X) = 0.15$	

- Test 1: belief and plausibility values:

$$\mathrm{Bel}(\{X_1, X_2\}) = m_1(\{X_1, X_2\}) = 0.4$$

$$\mathrm{Bel}(\{X_2, X_3\}) = m_1(\{X_2, X_3\}) = 0.3$$

$$\mathrm{Bel}(\{X_4\}) = m_1(\{X_4\}) = 0.1$$

$$\mathrm{Pl}(\{X_1, X_2\}) = m_1(\{X_1, X_2\}) + m_1(\{X_2, X_3\}) + m_1(X) = 0.9$$

$$\mathrm{Pl}(\{X_2, X_3\}) = m_1(\{X_1, X_2\}) + m_1(\{X_2, X_3\}) + m_1(X) = 0.9$$

$$\mathrm{Pl}(\{X_4\}) = m_1(\{X_4\}) + m_1(X) = 0.3$$

- Test 2: belief and plausibility values:

$$\mathrm{Bel}(\{X_1\}) = m_2(\{X_1\}) = 0.3$$

$$\mathrm{Bel}(\{X_2, X_3\}) = m_2(\{X_3\}) + m_2(\{X_2, X_3\}) = 0.35$$

$$\mathrm{Bel}(\{X_1, X_2, X_4\}) = m_2(\{X_1\}) + m_2(\{X_1, X_2, X_4\}) = 0.5$$

$$\mathrm{Bel}(\{X_3\}) = m_2(\{X_3\}) = 0.15$$

$$\mathrm{Pl}(\{X_1\}) = m_2(\{X_1\}) + m_2(\{X_1, X_2, X_4\}) + m_2(X) = 0.65$$

$$\mathrm{Pl}(\{X_2, X_3\}) = m_2(\{X_2, X_3\}) + m_2(\{X_1, X_2, X_4\}) + m_2(\{X_3\}) + m_2(X) = 0.7$$

$$\mathrm{Pl}(\{X_1, X_2, X_4\}) = m_2(\{X_1\}) + m_2(\{X_2, X_3\}) + m_2(\{X_1, X_2, X_4\})$$
$$+ m_2(X) = 0.85$$

$$\mathrm{Pl}(\{X_3\}) = m_2(\{X_2, X_3\}) + m_2(\{X_3\}) + m_2(X) = 0.5$$

- Test 3: belief and plausibility values:

$$\mathrm{Bel}(\{X_1, X_2, X_3\}) = m_3(\{X_1, X_2, X_3\}) + m_3(\{X_1, X_3\}) = 0.6$$

$$\mathrm{Bel}(\{X_1, X_3\}) = m_3(\{X_1, X_3\}) = 0.2$$

$$\mathrm{Bel}(\{X_4\}) = m_3(\{X_4\}) = 0.2$$

$$\mathrm{Pl}(\{X_1, X_2, X_3\}) = m_3(\{X_1, X_2, X_3\}) + m_3(\{X_1, X_3\}) + m_3(X) = 0.8$$

$$\mathrm{Pl}(\{X_1, X_3\}) = m_3(\{X_1, X_2, X_3\}) + m_3(\{X_1, X_3\}) + m_3(X) = 0.8$$

$$\mathrm{Pl}(\{X_4\}) = m_3(\{X_4\}) + m_3(X) = 0.2$$

Interpreting Belief and Plausibility Values. The belief and plausibility values between tests can be compared. The higher the number computed for a belief or plausibility on an observation, the more confidence in the truth of that observation; however, unlike probability values, fuzzy measures do not give predictions of frequency of occurrence (a 0.50 value for belief does not express that in 10 similar situations one expects this event will occur 5 times). Still, the relative sizes of the fuzzy measures can be used to express the relative likelihood. These three tests have been chosen to show that there is no clear indication of fault type based on any individual test, although there appears to be strong evidence that a fault exists.

EXAMPLE 3.2: Combining Evidence with Dempster-Shafer Theory
In Example 3.1, there is conflicting evidence between the tests; for example, compare $m(X_4)$ of tests 1 and 3. Further, there is an incompleteness to the tests. Notice none of the tests assign evidence values to X_2 (a thermal fault), which means that $\text{Bel}(X_2) = 0$ in all cases. In order to resolve these problems and allow a single coherent conclusion, Dempster-Shafer theory is used to combine the bodies of evidence. Using Eqs. (3.14) and (3.15) with the data from the previous example, the combined evidence values between tests can be found as follows:

   i. First compute the K value for tests 1 and 2:

$$K_{12} = m_1(\{X_1, X_2\})m_2(\{X_3\}) + m_1(\{X_2, X_3\})m_2(\{X_1\})$$
$$+ m_1(\{X_4\})m_2(\{X_1\}) + \cdots + m_1(\{X_4\})m_2(\{X_2, X_3\})$$
$$+ m_1(\{X_4\})m_2(\{X_3\}) = 0.215$$

   ii. Compute the combined evidence values:

$$m_{12}(X_1, X_2) = \frac{m_1(\{X_1, X_2\})m_2(\{X_1, X_2, X_4\}) + m_1(\{X_1, X_2\})m_2(X)}{1 - K_{12}} = 0.178$$

Similarly the following values are found:

$$m_{12}(\{X_2, X_3\}) = 0.185$$
$$m_{12}(\{X_4\}) = 0.045$$
$$m_{12}(\{X_1\}) = 0.229$$
$$m_{12}(\{X_3\}) = 0.096$$
$$m_{12}(X) = 0.038$$
$$m_{12}(\{X_1, X_2, X_4\}) = 0.051$$

Note that while there are no tests done on X_2 alone, Dempster's rule of combination assigns evidence on X_2 based on resolving conflicts between tests 1 and 2. Thus, the following is also found:

$$m_{12}(\{X_2\}) = 0.178$$

A good check on these calculations is to ensure that the summation of the observations in the new body of evidence equals 1. The next step is to use Dempster combination to combine the new body of evidence with the evidence from the third and final test:

iii. Again, compute K for the combination of the tests:

$$
\begin{aligned}
K_{123} = {} & m_{12}(\{X_1, X_2\})m_3(\{X_4\}) + m_{12}(\{X_2, X_3\})m_3(\{X_4\}) + m_{12}(\{X_4\}) \times \cdots \\
& \times (m_3(\{(X_1, X_2, X_3\}) + m_3(\{X_1, X_3\})\}) + m_{12}(\{X_1\})m_3(\{X_4\}) + \cdots \\
& + m_{12}(\{X_3\})m_3(\{X_4\}) + m_{12}(\{X_2\})(m_3(\{X_4\}) + m_3(\{X_1, X_3\})\}) = 0.2358
\end{aligned}
$$

iv. Now compute the evidence values for the combination of the tests:

$$
m_{123}(\{X_1, X_2\}) = \frac{\begin{aligned} & m_{12}(\{X_1, X_2\})m_3(\{X_4\}) + m_{12}(\{X_1, X_2\})m_3(\{X_1, X_2, X_3\}) \\ & \qquad\qquad + m_{12}(\{X_1, X_2, X_4\})m_3(\{X_1, X_2, X_3\}) \end{aligned}}{1 - K_{123}}
$$

$$
= 0.167
$$

Similarly the following values are found:

$$
m_{123}(\{X_2, X_3\}) = 0.145
$$
$$
m_{123}(\{X_4\}) = 0.047
$$
$$
m_{123}(\{X_1\}) = 0.299
$$
$$
m_{123}(\{X_1, X_2, X_4\}) = 0.0133
$$
$$
m_{123}(\{X_3\}) = 0.149
$$
$$
m_{123}(X) = 0.01
$$
$$
m_{123}(\{X_2\}) = 0.14
$$

Combination with the third test also results in evidence distributed among several other sets:

$$
m_{123}(\{X_1, X_2, X_3\}) = 0.02 \qquad m_{123}(X_1, X_3) = 0.01
$$

The above calculations have merely followed the rules of combination. It is useful to view these results from the other perspective. If one desires evidence of a particular condition, a test can be designed that will clarify the evidence. For example, DGA that distinguishes between two possible faults by looking at gas ratios could be combined with a test that looked at total gas concentrations. Dissolved-gas analysis would reveal the particular fault type, but that would need to be backed up by a test indicating the presence of a fault.

EXAMPLE 3.3: Calculation of Fuzzy Measures for Combined Evidence
In order to compare the combined test with the other tests, the resulting belief and plausibility measures are computed. In the following, evidence values are for the basic assignment m_{123}. A few of the required computations are shown. The entire set of values is presented in Table 3.2.

TABLE 3.2 SUMMARY OF FUZZY MEASURE COMPUTATIONS FOR COMBINING EVIDENCE

(a) Test 1

Set	Evidence (m)	Belief	Plausibility
X_1, X_2	0.4	0.4	0.9
X_2, X_3	0.3	0.3	0.9
X_4	0.1	0.1	0.3
X	0.2	1.0	1.0

(b) Test 2

Set	Evidence (m)	Belief	Plausibility
X_1	0.3	0.3	0.65
X_2, X_3	0.2	0.35	0.7
X_1, X_2, X_4	0.2	0.5	0.85
X	0.15	1.0	1.0
X_3	0.15	0.15	0.5

(c) Test 3

Set	Evidence (m)	Belief	Plausibility
X_1, X_2, X_3	0.4	0.6	0.8
X_1, X_3	0.2	0.2	0.8
X_4	0.2	0.2	0.4
X	0.2	1.0	1.0

(d) Combined Tests

Set	Evidence (m)	Belief	Plausibility
X_1, X_2	0.167	0.606	0.8043
X_2, X_3	0.145	0.434	0.6543
X_4	0.047	0.047	0.0703
X_1	0.299	0.299	0.5193
X_1, X_2, X_4	0.0133	0.6663	0.851
X_3	0.149	0.149	0.334
X	0.01	1.0	1.0
X_2	0.14	0.14	0.4953
X_1, X_2, X_3	0.02	0.93	0.953
X_1, X_3	0.01	0.458	0.8133

- Belief values:

$$\text{Bel}(\{X_1\}) = m(\{X_1\}) = 0.299$$
$$\text{Bel}(\{X_2\}) = m(\{X_2\}) = 0.14$$
$$\text{Bel}(\{X_3\}) = m(\{X_3\}) = 0.149$$

$$\text{Bel}(\{X_4\}) = m(\{X_4\}) = 0.047$$

$$\text{Bel}(\{X_1, X_2\}) = m(\{X_1\}) + m(\{X_2\}) + m(\{X_1, X_2\}) = 0.606$$

$$\text{Bel}(\{X_2, X_3\}) = m(\{X_2\}) + m(\{X_3\}) + m(\{X_2, X_3\}) = 0.434$$

$$\vdots$$

- Plausibility values:

$$\text{Pl}(\{X_1\}) = m(\{X_1, X_2\}) + m(\{X_1\}) + m(\{X_2\}) + m(\{X_1, X_2, X_4\}) + \cdots$$
$$+ m(\{X_1, X_2, X_3\}) + m(\{X_1, X_2\}) + m(X) = 0.519$$

$$\text{Pl}(\{X_2\}) = m(\{X_1, X_2\}) + m(\{X_2, X_3\}) + m(\{X_1, X_2, X_4\}) + m(X) + \cdots$$
$$+ m(\{X_2\}) + m(\{X_1, X_2, X_3\}) = 0.4953$$

$$\text{Pl}(\{X_3\}) = m(\{X_2, X_3\}) + m(\{X_3\}) + m(X) + m(\{X_1, X_2, X_3\})$$
$$+ m(\{X_1, X_3\}) = 0.334$$

EXAMPLE 3.4: Information Measures—Nonspecificity, Confusion, and Dissonance

While belief and plausibility values give a measure of the confidence of the end result, information measures show how well a test is structured in order to reach its conclusions. In probability, erroneous data can be identified by large deviations from expected values. Similar techniques do not exist for fuzzy set approaches. On the other hand, information measures similar to entropy in classical communication theory can give a sense of the quality of data and conclusions.

There are three commonly used methods for measuring uncertainty, as discussed in Section 3.3. One method is called nonspecificity, commonly represented by $V(m)$. Nonspecificity measures the uncertainty associated with a location of an element within an observation or set. The other measures of uncertainty are dissonance, commonly represented by $E(m)$, and confusion, designated by $C(m)$. Their difference lies in that dissonance is defined over conflicts in plausibility values and confusion over conflicts in belief values. Both dissonance and confusion arise when evidence supports two observations that are disjoint (cannot occur at the same time). For example, if one assumes that the transformer has only one type of fault, then evidence on different types of faults is conflicting and can be said to add to the "dissonance," or "confusion." In the following, these information measures are applied to the transformer diagnosis example. These calculations illustrate the relation between uncertainty and information.

For test 1:

$$V(m_1) = m_1(\{X_1, X_2\}) \log_2(|\{X_1, X_2\}|) + m_1(\{X_2, X_3\}) \log_2(\{|X_2, X_3|\}) + \cdots$$
$$+ m_1(\{X_4\}) \log_2(|X_4|) + m_1(X) \log_2(|X|) = 1.100$$

$$E(m_1) = -m_1(\{X_1, X_2\}) \log_2(\text{Pl}(\{X_1, X_2\})) - m_1(\{X_2, X_3\}) \log_2(\text{Pl}(\{X_2, X_3\})) - \cdots$$
$$- m_1(\{X_4\}) \log_2(\text{Pl}(\{X_4\})) - m_1(X) \log_2(\text{Pl}(X)) = 0.280$$

$$C(m_1) = -m_1(\{X_1, X_2\}) \log_2(\text{Bel}(\{X_1, X_2\})) - m_1(\{X_2, X_3\}) \log_2(\text{Bel}(\{X_2, X_3\})) - \cdots$$
$$- m_1(\{X_4\}) \log_2(\text{Bel}(X_4)) = 1.382$$

Computations are similar for tests 1 and 2. For the combined test, it is instructive to take a closer look at the computation of confusion (boldface quantities identify significant contributions to the confusion values):

$$C(m_{123}) = -m(\{X_1, X_2\}) \log_2(\text{Bel}(\{X_1, X_2\})) - m(\{X_2, X_3\}) \log_2(\text{Bel}(\{X_2, X_3\})) - \cdots$$
$$- m(\{X_4\}) \log_2(\text{Bel}(\{X_4\})) - \boldsymbol{m(\{X_1\})} \log_2(\boldsymbol{\text{Bel}(\{X_1\})}) - \cdots$$
$$- m(\{X_1, X_2, X_4\}) \log_2(\text{Bel}(\{X_1, X_2, X_4\})) - \boldsymbol{m(\{X_3\})} \log_2(\boldsymbol{\text{Bel}(\{X_3\})}) - \cdots$$
$$- m(X) \log_2(\text{Bel}(X)) - \boldsymbol{m(\{X_2\})} \log_2(\boldsymbol{\text{Bel}\{X_2\}}) - \cdots$$
$$- m(\{X_1, X_2, X_3\}) \log_2(\text{Bel}(\{X_1, X_2, X_3\})) - \cdots$$
$$- m(\{X_1, X_3\}) \log_2(\text{Bel}(\{X_1, X_3\})) = 1.8508$$

A summary of the resulting information measures is given in Table 3.3.

Note that by carefully observing the calculations term by term, it is possible to see the largest contributions to the information uncertainty. For example, in computing the confusion of the combined test, the $-m(\{X_1\})\text{Bel}(\{X_1\})$ term is 0.521, which is about 33% of the entire measure. Table 3.4 shows that three terms in that summation account for over 80% of the confusion of the entire set of observations. This illustrates the fact that in practice a few results of a test tend to dominate the overall uncertainty. On the other hand, note that these same three terms would not contribute to the nonspecificity of the observation.

TABLE 3.3 INFORMATION MEASURES FOR EXAMPLE 3.4

Value	Test 1	Test 2	Test 3	Combined Tests
$E(m)$	0.280	0.487	0.458	0.989
$C(m)$	1.382	1.435	1.224	1.856
$V(m)$	1.100	0.817	1.234	0.395

TABLE 3.4 CONTRIBUTIONS TO CONFUSION FOR COMBINED TESTS

Confusion Term	Calculated Value	Percent of Total
$m(\{X_1, X_2\}) \log_2(\text{Bel}(\{X_1, X_2\}))$	0.1206	7.63
$m(\{X_2, X_3\}) \log_2(\text{Bel}(\{X_2, X_3\}))$	0.1746	11.05
$m(\{X_4\}) \log_2(\text{Bel}(\{X_4\}))$	0.2073	13.11
$\boldsymbol{m(\{X_1\})\log_2(\text{Bel}(\{X_1\}))}$	**0.5207**	**32.94**
$m(\{X_1, X_2, X_4\}) \log_2(\text{Bel}(\{X_1, X_2, X_4\}))$	0.0078	0.49
$\boldsymbol{m(\{X_3\})\log_2(\text{Bel}(\{X_3\}))}$	**0.4093**	**25.89**
$m(\{X\}) \log_2(\text{Bel}(\{X\}))$	0	0
$\boldsymbol{m(\{X_2\})\log_2(\text{Bel}(\{X_2\}))}$	**0.3972**	**25.13**
$m(\{X_1, X_2, X_3\}) \log_2(\text{Bel}(\{X_1, X_2, X_3\}))$	0.0021	0.13
$m(\{X_1, X_3\}) \log_2(\text{Bel}(\{X_1, X_3\}))$	0.0112	0.71

3.5. A PROPOSED IMPLEMENTATION

In the preceding sections, a foundation has been laid for processing uncertain diagnostic and monitoring information. In this section, implementation issues are discussed. There are essentially two concerns: the rule structure and the inference process. The rule structure must provide an intuitive representation of knowledge as well as a complete description of the uncertainty. As described in the last section, the inference process requires propagation of uncertainties that depend on the situation. Furthermore, the inference process cannot ignore computational efficiency issues when considering larger domains. In this section, a rule structure and an inference engine are proposed. Within this structure, techniques are described for defining fuzzy membership functions and assigning degrees of confidence to relations.

3.5.1. Rule Structure and Assigning Values to Uncertainties

In Section 3.2.1, requirements on the rule structure were identified. The rule structure is more fully explored here, for rule R_i:

IF fuzzy condition A_j *THEN* equipment condition is B_k
AND this relation is necessary to degree C_m

An example rule in DGA of power transformers is

IF acetylene (C_2H_2) concentration is high
THEN arcing is indicated
AND the presence of this gas is very important to conclude arcing

In this rule, the expert or knowledge engineer must determine what constitutes a high concentration of acetylene (a fuzzy set) and the degree of importance of this evidence (a fuzzy measure). There are certainly many other forms that could be used to represent uncertain diagnostic knowledge; however, this structure strikes a good balance between representational power and complexity. Establishing the required uncertainty values will be discussed next before continuing with computational aspects.

Establishing Membership Functions. One of the primary difficulties faced in applying fuzzy sets is the rational assignment of membership values. The following approaches are considered here:

1. As the ordering of set elements, rather than absolute fuzzy values, is most important, design should emphasize consistency in assignment of values. This should be maintained within a particular fuzzy set definition as well as between fuzzy sets. For example, the fuzzy set "high acetylene concentration" used above should have a monotonically increasing membership function and strive for consistency with other similar fuzzy sets (e.g., "high methane concentration").
2. Standard membership function forms can be used. Trapezoidal and triangular forms are widely used in the literature. These standard forms should have

parameters that correspond to entities familiar to experts, and they should not rely on involved computations. Curve-fitting algorithms can be used to define functions if data points are available.

3. Program design should ensure that the solution is not highly sensitive to the fuzzy values. It is logically inconsistent to require extreme accuracy of fuzzy values when their purpose is to express approximations. A design that requires high accuracy in measurement values should be reconsidered. One method of avoiding such sensitivity is to build redundancy into the rule base.

Issue 2 above is taken up first. A standard membership function proposed by Dombi [18] is used in this work, as given below with $x \in [a, b]$:

$$\mu_+(x) = \frac{(1 - \nu)^{\lambda-1}(x - a)^{\lambda}}{(1 - \nu)^{\lambda-1}(x - a)^{\lambda} + \nu^{\lambda-1}(b - x)^{\lambda}} \tag{3.19}$$

where this membership function requires the specification of four parameters: a, the lower limit; b, the upper limit; λ, the transition rate; and ν, the inflection point. The subscript $+$ indicates this is a monotonically increasing function. Decreasing functions can be represented by

$$\mu_-(x) = \frac{\nu^{\lambda-1}(b - x)^{\lambda}}{(1 - \nu)^{\lambda-1}(x - a)^{\lambda} + \nu^{\lambda-1}(b - x)^{\lambda}} \tag{3.20}$$

and more complex functions can be constructed from these forms. One of the principal advantages of this form is that logarithmic transformation allows linear regression for parameter estimation. In order to describe this procedure, define the following [18]:

$$y_i = \ln\left(\frac{1 - \mu_+(x_i)}{\mu_+(x_i)}\right) \tag{3.21}$$

$$z_i = \ln\left(\frac{b - x_i}{x_i - a}\right) \tag{3.22}$$

$$d = (\lambda - 1)\ln\left(\frac{\nu}{1 - \nu}\right) \tag{3.23}$$

Then (3.19) can be rewritten as

$$y_i = \lambda z_i + d \tag{3.24}$$

Two points are required in order to specify λ and d. If more than two points are available, then linear regression techniques can be applied [19]. In summary, the procedure followed to establish the membership requires the expert to specify an upper and lower limit for each measurement condition and then specify at least two intermediary values. Figure 3.1 shows a membership function generated given the following specifications supplied by an engineer for the fuzzy set "high methane

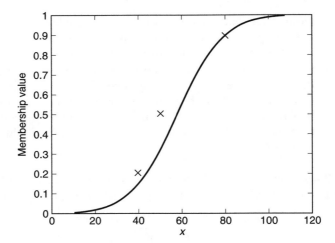

Figure 3.1 Fuzzy set "high methane concentration."

concentration":

> Range: [0, 120] ppm.
> Intermediary values: 80 ppm (relatively high, $\mu = 0.90$), 50 ppm (caution level, $\mu = 0.50$), and 40 ppm (relatively low, $\mu = 0.20$).

Establishing Fuzzy Measures of Rule Importance. There are similar problems in determining values for fuzzy measures as in establishing membership functions. Again, it is prudent to design a system that is not highly sensitive to fuzzy values. Some guidance is found in the relationships between fuzzy measures and probability. Recall that belief and plausibility measures provide upper and lower bounds on the probability. Thus, a belief value is chosen as a "conservative" estimate on the probability and a plausibility value is chosen as a "liberal" estimate of that probability. In the proposed rule structure, only a belief measure is specified for each rule. A trial-and-error approach has been used that increases the belief values associated with a rule incrementally as experience is gained with the rule base and the reliability of a rule is established. More sophisticated methods are discussed in Section 3.6.

3.5.2. Computation and Propagation of Uncertainties

The initial state of knowledge of the problem is complete ignorance. That is, nothing can be stated about any conclusion except that it is possible. As indicated earlier, the fuzzy measure representation of ignorance is all conclusions have plausibility 1 and belief 0. Application of any rule acts to decrease the plausibility of a conclusion or increase the belief in that conclusion. Consider applying rule R_i to a conclusion B_k that has some initial plausibility $\text{Pl}(B_k)$; the resulting plausibility is given by

$$\text{Pl}^*(B_k) = f(\text{Pl}(B_k), \text{Pl}(R_i)) \tag{3.25}$$

where intuitively the function f must satisfy properties similar to set intersection. For example, the resulting plausibility must be less than the initial plausibility. A more theoretical discussion can be found elsewhere [14]. Then, (3.25) can be written as

$$\text{Pl}^*(B_k) = \text{Pl}(B_k) \cap \text{Pl}(R_i) \tag{3.26}$$

The plausibility of the rule is determined by the degree to which a rule condition is satisfied and the importance of the rule. To begin, the fuzzy set condition must be related to a fuzzy measure. Consider a sequence of α-cuts [Eq. (3.6)] on a fuzzy set A so that a sequence of embedded sets is generated:

$$A_{\alpha_1} \subseteq A_{\alpha_2} \subseteq \cdots \subseteq A_{\alpha_n}$$

with

$$\alpha_1 \geq \alpha_2 \geq \cdots \geq \alpha_n$$

Then from the fuzzy measure axioms the following must hold:

$$\text{Bel}(A_{\alpha_1}) \leq \text{Bel}(A_{\alpha_2}) \leq \cdots \leq \text{Bel}(A_{\alpha_n})$$

and a consistent association between belief values and α-cuts can be given in general as

$$\text{Bel}(A_\alpha) = 1 - \alpha \tag{3.27}$$

or

$$\text{Pl}(\overline{A_\alpha}) = \alpha \tag{3.28}$$

Now there is sufficient background to state the plausibility of some conclusion based on a given rule as

$$\text{Pl}(R_i) = \text{Pl}(\overline{A_{\alpha_i}}) \cup \overline{C_m} \tag{3.29}$$

where $\text{Pl}(\overline{A_{\alpha_i}})$ is taken to be the degree to which the rule condition is satisfied and C_m is the necessity or importance of the rule being satisfied. The importance is complemented since an unimportant rule has less impact on the plausibility of a conclusion. Finally, combining (3.25) and (3.29) yields

$$\text{Pl}^*(B_k) = \text{Pl}(B_k) \cap [\text{Pl}(\overline{A_{\alpha_i}}) \cup \overline{C_m}] \tag{3.30}$$

The above formulation allows uncertainties to be calculated independently, yet it does not guarantee consistency among rules if the rule conclusions B_k are related. Practically, there must be a method of ensuring consistency and resolving conflicts when they do arise. Specifically, the following constraints must be enforced:

C1: $\text{Pl}(B_i \cap B_j) \leq \min(\text{Pl}(B_i), \text{Pl}(B_j))$.

C2: $\text{Bel}(B_i \cap B_j) \leq \min(\text{Bel}(B_i), \text{Bel}(B_j))$.

C3: $\text{Pl}(B_i \cup B_j) \geq \max(\text{Pl}(B_i), \text{Pl}(B_j))$.

C4: $\text{Bel}(B_i \cup B_j) \geq \max(\text{Bel}(B_i), \text{Bel}(B_j))$.

C5: $\text{Bel}(A) \leq \text{Pl}(A)$.

For example, consider the transformer diagnostic problem. If one assumes only a single type of fault, then the transformer cannot have both an electrical and a thermal fault; however, in most cases, the rules will indicate some evidence for both types of faults. There are several ways to manage such conflicts: (1) maximum assignment of conflicting evidence to an unknown, (2) minimum assignment of conflicting evidence to the unknown, or (3) Dempster's rule of combination. The next example highlights the first two approaches.

EXAMPLE 3.5: Assignment of Evidence

This example explores two ways in which evidence can be reassigned to manage conflict between rules. Consider the transformer problem again, which has been implemented with only two rules and three possible conclusions (thermal fault, electrical fault, or no fault) and these conclusions are mutually exclusive. The evidence given is as follows:

Rule 1 concludes Pl(Thermal) = 0.6.
Rule 2 concludes Pl(Electrical) = 0.5.

Approach 1: Unknown Evidence Maximized. Evidence of 0.5 is assigned to the unknown so that the resulting basic assignment is

$$m(\text{Thermal}) = 0.1$$
$$m(\text{Electrical}) = 0$$
$$m(\text{Not faulted}) = 0$$
$$m(\text{Unknown}) = 0.5$$

and fuzzy measures are

Pl(Thermal) = 0.6	Bel(Thermal) = 0.1
Pl(Electrical) = 0.5	Bel(Electrical) = 0.0
Pl(Not faulted) = 0.5	Bel(Not faulted) = 0.0

Approach 2: Unknown Evidence Minimized. Evidence of 0.1 is assigned to the unknown so that the resulting assignment is

$$m(\text{Thermal}) = 0.5$$
$$m(\text{Electrical}) = 0.4$$
$$m(\text{Not faulted}) = 0$$
$$m(\text{Unknown}) = 0.1$$

and fuzzy measures are

Pl(Thermal) = 0.6	Bel(Thermal) = 0.5
Pl(Electrical) = 0.5	Bel(Electrical) = 0.4
Pl(Not faulted) = 0.1	Bel(Not faulted) = 0.0

By minimizing the evidence assigned to the unknown, there is more clarity in the conclusions. The difference between the belief and plausibility values is smaller and there is a clearer distinction between those faulted and those not faulted. In this work, evidence assigned to the unknown is minimized within a particular diagnostic method. Between diagnostic methods (e.g., DGA and acoustic measurements) the Dempster rule of combination is applied. This approach assumes that different diagnostic methods present independent sources of information while within a particular method the rules are closely related and evidence assignments must be as consistent as possible.

3.5.3 Considerations of Efficiency

One of the principal objections to the evidence methods presented here is computational complexity. If there are n possible conclusions, then a naive implementation of these methods must operate on 2^n sets. There are two methods used to reduce these computations:

1. Most of the 2^n sets are not physically interesting. A number of conclusions will be mutually exclusive; for example, one need not be concerned with the possibility that a transformer is both operating normally and experiencing arcing faults. The interesting sets can be represented as a fault tree, as in ref. 9, or simply delineated for a particular problem.
2. For an informative test, one does not expect evidence for a large number of possibilities. Only nonzero evidence values and the corresponding sets contribute to the uncertainty calculations. Thus, only nonzero values are stored for problems with a large number of possible conclusions.

An expert system shell has been implemented in $C++$ on an IBM-PC using object-oriented programming techniques. For the transformer problem, computational time has not been a problem. In fact, the time required to analyze a particular case is less than the time required to access all the relevant data from the data base.

3.6. EVALUATION, LEARNING, AND INFORMATION MEASURES PERFORMANCE

Performance evaluation and subsequent performance improvement are on-going research topics in expert system development [20]. The eventual goal of such research is to develop automated systems that can be said to learn. Unfortunately, traditional expert system approaches do not lend themselves easily to learning. In contrast, artificial neural nets (ANNs) incorporate very powerful methods of learning from data. In the diagnostic problem, the ANN approach faces one major difficulty: the acquisition of interesting data. For example, Tomsovic et al. [9] observed around 20 transformer faults in over 2000 gas samples. Such a sparsity of fault cases complicates learning from data. Still, there have been reports of some limited success [11, 21]. In this section, these problems are discussed first in terms of evaluating performance and second in terms of improvements.

3.6.1 Performance Evaluation and Tuning

The difficulty of assessing expert system performance arises from the complexity of the problem domain and the lack of any clear optimal solution. One approach is to evaluate the consistency of the rules independently of data. Marathe et al. [22] used this approach to identify conflicts and redundancies in the knowledge. By itself, such an approach is not adequate for systems with uncertainty, where the consistency of rule relations varies greatly with data. Tomsovic et al. [10] took an approach to tuning the knowledge base by analyzing interesting cases and evaluating performance based on the correct classification of the fault and an analysis of the information content in the solution. The following steps were performed on the transformer diagnostic system:

1. A set of 20 interesting case studies was chosen. These cases were selected based on the difficulty of classification. A transformer condition was deemed difficult to classify either because measurements were close to threshold levels or because the measurements gave conflicting evidence.
2. A prototype knowledge base with no tuning of parameters was applied to these case studies.
3. The fuzzy membership functions were redefined based on the process in Section 3.5. The importance of each rule was determined by a trial-and-error adjustment of the confidence in each rule. Confidence values were adjusted until correct classification of all cases was obtained.
4. Information measures for the systems in steps 2 and 3 were computed.

The results of this study are shown in Table 3.5. The tuning significantly improved the correct classification of the results with some minor improvement in the information measures. From observing the effects of tuning, the following general conclusions can be made:

- Tuning using the fuzzy set techniques does allow for improvement in the classification of faults. This in part justifies the use of an uncertainty model as strict logic would not have been able to separate the normal and faulted cases based only on acceptable ranges for measurements.

TABLE 3.5   CASE STUDY FOR PERFORMANCE IMPROVEMENT

	Cases Classified as Faulted	Cases Classified as Unfaulted	Average Confusion (% of maximum)	Average Vagueness (% of maximum)
Actual	6	14	—	—
Untuned system	15	5	35.9	12.0
Tuned system	6	14	40.7	16.2

- In general, the information measures are affected by tuning of the parameters. Although improvement for the selected cases was not great, a measurement of the improvement in the clarity of the rule base for these cases was obtained. The small improvement is partly due to the fact that difficult case studies were chosen. The information measures may provide more useful information for typical cases. This was validated by a further experiment that showed that, on average, information measures improved for more typical data that included easy-to-classify cases [10].

3.6.2. ANN Approaches

Artificial neural nets can be trained to represent arbitrary nonlinear mappings. The principal benefit of ANNs over expert systems is providing a systematic approach to learning. On the other hand, the principal drawback of ANNs is that they implement their knowledge in terms of the weights and connections in a network with no explicit method to incorporate a priori knowledge. A straightforward ANN approach to diagnostics would implement the relationship between measurements and equipment condition as a nonlinear mapping. That is, the measurements would be inputs and the various equipment states would be outputs. The input-output relationship would be found by "training" the network on a set of data for which the conclusions are known.

Over the years, a utility or testing firm will develop a large data base of equipment tests. These data can provide a convenient test bed for analyzing an expert system approach. It is tempting then to develop learning-based methods to utilize these data. Often, these data have not been carefully analyzed and are quite difficult to interpret. Note that most power system equipment is highly reliable; therefore, there are relatively few field failures of equipment. The development of a diagnostic technique is based not only on field experience but also on theoretical approaches and laboratory data. For example, knowledge of temperature and gas solubility relations is the foundation for DGA. Such information is difficult to incorporate into the neural net. Three methods of combining ANNs and fuzzy methods are discussed here:

1. Artificial neural nets can implement fuzzy membership functions. As the ANN can represent arbitrary functions, it can obviously be trained to model membership functions. This is a common approach in control applications where extensive data can be generated by simulation. In diagnostics, this would require even a greater amount of data than simply modeling the overall input-output relationship. Thus, this approach appears of little use in this work.

2. Artificial neural nets can process outputs of the fuzzy rules. In this approach, each fuzzy relation is represented as an input to the neural net. This ensures that some preprocessing of the data is performed. The underlying assumption is that such a formulation will require less training. Experiments in ref. 11 showed that this is still a difficult task for the ANN owing to severely

limited failure data. In that work, the ANN was unable to approach the performance of the fuzzy set methods.

3. The developed fuzzy expert system is used to generate input data for the ANN. In this approach, one can view the ANN as a method of implementation for the expert system. The ANN is trained to match the expert system performance. As experience with a technique is gained, the ANN can be trained on new data and presumably improve on the expert system performance. This approach appears to show promise in that it requires the least amount of failure data for training.

3.7. SUMMARY AND DISCUSSION

This chapter has explored a number of methods for incorporating uncertainty into expert systems for power systems equipment diagnosis and condition monitoring. The application of these techniques is summarized below:

1. Uncertainty in acceptable ranges for measurements is modeled by fuzzy sets.
2. A systematic approach for establishing membership functions is described.
3. The importance of a relation between a measurement or observation and equipment condition is modeled by a fuzzy measure.
4. Uncertainties are initially propagated by the application of individual rules without global information.
5. After application of all rules within a particular diagnostic method, evidence is distributed to minimize uncertainty assigned to the unknown such that the evidence assignment is consistent with the fuzzy measure axioms. This is a global computation.
6. The Dempster rule of combination is used to combine different diagnostic measurements based on the assumption that each method gathers independent information. Again, this is a global computation.
7. Information measures are proposed for evaluating the consistency of conclusions in a rule base including uncertainty. It was shown that these measures can be used to determine appropriate tests to clarify analysis when there is conflicting information.
8. Information measures can be used to assess the performance of an expert system implemented using the proposed techniques. The system that provides the most consistent and specific conclusions performs best. This can be extended to provide a means of tuning fuzzy parameters in order to improve performance.

Acknowledgments

This research has been supported in part by Washington State University. The transformer diagnostic system used for testing was initially developed with support from Vattenfall AB (formerly Swedish State Power Board) under the direction of K. Tomsovic , M. Tapper, and T. Ingvarsson. Discussions with T. Haupert, F. Jakob, and D. Hanson of Analytical Associates were very helpful in gaining a further understanding of transformer testing.

References

[1] Z. Z. Zhang, G. S. Hope, and O. P. Malik, "Expert Systems in Electric Power Systems: A Bibliographic Survey," *IEEE Transactions on Power Systems*, Vol. 4, No. 4, 1989, pp. 1355–1362.

[2] R. Levi and M. Rivers, "Substation Maintenance Testing Using an Expert System for On-Site Equipment Evaluation," *IEEE Transactions on Power Delivery*, Vol. 7, No. 1, 1992, pp. 269–275.

[3] T. H. Crowley, "Automated Diagnosis of Large Power Transformers Using Adaptive Model-Based Monitoring," M.S. Thesis, Massachusetts Institute of Technology, Cambridge, MA, June 1990.

[4] S. Tanimoto, *Elements of Artificial Intelligence*, Computer Science Press, New York, 1987.

[5] J. Momoh, X. Ma, and K. Tomsovic, "Overview and Literature Survey of Fuzzy Set Theory in Power Systems," *IEEE Transactions on Power Systems*, Vol. 10, No. 3, 1995, pp. 1676–1690.

[6] C. Lin, J. M. Ling, and C. L. Huang, "An Expert System for Transformer Fault Diagnosis and Maintenance Using Dissolved Gas Analysis," *IEEE Transactions on Power Delivery*, Vol. 8, No. 1, 1993, pp. 231–238.

[7] H. E. Dijk, "Exformer an Expert System for Transformer Faults Diagnosis," paper presented at the Ninth Power Systems Computation Conference, Lisbon, 1987, pp. 715–721.

[8] G. Tangen, L. E. Lundgaard, and K. Faugstad, "A Knowledge Based Diagnostic System for SF_6 Insulated Substations," paper presented at the Nordic Insulation Symposium, Nord-IS 90, Denmark, 1990, pp. 6.2:1–6.2:11.

[9] K. Tomsovic, M. Tapper, and T. Ingvarsson, "A Fuzzy Information Approach to Integrating Different Transformer Diagnostic Methods," *IEEE Transactions on Power Delivery*, Vol. 8, No. 3, 1993, pp. 1638–1646.

[10] K. Tomsovic, M. Tapper, and T. Ingvarsson, "Performance Evaluation of a Transformer Condition Monitoring Expert System," in *Proceedings of the 1993 (CIGRÉ) Symposium on Diagnostic and Maintenance Techniques*, Berlin, Germany, April, 1993.

[11] K. Tomsovic and A. Amar, "On Refining Equipment Condition Monitoring Using Fuzzy Sets and Artificial Neural Nets," paper presented at the 1994 International Conference on Intelligent System Applications to Power Systems, Montpellier, France, Sept. 1994, pp. 363–370.

[12] R. S. Gorur, "Aging of Power System Hardware, Part 1: Mechanisms and Laboratory Simulation," in *Proceedings of the NSF Workshop on Electric Power System Infrastructure Issues*, to appear.

[13] IEC Publication 599, *Interpretation of the Analysis of Gases in Transformers and Other Oil-Filled Electrical Equipment in Service*, 1st ed., 1978.

[14] G. J. Klir and T. A. Folger, *Fuzzy Sets, Uncertainty, and Information*, Prentice-Hall, Englewood Cliffs, NJ, 1988.

[15] H. Prade, "A Computational Approach to Approximate and Plausible Reasoning with Applications to Expert Systems," *IEEE Transactions on Pattern Analysis and Machine Intelligence*, Vol. PAMI-7, No. 3, 1985, pp. 260–283.

[16] M. M. Gupta and J. Qi, "Theory of *T*-Norms and Fuzzy Inference Methods," *Fuzzy Sets and Systems*, Vol. 40, 1991, pp. 431–450.

[17] J. Dombi, "A General Class of Fuzzy Operators, the De Morgan Class of Fuzzy Operators and Fuzziness Induced by Fuzzy Operators," *Fuzzy Sets and Systems*, Vol. 11, 1983, pp. 115–134.

[18] J. Dombi, "Membership Function as an Evaluation," *Fuzzy Sets and Systems*, Vol. 35, 1990, pp. 1–21.

[19] W. H. Press, *Numerical Recipes in C*, 2nd ed., Cambridge University Press, 1992, pp. 517–565.

[20] R. M. O'Keefe, O. Balci, and E. P. Smith, "Validating Expert System Performance," *IEEE Expert*, Vol. 2, No. 4, 1987, pp. 81–90.

[21] S. K. Bhattacharya, R. E. Smith, and T. A. Haskew, "A Neural Network Based Approach to Transformer Fault Diagnosis Using Dissolved Gas Analysis," in *Proceedings of the 1993 NAPS*, Washington, DC, Oct. 1993, pp. 125–129.

[22] H. Marathe, T. K. Ma, and C. C. Liu, "An Algorithm for Identification of Relations Among Rules," in *Proceedings of the 1989 Workshop on Tools for AI*, Oct. 1989, pp. 360–367.

M. A. El-Sharkawi
R. J. Marks II
R. J. Streifel
Department of Electrical Engineering
University of Washington
Seattle, WA 98195

I. Kerszenbaum
Southern California Edison Company
Research Center
Irwindale, CA 91702

Chapter 4

Detection and Localization of Shorted Turns in the DC Field Winding of Turbine-Generator Rotors Using Novelty Detection and Fuzzified Neural Networks

4.1. INTRODUCTION

One of the most difficult problems in the operation of large synchronous turbine-generators is the detection of shorted turns in the DC field of the rotor. Not only is the existence of a shorted turn in the field winding hard to detect, its correction may result in an expenditure of several hundred thousand dollars when including the cost of replacing the lost power generation with more expensive sources, such as large nuclear-powered machines. Unfortunately, this expense is incurred even in the case of a wrong diagnosis. This is because the major expense results from the disassembly and assembly of the machine and in the added cost of alternative production. Proper localization, and more important, accurate determination of the actual existence of a shorted turn, is therefore essential to avoid huge unnecessary monetary losses. A general solution to this problem has so far remained elusive [1–3].

The type of machines under consideration are two- or four-pole cylindrical rotors. Slower machines are almost invariantly of the salient-pole construction. Shorted turns in salient-pole rotors are less detrimental to the operation of the machine and, at the same time, easier to detect through the pole-drop test. This test simply requires measurement of the voltage across each pole of the machine while the entire rotor field winding is being fed from a DC source. Since each pole has the same number of turns, all poles will experience the same voltage drop. If any pole has one or more turns shorted, a lower voltage will exist. Normally, a voltage difference of 5% warrants further investigation. On the other hand, the pole windings in a cylindrical rotor are totally inaccessible for this type of test. The windings are located underneath large metallic retaining rings and the slot portion of the coil is contained underneath metallic wedges that can only be removed by removing the retaining springs. Removing and remounting the retaining rings is a complicated and costly operation.

There are various mechanisms by which shorted turns develop in a synchronous machine. For instance, when not in operation, turbine-generators are often kept rotating at low speeds (several turns per minute). This operational status is referred to as a *turning-gear* mode. This practice eliminates the natural bowing of the rotor due to its own weight when stationary. Therefore, if required, the machine can be brought rapidly into full operation without the excessive vibration produced by a rotor that has developed a bow. Unfortunately, turning-gear operation has a negative side effect on the integrity of the rotor winding. The rotor field windings consist of heavy copper conductors. During the life of the machine, small clearances between layers of conductors are created in the radial direction. These are the product of centrifugal forces and thermomechanical induced aging and drying of the insulation. These clearances allow the heavy conductors to move slightly in the slot. During normal operation, the large centrifugal forces developed in the rotor keep the conductors pressing toward the wedge. However, during turning-gear operation, the conductors fall and rise with the rotation of the machine. This incessant pounding results in the creation of copper dust that accumulates in the slot. Occasionally, sufficient accumulation creates a shorted turn between contiguous conductors.

Another process by which shorted turns are generated is the relentless expansion and contraction of the conductors in machines subjected to varying load. When expanding and contracting, the motion tends to damage the insulation between the conductors or between a conductor and the rotor forging (wall insulation). In machines subjected to many such load changes and/or machines with many years of operation, the insulation may tear in some places, allowing contact between contiguous turns.

A third mode of failure is caused by broken DC field conductors puncturing or tearing the insulation layers. The copper conductors may break because of metal fatigue, insufficient support, or overheating during improper operation.

4.2. EXISTING METHODS OF DETECTION

To better understand the existing methods for detection of shorted turns, a brief description of the effect shorted turns have on the operation of the machine is appropriate. A shorted turn or a number of shorted turns in the DC field of a cylindrical rotor affect the operation of the machine by way of two separate phenomena:

1. *Unbalanced magnetic pull*: The shorted turns introduce an asymmetry in the distribution of rotor-originated magnetic vector potential. The resulting flux in the gap of the machine will thus be asymmetric [4, 5]. Assuming the armature is balanced, the resultant air/gas gap flux density degree of asymmetry will depend on the level of excitation; that is, the higher the DC excitation,

the stronger the asymmetry. Asymmetric flux distribution causes unbalanced magnetic pull (UMP) between the rotor and the stator core. The end result is augmented vibrations at one and two times the synchronous speed. If the time constants of the electromagnetic phenomena are very short, any change in the excitation results in immediate change in the level of vibration. Therefore, monitoring the vibration as a function of the excitation is used as one indicator of a possible shorted turn. Unfortunately, there are many other variables in play during the process of changing the excitation, making reliable discrimination based solely on this test almost impossible.

2. *Thermal bow*: When shorted turns are present in the field winding of a cylindrical rotor, a thermal unbalance is created by the uneven flow of DC current along the coils. This unbalance, though small, bows the rotor sufficiently to produce an abnormal level of vibration due to the large centrifugal forces present in high-speed turbine-generators. The thermal time constant of the rotor is significantly long. Therefore, changes in the excitation current will result in delayed changes in the vibration levels at synchronous speed. It is interesting to note that the thermal bow of the rotor results not only in higher vibrations due to centrifugal forces but also in higher vibrations due to added UMP (by adding eccentricity, a bowed rotor will further increase the unbalanced magnetic pull). In addition, the magnetic axis of asymmetry may not coincide with the axis of the thermal bow. This phase shift results in additional harmonics of the vibration, making a determination of a shorted turn based only on vibration very uncertain.

Given the difficulty in reaching a reliable detection decision based on monitoring the vibration of the machine, other methods of detecting shorted turns have been employed over the years. One such method relies on the indirect measurement of the impedance (resistance) of the rotor field winding during operation. The DC value of the resistance of the winding is calculated from the measurement of the current flowing into it (the excitation current) and the voltage applied at the slip rings (collector rings). This resistance also varies according to the temperature of the machine, that is, according to the load. Plots of resistance versus load are generated. These plots can then be used to estimate the temperature of the field winding for a particular load condition. It is obvious that any number of shorted turns in the field winding will result in a lower resistance than expected for a particular load and excitation condition. Unfortunately, by itself, this technique yields dubious results unless the number of shorted turns is significant. One positive characteristic of this technique is the possible detection of a shorted turn if this condition disappears at certain speeds. Shorted turns are in many cases speed dependent; that is, when the speed of the machine decreases (e.g., during coastdown of the machine), the centrifugal forces acting on the conductors relax, and, in many cases, the shorted turn disappears. Continuous monitoring of the field resistance during the coastdown operation may reveal an abrupt change in value. This most certainly can be related to an intermittent shorted turn. However, this technique will not provide any help when a constant short is present.

Some methods detect the flux asymmetry created by a shorted turn by applying AC current to the field through the collectors and holding a C-shaped pick-up coil across the slot. This method is accurate but can only be performed after removing the rotor from the bore. This is an expensive exercise. In addition, detection of all shorts that tend to disappear when the rotor is brought to a stand-still is precluded.

Other methods rely on special design of the stator winding. The special design includes two parallel sections in each phase. Flux asymmetries generate circulating currents that can be measured. Although the technique has the advantages of being applied to the machine under operation and not being intrusive, it also presents some serious disadvantages. For instance, many machines presently in operation do not have a winding design that lends itself to the application of this method. Redesigning a machine for the sole purpose of detecting shorted turns is not practical.

One of the most reliable methods developed to date is based on the direct measurement of the air/gas gap magnetic flux with the machine in operation. The flux is measured by a pick-up coil installed in the gap. Two such methods for installing the probes exist: the wedge probe and the case-mounted probe. Both systems require the machine be brought off-line for their installation. This results in substantial down time. The advantages of this method are the high degree of discrimination, the fact that shorted turns are detected with the machine in normal operation, and the ability to detect shorts in self-excited machines (i.e., machines without collector rings). Unfortunately, the presence of these coils in existing machines (and new ones) is rare.

It is evident there is room for improved methods to detect shorted turns in the rotor of a turbogenerator. The method described here is nonintrusive; that is, no installation work is required. This eliminates installation, disassembly, and down-time costs. The method can be applied to both a running and a still rotor, eliminating the problem of missed detection of intermittent shorted turns. The equipment is very portable and the test requires little time. Tests performed so far have indicated reliable detection of shorted turns. Faults not only are detected but also are located with a reasonable degree of certainty. The method's major disadvantages are that it is only viable on machines with externally excited rotors and a neural network must be trained. The first disadvantage is not critical since the vast majority of machines in operation today are externally excited; that is, they have slip rings (collector rings). The second disadvantage will slowly become less important with the establishment of a data bank accumulated over repeated use of the instrument.

4.3. LOCATION OF SHORTED WINDINGS IN STANDSTILL ROTORS

An effective method of localizing shorts in rotors makes use of traveling waves. Due to reactive coupling between windings, conventional time-domain reflectometry will not work. Neural networks with fuzzy logic output, however, can be used to locate shorted turns. The method is quite general and can also be used for locating shorted turns in power devices such as transformers and motors.

Figure 4.1 shows the basic concept of the traveling-wave method for fault detection. Two identical signals are injected into the winding from either side and are received on the opposite end. The receiving signals are subtracted to form the *signature signal*, $A - B$. The frequency of the injected signals should be selected at a rate no greater than about $1/10\tau$, where τ is the traveling time of pulses through the field winding, which is dependent on winding parameters. The interference between the falling edge of the injected signal and the reflected wave is then essentially eliminated.

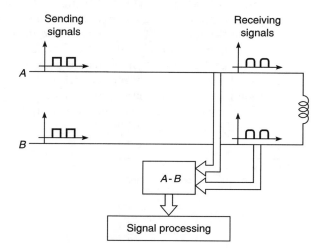

Figure 4.1 Traveling-wave signature signal acquisition.

The method's circuitry is depicted in Figure 4.2. A pulse generator is used to send two signals to the symmetrical circuit. The frequency of the signals is dependent on the winding parameters. The symmetrical circuit receives the two reflected signals and provides the signature signal to a PC-based computer through a general-purpose interface bus (GPIB) circuit board (the CompuScope LITE 220 from Omega). Several sampled signals for different short locations are collected for feature extraction and subsequent neural network training. The neural network used in this study

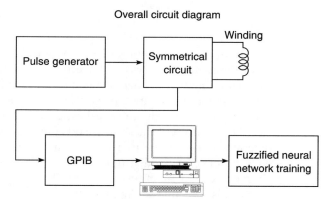

Figure 4.2 Signature signal circuitry block diagram.

is the layered perceptron [6–8]. The output of the neural network is fuzzified to increase the accuracy and the dynamic range of the neural network's output and also to diminish the effect of noisy measurements [9–12].

The shape of the signature signal is used to perform two functions: (1) detect the existence of a shorted turn and (2) localize the short. A high-frequency sampling device is used to ensure that the entire signature signal is captured. This results in a vector whose high dimension cannot be easily processed by a neural network. Hence, the cardinality of the training data must be reduced without destroying the data's information content [13–17]. In phase 1, the features are extracted from signature signals by computing the area under the waveform in fixed time intervals (see Figure 4.6 in Section 4.4). The process is described below in more detail.

Figure 4.3 outlines the general procedure for shorted-turn detection and localization. The training data acquired by the setup for acquisition of the signature signal are used for neural network training. Extracted features from the signature signal are used as inputs to train a standard feed-forward layered perceptron artificial neural network. The network output is defuzzified to provide a number to identify the short location.

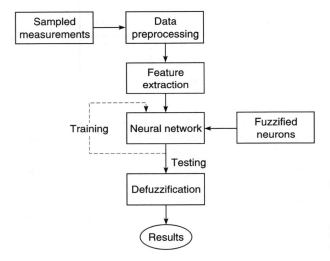

Figure 4.3 Shorted-turn localization procedure.

With the introduction of new neural network topologies and efficient training algorithms, neural networks have proven useful in several power applications [7]. The neural network, when adequately designed and trained, can synthesize a useful nonlinear mapping between input and output patterns. This is a key property for shorted-turn detection and localization [18–21].

The location of the short is coded into a number of fuzzy membership functions determined by the desired resolution of the short location. For example, a field winding may be divided into coils and the coils divided into turns. In this study, six membership functions are used. The number of output neurons of the neural network is the same as the number of the fuzzy membership functions, as illustrated

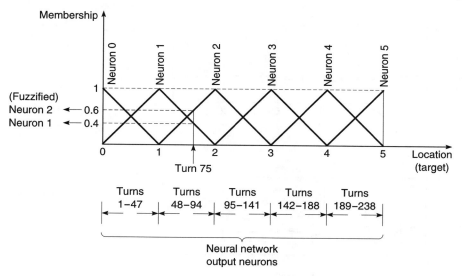

Figure 4.4 Membership function for the fuzzification mappings.

in Figure 4.4. The turbogenerator used to test the concept has 14 coils with 17 turns in each coil for a total of 238 turns. These turns are partitioned into six groups. Each output neuron corresponds to the value of the corresponding membership function. For a short at turn 75, the membership (neural network output) is $[0\ 0.4\ 0.6\ 0\ 0\ 0]^T$.

During testing, the output of the neural net is *defuzzified* where each membership function is weighted by the state of the corresponding output neuron [9–12]. The weighted membership functions are then added and the center of mass (first moment) of the sum is the short location. If, for example, each membership function is of identical shape and has a center of mass C_i, then the defuzzified output is given by

$$\text{Short location} = \frac{\sum_i \beta_i C_i}{\sum_i \beta_i} \qquad (4.1)$$

where β_i is the output of neuron i. Other defuzzification methods can also be used [11].

4.4. RESULTS

The proposed detection method was tested in the Southern California Edison Company facilities on a 60-MVA two-pole turbogenerator. The generator has 14 coils with 17 turns in each coil. The shorts between windings were intentionally introduced to verify the proposed technique. The loss of one turn reduces the ampere-turns of that pole by about 0.85%.

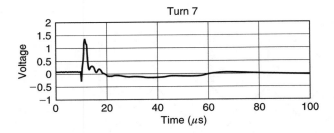

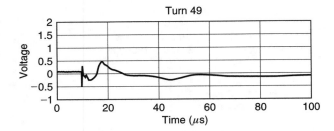

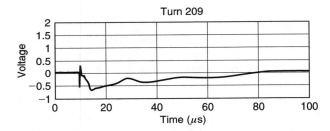

Figure 4.5 Samples of signature signals.

To train and test the neural network, temporary shorts were introduced between adjacent windings. Two simultaneous signals were then injected from both sides of the field winding. The difference between the two receiving signals is the signature signal. This signal is used to detect and localize the shorted turn. Examples of several sampled waveforms are shown in Figure 4.5. The horizontal axis represents the time in microseconds and the vertical axis is the magnitude of the signature signals in volts. The signature signal is the difference $A - B$.

The signature signals are sampled at 5 MHz and a total of 500 samples are collected. If the entire signal were used to train the neural net, the network would certainly suffer from scaling problems and the curse of dimensionality. Feature extraction, rather, must be used to capture the information content of the signal and reduce the dimension of the sampled signal vector. Some feature extraction methods are based on mathematical techniques [16, 17]. Others are based on engineering judgment and heuristics.

Figure 4.6 shows an expansion of a signature signal. Dispersion causes the signature signal to have large differences for different short locations near the initial

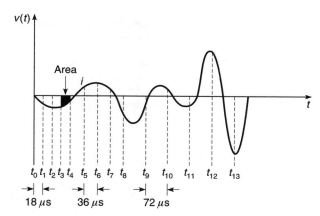

Figure 4.6 Features extracted from signature signal.

time. Therefore, the signature signal is divided into two sections: initial and extended. In the initial section, the waveform is partitioned into four intervals of 18 μs each. The extended section is composed of four intervals of 36 μs and five intervals of 72 μs. The area of each interval is given by

$$\text{Area } i = \int_{t_{i-1}}^{t_i} v(t) \, dt \tag{4.2}$$

Area i is the area of segment i, t_i is the end time of segment i, and $v(t)$ is the voltage waveform. Of course, the signal has been sampled and the area is merely taken as the discrete voltage sample multiplied by the width of segment i. A total of 13 areas are obtained for each signature signal. These areas are the features used as the input to the neural network. A total of 67 training patterns were collected by shorting adjacent turns at several locations within the field winding. A neural network with 1 hidden layer, 13 input neurons, 4 hidden neurons, and 6 output neurons was used. This architecture gave a lower test error than other architectures. The network was trained by using the standard back-error propagation method.

After training, the neural network was tested for 60 short locations taken at random points. None of the test data was used during training. The test results are listed in Table 4.1. As seen in the table, the proposed technique is highly accurate and very robust. In all test cases, the coil with shorted turns was accurately identified. Moreover, the shorted turns were all located to within a few turns of the actual short location.

The technique for locating shorted windings requires the training of a neural network with input-output pattern pairs. Thus, the ability to introduce shorts for acquiring training data is essential. The expense of dismantling and reassembling a large turbogenerator prevents the introduction of shorts for training sample acquisition when shorts are to be detected in an operational rotor. Therefore, an alternate technique is required to detect shorted turns without the need to introduce shorts into the field windings.

TABLE 4.1   FIELD TEST RESULTS

NN Structure	
Input neurons	13
Hidden neurons	4
Output neurons	6
Patterns	
Training patterns	67
Testing patterns	60
Test Results	
Percentage of accurate identification of coil	100%
Maximum error in short localization	±7 turns (±3%)

4.5. DETECTION OF SHORTED TURNS IN OPERATIONAL GENERATORS

The detection of shorted turns in operational rotors can be accomplished uing the traveling-wave technique and novelty detection. The detection of faults for a rotor that is in operation is made much more difficult due to the presence of noise caused by the brushes, the mechanical vibrations during rotation, and the presence of the excitation source in the rotor circuit.

The brushes of a turbogenerator connect the excitation source to the slip rings and supply power to the rotor. The motion of the brushes and the current flowing through the brushes cause arcing. The arcs produce high-energy spikes that appear as noise in the signal. The noise will be measured by the signature signal acquisition hardware and will corrupt the signal. A study of brush noise was performed on a small laboratory machine and the results are shown below.

The rotation of the rotor causes the windings to move and shift as the rotor turns. The signature signal is dependent on the physical characteristics of the rotor windings, and thus the signature signal will not be identical when measured at different points in the rotational cycle.

The excitation source provides another path for the pulses of the signature signal acquisition system. The added path will greatly change the reflected signals and thus change the signature signal. As long as the excitation windings are symmetrical from the perspective of the injected pulses, the ability to detect faults will not be hindered.

All of the modifying effects described prevent the use of training examples collected while the rotor is dismantled to detect shorted turns while the rotor is operating. Since inducing shorted turns in an operational rotor is expensive, the concept of a novelty filter must be used for the detection of shorted turns.

4.6. HARDWARE FOR SIGNATURE MEASUREMENT

The acquisition of signature signals for operational rotors is similar to the acquisition of signature signals for dismantled rotors. The addition of blocking capacitors in series with each pulse generation path is required to protect the measurement circuits from the high excitation source voltage present on the brushes during operation. The excitation source required by most turbogenerators supplies a high DC voltage and the voltage can easily be blocked by the capacitors. The pulses are not significantly affected by the capacitors, and thus measurement of the signature signals is not seriously affected.

Since the signature signals will be corrupted by noise, the signature signals are averaged over many collected waveforms. Averaging of the signals will remove any noise with zero mean such as brush noise. The number of signals averaged depends on the standard deviation of the noise and the degree to which the noise is to be removed. For noise outside the bandwidth of the signature signals, a low-pass frequency-selective filter is used. Different rotors will produce signature signals with different frequency content. Therefore, the signature signals and the undesired noise frequency spectra must be analyzed for each rotor. The spectrum of the noise is easy to obtain using the signature signal acquisition hardware. The pulses should be disabled and many samples of the noise signal gathered. The average power spectral density can then be computed. The determination of the frequency content of the signature signal is not possible due to the presence of noise.

4.7. DETECTION OF SHORTED WINDINGS

Use of the signature signals and a novelty filter allows detection of shorted turns in rotor windings. The procedure for detecting shorted windings is described below. Difficulties arise in determining whether a fault exists when noise and other random effects are present in the signal. The detection algorithm can be designed to be conservative, which allows for the possibility that many faults will not be detected (missed detections). The detection algorithm can also be designed to detect all faults but may then allow faults to be detected when no fault is present. The cost of servicing a machine that does not have a shorted turn must be compared to the cost of allowing a machine to operate with a fault present. The cost of dismantling a healthy rotor is assumed to be much greater than the cost of allowing a rotor with shorted turns to remain in operation. If signature signals are observed for a period of time, the probability of making a correct decision increases.

4.8. NOVELTY DETECTION

A novelty filter, as defined by Kohonen [22], is a system that extracts the new, anomalous, or unfamiliar part of the input data. Neural networks have also been applied to novelty filtering [23–24]. The novelty detector neural network operates as an associative memory where the neural network, normally a two-layer feed-forward network, is trained to identify the inputs. During training of the network by back propagation, the output of the network is forced to repeat the input. The output layer must thus have the same number of units as the number of input values. During operation of the network, the inputs are presented and the output produces the values corresponding to the training input, which resembles the current input. The novelty is then the difference between the current input and the output produced by the network. The difference is then compared with a threshold using some form of distance function. Use of neural networks for detection of shorted windings is difficult due to the high dimension of the signature signal. The neural network must be able to memorize all signature signals for healthy operational rotors. For this reason, other forms of novelty filtering are used.

Many measurements can be made for healthy rotors. Data are collected when the machine is new or immediately after maintenance or cleaning, increasing the chances that data for a healthy machine are being gathered. The presence of a fault should be indicated when the signature signals differ significantly from healthy signature signals. Since many "good" signals can be measured, a statistical view can be obtained about the signal. The basic assumption is that the signal will vary significantly when a fault is present.

4.9. COMPUTATION OF DETECTION THRESHOLD

The first operation on the signature signals is the removal of the mean signal. Since the signature signals are sampled and converted to an array of digital words, each signal can be considered a discrete vector or a point in the signal space. The average signal is computed by summing all healthy rotors as vectors and dividing each component by the number of healthy signature signals used in the sum. The average vector is subtracted from each signature signal to translate the signature signals toward the signal space origin. The average signal is called the prototype.

The signal space is then partitioned into two regions: one for healthy rotors and one for faulted rotors. The magnitude of a translated signature signal then corresponds to the size of the signal novelty. The standard Euclidean distance is used as a measure of vector magnitude in this work. Due to the random character of the signature signal, the region corresponding to healthy rotors will extend away from the origin. A surface separating the two regions must be defined to enclose the healthy region. Any translated signature signals outside the detection surface will be assumed to represent a rotor with a fault.

The simplest detection surface is a hypersphere. The largest Euclidean length of the translated healthy signature signals is used as the threshold. The region for healthy rotors is inside the resulting hypersphere. In theory, any signature signal outside the hypersphere will represent a faulted rotor.

$$\text{Fault} = \begin{cases} \text{True} & \text{if } \|x - \bar{x}\| > T \\ \text{False} & \text{if } \|x - \bar{x}\| < T \end{cases} \tag{4.3}$$

where x is a signature signal, $\bar{x}$ is the prototype vector, and T is the detection threshold or radius of the hypersphere. This method could potentially produce a detection algorithm with a high rate of missed detections. For example, if the translated healthy signature signals all lie on a line, the sphere will enclose all the points but will have a great deal of space with no healthy signature signals nearby. A signature signal for a faulted rotor may lie within the hypersphere but not on the line and thus no fault would be detected.

An improvement in detection performance will result from using hyperboxes rather than hyperspheres. The minimum and maximum coordinates of each component of all translated healthy signature signals are recorded. The vector components of translated signature signals are then compared to the stored bounds. If any vector component lies outside the corresponding bound, the signature signal is outside the hyperbox and is determined to represent a faulted rotor. If the translated healthy signature signals again lie along a line but the line is not along one of the vector coordinate axes, missed detections will again result.

An even more accurate detection boundary can be defined using hyperellipses. The hyperellipse can be oriented in any direction and thus would not contain large empty spaces unless the healthy region consisted of multiple disjoint subregions. A signature matrix X is formed by using the healthy signature signals as the columns. The correlation matrix C is formed by

$$C = XX^T \tag{4.4}$$

The eigenvectors of the correlation matrix C define the principal axes of the hyperellipse. Each translated healthy signature signal is rotated into a new space by vector multiplication by the eigenvectors. The hyperellipse detection boundary can then be defined using the healthy signature signals.

The advantage of using the hypersphere as a detection boundary is the ease with which the threshold can be calculated and the limited computer resources required to detect faults. The hyperellipse detection boundary computation will require a great deal of computer memory and computing time. The correlation matrix for signature signals with 1000 samples will require 1,000,000 elements. The number of nonzero eigenvalues determines the number of eigenvectors that must be stored for the rotation operation. Fortunately, the correlation matrix will typically be singular and many of the eigenvalues will be zero. Once the eigenvectors are computed, the memory requirement is reduced. The hyperbox detection boundary is a compromise between the hypersphere and hyperellipse boundaries.

4.10. RESULTS

The following paragraphs describe tests and measurements related to detection of shorted rotor windings. The tests include measurements and characterizations of brush noise, detection experiments with laboratory machines, and a field test involving two large generators.

4.10.1. Brush Noise Characterization

The feasibility of detecting faults in a dismantled rotor is easily demonstrated. However, noise is introduced by the brushes when the signal is applied to a rotor in operation. The effect of brush noise on signals used to detect rotor winding faults while the machine is operating must be determined.

The noise on the brushes of a DC machine operating at full load was measured and collected using a Tektronix TDS540 oscilloscope and a GPIB interface to a personal computer. The power spectrum of the stored noise data was computed using the following procedure:

1. The mean of each signal was removed. This acts as a high-pass filter, especially at the high sampling frequency relative to the power supply ripple frequency.
2. The autocorrelation function was estimated by computing the magnitude squared of the Fourier transform of the signal and inverse transforming. The inverse transform was then scaled to produce an unbiased estimate of the autocorrelation function.
3. The estimated autocorrelation functions of several (usually 10) collected signals were averaged and windowed to remove edge transitions. A Hamming window was used in this study.
4. The magnitude of the Fourier transform of the windowed autocorrelation average was computed to produce the estimated power spectral density of the noise signal.

Figure 4.7 shows the measured signal with the machine rotating at rated speed (1800 rpm) under rated load (field excitation of 90 V). The slow oscillation is the full-wave rectified power supply ripple at 120 Hz. The noise at the sampling frequency used for this measurement is not high enough to prevent aliasing. The sampling frequency at which aliasing did not occur was determined by progressively increasing the sampling rate until the power spectrum did not change. The step increases in

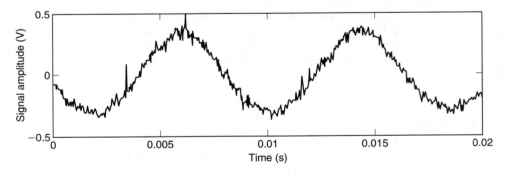

Figure 4.7 Brush noise measurement showing noise spikes and power supply ripple.

sampling rate required that the data be collected using 25×10^6 samples per second to avoid aliasing.

One further observation about the data in Figure 4.7 is the periodic noise spikes. These spikes occur at approximately 1500 Hz. The periodic nature is more evident in the expanded signal of Figure 4.8. Unfortunately, hardware limitations prevent simultaneous collection of the spikes while using a sampling rate that avoids aliasing. If a high sampling rate is used, an insufficient number of samples can be collected to obtain more than one of these spikes. A detailed analysis of the spike frequency is a potential topic of future work.

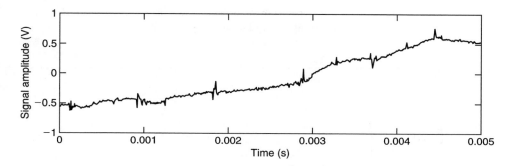

Figure 4.8 Brush noise measurement showing the periodic nature of the noise spikes.

The analysis of the power spectrum of the brush noise in general is hindered by this limitation. The oscilloscope trigger is required to capture the noise spikes within the displayed trace. However, the spikes are then overrepresented in the data. For example, it would be desirable to remove the brush noise by averaging many sweeps of the data. If the oscilloscope trigger is used to catch the spikes, the average is not zero since the spikes have a similar shape from one sweep to the next. But if the oscilloscope trigger is not used, the probability of collecting a spike becomes small. The ability to remove the brush noise was thus verified visually by noting the absence of spikes flashing across the oscilloscope screen. Data are collected and analyzed by repeatedly triggering the oscilloscope at random points in an attempt to capture the noise character. The resulting power spectrum is shown in Figure 4.9 for averaging of 100 sweeps before collection. It should be noted that although the probability of capturing a noise spike in the oscilloscope trace is very low, the probability that a fault detection pulse will overlap with a noise spike is much higher. The time between oscilloscope samples is very small but the time between sweeps is quite large. As described below, the noise spike frequency is relatively close to the pulse frequency of the fault detection system.

Figure 4.10 shows the power spectrum of the voltage across the brushes of a motor operating at rated speed and field excitation using the oscilloscope trigger to capture the noise spikes. Averaging was not used during this measurement. Note the difference in the scales between Figures 4.9 and 4.10. The frequency content appears

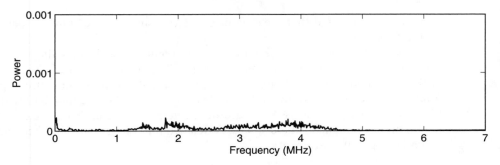

Figure 4.9 Power spectrum of brush noise where each measurement was the average of 100 oscilloscope sweeps.

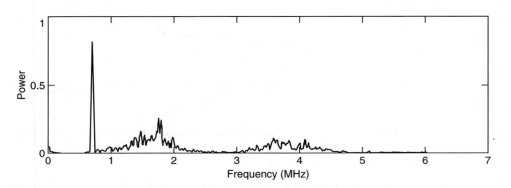

Figure 4.10 Power spectrum of brush noise.

to be roughly the same except the power scale is greatly expanded. The noise has been essentially removed by the averaging process.

Figure 4.11 shows the power spectral density for the brush noise measured on a different machine of the same type. The major difference between the two machines is the absence of the large spike at approximately 700 kHz in the spectrum of the second machine. The cause of this large spike is unknown. The distribution of power in the second machine is also shifted more toward the higher end of the spectrum. The range of frequencies contained in the noise for both machines appears to be quite similar.

The character of the brush noise was also measured under several other conditions. Figure 4.12 shows the power spectrum of the noise measured on a machine operating at rated speed without a load (field excitation was 0). Compared to the spectrum shown in Figure 4.11, the power spectrum shows a slight compression of the frequencies present and a significant reduction in the overall power at the higher frequencies. This would imply that any method that will effectively remove the effects of brush noise for a machine under load would also function for an unloaded machine.

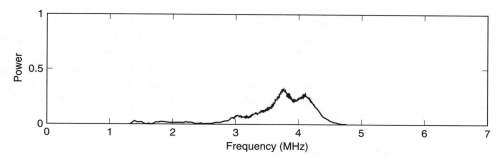

Figure 4.11 Power spectrum of noise spikes for another similar machine.

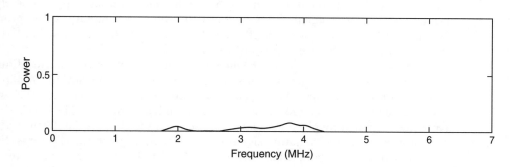

Figure 4.12 Power spectrum for a rotating machine without field excitation.

The power spectrum of the signature signal collected from a rotor is displayed in Figure 4.13. The signals were applied to a rotor with shorts in eight arbitrary locations and once to an unshortened rotor. The nine signals were used to compute the power spectrum as described above. The frequency axis has been greatly expanded to show the frequency content more clearly. The majority of the signal power lies in a range less than 100 kHz and the frequencies extend out to close to 5 MHz. It appears

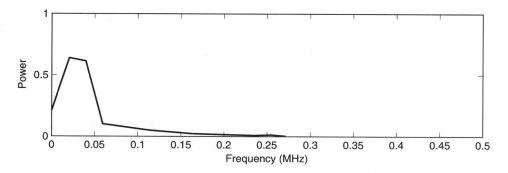

Figure 4.13 Power spectrum of signature signal collected from a dismantled rotor. The scale has been expanded to show the variation.

that the signal spectrum and the noise spectrum overlap to a large extent. Recall, however, that the oscilloscope trigger was used to capture the brush noise spikes, and thus the noise spikes are overrepresented in the data. This may significantly alter the relative power of the noise, since in normal operation, the spikes are absent for a majority of the time. It is also postulated that the higher frequencies of the signal are not needed to detect the rotor faults, and thus a low-pass filter may be sufficient to remove the brush noise.

4.10.2. Laboratory Tests

Signature signals were collected for several small laboratory machines in an attempt to determine if the signature signals for similar machines are approximately the same. The signature signals were analyzed to estimate the ability of the proposed system to detect rotor faults. The effects of the level of excitation applied to the rotor on the signature signal were also investigated.

The first test performed was similar to the dismantled rotor fault location test. The signal was injected into the rotor winding by direct connection to the slip rings of the laboratory machine. The machine is small compared to machines that will be tested in the field but should represent a more difficult problem. The brushes of the rotor were isolated by placing an insulator between the slip rings and brushes. With the rotor stopped, the signature signal was measured between the slip rings. For all tests the signal was sampled at a rate of 25×10^6 samples per second for 1000 samples per signal. The Tektronix TDS540 oscilloscope was set to collect 50 samples (5%) prior to the trigger event. The oscilloscope was triggered on the rising edge of the A signal and a 20-MHz bandwidth was selected.

A typical signature signal appears in Figure 4.14. Only 250 samples of the signal are plotted in Figure 4.14 since the remaining samples are of little interest. The greatest detail appears within this interval, but more samples were collected for completeness. The power spectral density as a function of frequency is shown in

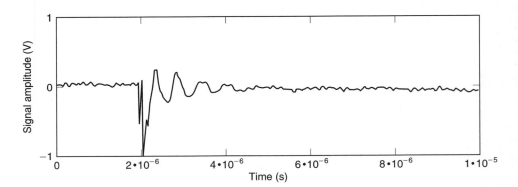

Figure 4.14 Typical signature signal for pulses applied directly to the slip rings. The excitation supply is not connected and the rotor is stationary.

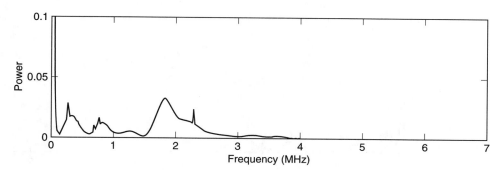

Figure 4.15 Power spectral density of signature signals applied directly to the slip rings of the rotor. The signals are filtered to remove the DC component and all frequencies above 4.0 MHz.

Figure 4.15. Most of the signal power is concentrated below 3.0 MHz with some power at approximately 5 MHz. The plot shows the power density spectrum after applying a bandpass filter that removes the AC ripple and all frequencies above 4.0 MHz. The circuit used to measure the signature signals has a limited bandwidth and contributes to the low-pass nature of the signature signals.

No shorted-turn data were gathered for a dismantled rotor of this type due to the difficult nature of introducing shorts. The windings of the rotor take complex paths, making determination of the short location difficult, and a rotor that could be destructively tested was not available. Thus a comparison of the measured signal with signals for a dismantled rotor is not possible. The change in the signature signal as shorts are introduced could aid the detection process by providing information relative to the characteristics of the signal in the presence of shorts.

Next, the insulators between the brushes and slip rings were removed and the fault detection pulses were applied directly to the brushes. The rotor excitation was not connected and the rotor was not rotating. A typical signature signal appears in Figure 4.16. The signal shown differs significantly from the signature signal shown in Figure 4.14.

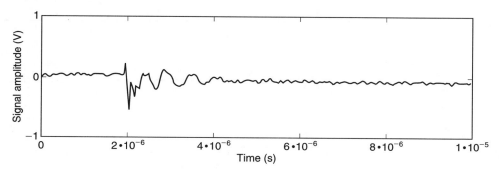

Figure 4.16 Typical signature signal when pulses are applied to a nonrotating machine through brushes. The excitation supply is not connected.

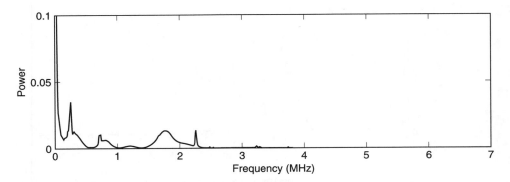

Figure 4.17 Power density spectrum for signals applied through the brushes of a nonrotating rotor.

The power spectral density is shown in Figure 4.17. This signal is very nearly the same as the signal applied directly to the slip rings. The brushes therefore have no adverse impact on the frequency content of the signature signal when no excitation source is connected and the rotor is not rotating. The power in the 2.0-MHz range is slightly reduced, but the general shape of the power density spectrum is the same.

Next, a three-phase AC induction motor was used to rotate the DC excitation source and the synchronous generator under test. The excitation was still not connected during the test. Figure 4.18 shows a typical signature signal. The signal now has a much larger positive pulse at the start of the signature signal. This indicates that the rotation of the machine introduces significant signal component variations.

The DC excitation source was then connected to the rotor field winding and used to supply various levels of excitation. The levels are specified in volts in the range from a minimum of approximately 2.5 V to a maximum of 145.0 V. The fault detection pulses were applied to the brushes of the rotor and the signature signals recorded. Numerous samples were taken using the averaging capability of the oscilloscope. Averaging the signature signals for 50 sweeps resulted in a very stable

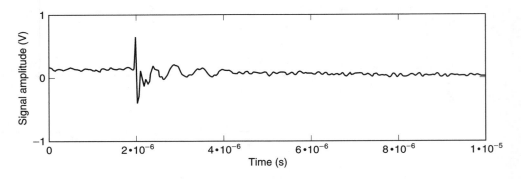

Figure 4.18 Typical signature signal when rotor is rotating without excitation connected.

waveform that seemed not to vary with excitation level. One of the difficulties of the measurement process is the rather large AC ripple present on the DC excitation. This ripple can be removed with a high-pass filter. The magnitude of the ripple is on the order of the measured signature signal and varies in proportion to the excitation level. This experiment was repeated on five similar rotors.

The excitation source significantly alters the shape of the signature signal. Thus, the signature signal from a dismantled rotor cannot be used as the signature for an excited rotating rotor. Rather, a prototype signature signal is formed by averaging the signature signals at different excitations. This approach does not require an operator to enter the excitation as a parameter of the detector. If, however, the excitation were made available, greater accuracy could possibly be achieved.

Figure 4.19 shows the variation of the signature signal distance from a prototype signature signal for different rotors. In this case, the prototype signature signal is the average signature signal collected at all excitation levels on all five rotors. The solid line shows the mean distance from the prototype for the different machines. The dotted line shows the standard deviation. Though the mean seems to vary considerably from one machine to another, the standard deviation is relatively low. This illustrates the consistency of the signature signals over different levels of excitation. The numbers at the bottom of the plot are for plotting purposes only and are not relevant to the data.

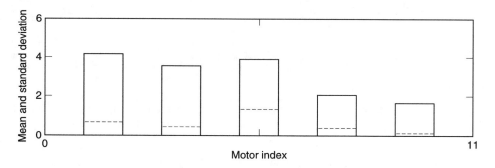

Figure 4.19 Mean (solid line) and standard deviation (dotted line) of signature signal distances from a prototype signature signal for different rotors.

Figure 4.20 shows the mean and standard deviation of the distance of signature signals from the prototype when signature signals at the same excitation level but different machines were used to compute the statistics. The same prototype was used here as in the paragraph above. The average distance as a function of excitation voltage is shown by the solid line. The dotted line shows the standard deviation of the samples. The large mean and standard deviation at 0 V excitation represent an extreme case. This signal, which came from machine 4 (as identified by the last digit of the bench serial number), is removed from consideration and the measurements repeated. The new mean is shown by the dashed line and the new standard

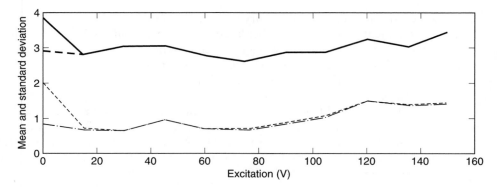

Figure 4.20 Mean (solid line) and standard deviation (dotted line) of distances as a function of excitation. The mean (dashed line) and standard deviation (dash-dot line) with the outlier signal removed.

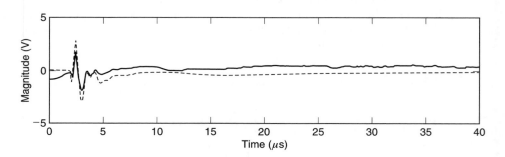

Figure 4.21 Outlier signature signal (solid line) and a typical signature signal (dotted line) from a different rotor.

deviation by the dash-dot line. Most of the statistics are virtually the same after removal of the outlier signature signal except for the values at zero excitation, as expected.

Figure 4.21 shows the outlier signature signal as a function of time. Another typical signature signal is also shown by the dashed line. The large offset over the full time span contributes to the large distance from the prototype signature signal.

Figure 4.22 shows the distance of signals from a prototype at each excitation voltage. The prototype consists of the average signal for a single machine at all excitation levels. The larger distances seen for two of the machines are correlated to observations during the data collection process. For the two machines in question, a much larger variation in the signal was noticed, perhaps as a low-frequency oscillation above the AC ripple frequency. The cause of the oscillation has not been determined. The very large distance for the zero excitation voltage signal shown by the dashed line is the same outlier signature signal mentioned above.

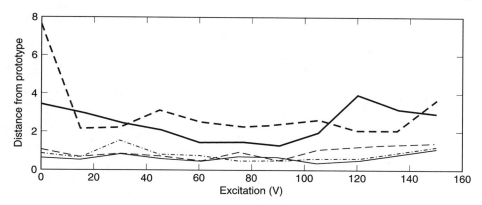

Figure 4.22 Distance of signature signals from prototype signature signal for five rotors.

The preceding experiments show that faults can be detected on an operating rotor by first recording the signal that represents a rotor with no faults. The introduction of the DC excitation source changes the character of the signature signal relative to a dismantled rotor, and thus a complete determination of the signal under fault conditions is not possible. If a very large number of signals are collected for a rotor without faults, the deviation of the signal required for fault detection can be determined.

4.10.3. Field Tests

Signature signals were recorded for two rotors of small generators in various modes of operation. Signature signals for the first machine were taken at stand-still and while rotating at various speeds, with and without the brushes contacting the rings and with and without the excitation source connected to the rotor. Signature signals for the second machine were only collected with the brushes removed from the slip rings and the rotor not rotating. In both cases the instrumentation circuit was not grounded directly at the rotor. The signature signals were recorded at 250×10^6 samples per second with a 20-MHz low-pass filter. The signature signals consist of 1000 samples of the reflected difference signal. Each signature signal is the average of 50 individual signals to remove the effects of noise. The averaging was performed by the TDS540 oscilloscope.

The second experiment was conducted with the brushes in contact with the slip rings and the signature signal measured through the brushes. Figure 4.23 shows the signature signal when connected directly to the slip rings (solid line) and through the brushes (dashed line). The signature signal obtained through the brushes has a significantly different shape from the signature signal applied directly to the slip rings. Though not shown in the figure, connection of the excitation source also significantly altered the signature signal, as expected.

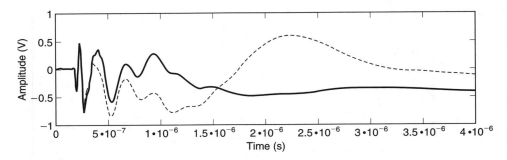

Figure 4.23 Plot of signature signal taken directly from the slip rings (solid
line) and the signature signal taken through the brushes
(dotted line).

The next experiment consisted of connecting the excitation source to the rotor.
In several stages, the rotor was brought up to full speed with the excitation connected, and thus providing power to the rotor. The amount of power applied to the
rotor is proportional to rotation speed. Figure 4.24 shows plots of the signature
signal at several different speeds. The solid line represents no rotation, the dotted
line represents a very slow rolling rate, the dashed line represents 1800 rpm, and the
dashed-dot line represents 3600 rpm. All of the signature signals are seen to be quite
similar, especially in the early portion. As the rotation speed increased, the signature
signals became more unstable, as if being modulated by a low frequency signal. The
same effect was noticed in a laboratory experiment where the low frequency modulation was at 120 Hz, the AC ripple on top of the DC excitation source.

Signature signals from a second machine were recorded by connecting the measurement leads directly to the slip rings. The rotor was not rotating during this test.

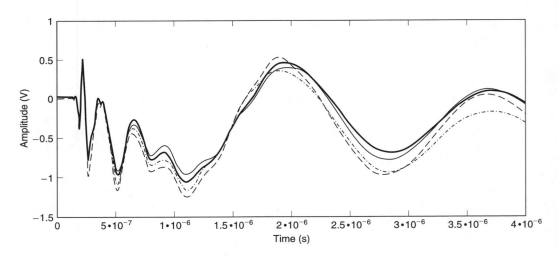

Figure 4.24 Signature signals for various rotation rates.

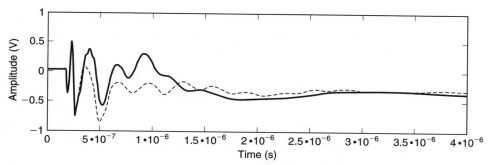

Figure 4.25 Signature signals for two different machines. The measurement
leads were connected directly to the slip rings

Figure 4.25 shows the signature signals for both machines are different or that one
machine has shorter turns. The resistance to ground for the second machine, measured before the experiment began, was found to be much lower than the resistance
for the first machine. It is not clear that the resistance to ground should have such an
impact and further studies are indicated.

The signature signals collected during the experiment demonstrate that detection of shorted rotor windings may indeed be possible while the rotor is in operation.
Several tests are required to improve the detection accuracy. Since the signature
signals for the two similar machines were different, a test should be conducted on
two machines that are known to have identical rotors and excitation sources, and
known not to have shorted rotor windings. The resulting signature signals will verify
whether the detection process will require signature signals for all machines before
shorted windings can be detected, or if a single signature signal for a given machine
type will be sufficient.

4.11. CONCLUSION

Use of neural networks with fuzzy logic outputs and traveling waves techniques has
been shown to be an accurate locator of shorted turns in turbo-generator rotors. The
technique is easy to apply not only to rotors but to transformers and other devices
containing symmetrical windings. The technique is extended to operational rotors by
the use of a novelty filter. These two forms of shorted-turn detection show great
promise in the monitoring of high-speed turbogenerators.

Though the novelty filter technique shows great promise, the detection capabilities have not been verified. Since introduction of shorts into on-line rotors is not
possible, the explicit verification of any fault detection technique is also not possible
without the availability of a machine with a suspected shorted turn. Signature signals
could be collected before the suspect rotor is dismantled for maintenance. After the
rotor is repaired and brought back on line, healthy signature signals can be collected.
The healthy regions and thresholds can then be established using the proposed

techniques. The signature signals recorded before correction of the fault can then be processed by the detection algorithm to determine the effectiveness of the proposed technique.

References

[1] D. R. Albright, "Interturn Short-Circuit Detection for Turbine Generator Rotor Winding," *IEEE Transactions on Power and Apparatus Systems*, Vol. PAS-90, 1971, pp. 478–483.

[2] J. Mulhaus, D. M. Ward, and I. Lodge, "The Detection of Shorted Turns Alternator Rotor Windings by Measurement of Circulating Stator Currents," *IEE Conference Publications*, Vol. 254, 1985, pp. 100–103.

[3] J. W. Wood and R. T. Hindmarch, "Rotor Winding Short Detection," *IEE Proceedings*, Vol. 133, Pt. B, No. 3, 1986, pp. 181–189.

[4] L. C. Shan and J. A. Kong, *Applied Electromagnetism*, PWS Publishers, Boston, 1987.

[5] S. Ramo, J. R. Whinnery, and T. Van Duzer, *Fields and Waves in Communication Electronics*, Wiley, New York, 1965.

[6] M. J. Damborg, M. A. El-Sharkawi, and R. J. Marks II, "Potential Applications of Artificial Neural Networks to Power System Operation," paper presented at the IEEE International Symposium on Circuits and Systems, Louisiana, May 1990.

[7] M. A. El-Sharkawi, R. J. Marks II, and S. Weerasooriya, "Neural Networks and Their Applications to Power Engineering," *Advances in Control and Dynamic Systems*, Vol. 41, Academic, 1992.

[8] P. K. Simpson, "Foundation of Neural Network Paradigms," in *Artificial Neural Networks*, IEEE Press, New York, 1992.

[9] L. A. Zadeh, "Fuzzy Sets," *Information and Control*, Vol. 8, 1965, pp. 338–353.

[10] L. A. Zadeh, "Outline of New Approach to the Analysis of Complex Systems and Decision Processes," *IEEE Transactions on Systems, Man and Cybernetics*, Vol. SMC-1, 1973, pp. 28–44.

[11] T. Terano, K. Asai, and M. Sugeno, *Fuzzy Systems Theory and Its Applications*, Academic, San Diego, 1992.

[12] G. J. Klir and T. A. Folger, *Fuzzy Sets, Uncertainty and Information*, Prentice-Hall, Englewood Cliffs, NJ, 1988.

[13] K. Fukunaga, *Introduction to Statistical Pattern Recognition*, Academic, New York, 1972.

[14] K. Fukunaga and W. L. G. Koontz, "Application of Karhunen-Loe've Expansion to Feature Selection and Ordering," *IEEE Transactions on Computers*, Vol. C-19, No. 4, April 1970.

[15] T. Y. Young and K. S. Fu, *Handbook of Pattern Recognition and Image Processing*, Academic, San Diego, 1986.

[16] S. Weerasooriya and M. A. El-Sharkawi, "Feature Selection for Static Security Assessment Using Neural Networks, paper presented at the IEEE International Symposium on Circuits and Systems, San Diego, May 1992, pp. 1693–1696.

[17] S. Weerasooriya and M. A. El-Sharkawi, "Use of Karhunen Loe've Expansion in Training Neural Networks for Static Security Assessment," paper presented at the First International Forum on Application of Neural Networks to Power Systems, Seattle, July 1991, pp. 59–64.

[18] M. Chow, G. L. Bilbro, and S.-O. Yee, "Application of Learning Theory to a Single Phase Induction Motor Incipient Fault Detector Artificial Neural Network," in *Proceedings of the First International Forum on Applications of Neural Networks to Power Systems*, El-Sharkawi and Marks (Eds.), Seattle, WA, 23–26 July 1991, pp. 97–101.

[19] M. Chow and S.-O. Yee, "Methodology for On-line Incipient Fault Detection in Single-Phase Squirrel-Cage Induction Motors Using Artificial Neural Networks," *IEEE Transactions on Energy Conversion*, Vol. 6, No. 3, 1991, pp. 536–545.

[20] M. Chow, P. M. Magnum, and S.-O. Yee, "A Neural Network Approach to Real-Time Condition Monitoring of Induction Motors," *IEEE Transactions on Industrial Electronics*, Vol. 38, No. 6, 1991, pp. 448–453.

[21] M. Chow, A. V. Chew, and S.-O. Yee, "Performance of a Fault Detector Neural Network Using Different Paradigms," *Proceedings of the SPIE, Applications of Artificial Neural Networks III*, Orlando, FL, Vol. 1709, The International Society for Optical Engineering, Bellingham, WA, 1992, pp. 973–981.

[22] T. Kohonen, *Associative Memory: A System Theoretic Approach*, Springer-Verlag, Berlin, 1997.

[23] H. Ko, R. Baran, and M. Arozullah, "Neural Network Based Novelty Filtering for Signal Detection Enhancement," in *Proceedings of the 35th Midwest Symposium on Circuits and Systems*, Washington, DC, August 1992, 9–12, The Institute of Electrical and Electronics Engineering, Piscataway, NJ, pp. 252–255.

[24] K. Simpson, *Artificial Neural Systems: Foundations, Paradigms, Applications and Implementations*, Pergamon, New York, 1993, pp. 80–82.

[25] S.-O. Yee and M. Chow, "Robustness of an Induction Motor Incipient Fault Detector Neural Network Subject to Small Input Perturbations," *IEEE Proceedings of SOUTHEASTCON*, Williamsburg, VA, April 1991, 7–10, The Institute of Electrical and Electronics Engineering, Piscataway, NJ, pp. 365–369.

O. P. Malik
Department of Electrical and
Computer Engineering
The University of Calgary
Calgary, Alberta
T2N 1N4 Canada

K. A. M. El-Metwally
Electric Machines and Power
Department, Cairo University
Cairo, Egypt

Chapter 5

Fuzzy Logic Controller as a Power System Stabilizer

5.1. INTRODUCTION

The structure and design of a fuzzy logic controller, and an algorithm to tune its parameters to achieve the desired performance are described. Two rule generation methods to automatically generate the fuzzy rule set are proposed. The application of the fuzzy logic controller as a power system stabilizer is investigated by simulation studies on a single-machine infinite-bus system and on a multi-machine power system. Implementation of the fuzzy logic based power system stabilizer on a micro-controller and results of experimental studies on a physical model of a power system illustrate the effectiveness of the fuzzy logic based controller.

Low-frequency oscillations are a common problem in large power systems. A power system stabilizer (PSS) can provide supplementary control signal to the excitation system and/or the speed governor system of the electric generating unit to damp these oscillations and to improve its dynamic performance [1]. Due to their flexibility, easy implementation, and low cost, PSSs have been extensively studied and successfully used in power systems for many years.

Most PSSs in use in electric power systems employ the classical linear control theory approach based on a linear model of a fixed configuration of the power system. Such a fixed-parameter PSS, called a conventional PSS (CPSS), is widely used in power systems and has made a great contribution in enhancing power system dynamics [2].

Power systems are dynamic systems and their operation is of a stochastic nature. The characteristics of the plant are nonlinear. For example, the gain of the plant increases with generator loading and AC system strength [2]. Also, the phase lag of the plant increases as the system becomes stronger. Thus controller parameters that are optimum for one set of operating conditions may not be optimum for another set of operating conditions. The system configuration also keeps changing either due to switching actions in the short term or system enhancements in the long term.

In the conventional fixed-parameter controllers, the gains and other parameters may not ideally suit the entire spectrum of operation. Developments in digital technology have made it feasible to develop and implement improved controllers based on modern, more sophisticated techniques. Power system stabilizers based on adaptive control [3, 4], artificial neural networks [5], and fuzzy logic [6, 7] are being developed. Each of these control techniques possesses unique features and strengths.

In the traditional bivalent logic of Aristotle [6] something either belongs to a set or does not, thereby leaving no room for ambiguities. However, ambiguities are common in the real world. Contrary to traditional logic, where boundaries are rigid, fuzzy logic not only tolerates but is based on the looseness of boundaries.

In recent years, fuzzy logic has emerged as a powerful tool and is starting to be used in various power system applications [7–13]. The application of fuzzy logic control technique appears to be the most suitable one whenever a well-defined control objective cannot be specified, the system to be controlled is a complex one, or its exact mathematical model is not available. Fuzzy logic controllers (FLCs) are robust and have relatively low computation requirements. They could be constructed easily using a simple microcomputer.

Development of a fuzzy-logic-based power system stabilizer to maintain stability and enhance closed-loop performance of a power system is described in this chapter. Simulation studies on a single-machine infinite-bus system and on a multi-machine power system model show very satisfactory performance.

The fuzzy-logic-based PSS (FLPSS) has been implemented on a low-cost micro-controller and tested in the laboratory on a physical model of a single-machine infinite-bus system. Experimental tests and results are also described.

5.2. FLC STRUCTURE

In conventional control, the amount of control is determined in relation to a number of data inputs using a set of equations to express the entire control process. Expressing human experience in the form of a mathematical formula is a very difficult task, if not an impossible one. Fuzzy logic provides a simple tool to interpret this experience into reality.

Fuzzy logic controllers are rule-based controllers. The structure of the FLC resembles that of a knowledge-based controller except that the FLC utilizes the principles of fuzzy set theory [14, 15] in its data representation and its logic. The basic configuration of the FLC can be simply represented in four parts, as shown in Figure 5.1 [16]:

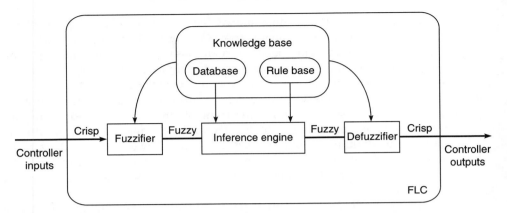

Figure 5.1 Schematic diagram of the FLC building blocks.

- *Fuzzification module*, the functions of which are, first, to read, measure, and scale the control variable (e.g. speed, acceleration) and, second, to transform the measured numerical values to the corresponding linguistic (fuzzy) variables with appropriate membership values.
- *Knowledge base*, which includes the definitions of the fuzzy membership functions defined for each control variable and the necessary rules that specify the control goals using linguistic variables.
- *Inference mechanism*, which is the kernel of the FLC. It should be capable of simulating human decision making and influencing the control actions based on fuzzy logic.
- *Defuzzification module*, which converts the inferred decision from the linguistic variables back to numerical values.

5.3. FLC DESIGN

The design process of an FLC may be split into the five steps described below.

5.3.1. Selection of Control Variables

The selection of control variables (controller inputs and outputs) depends on the nature of the controlled system and the desired output. It is more common to use the output error (e) and the rate or derivative of the output (e') as controller inputs. Some investigators have also proposed the use of error and the integral of error as an input to the FLC [16, 17].

5.3.2. Membership Function Definition

Each of the FLC input signals and output signals, fuzzy variables ($X_j = \{e, e', u\}$), has the real line R as the universe of discourse. In practice, the universe of discourse is restricted to a comparatively small interval $[X_{\min_j}, X_{\max_j}]$. The universe of discourse of each fuzzy variable can be quantized into a number of overlapping fuzzy sets (linguistic variables). The number of fuzzy sets for each fuzzy variable varies according to the application. A common and reasonable number is an odd number $(3, 5, 7, \ldots)$. Increasing the number of fuzzy sets results in a corresponding increase in the number of rules.

A membership function is assigned to each fuzzy set. The membership functions map the crisp values into fuzzy values. A set of membership functions defined for seven linguistic variables NB, NM, NS, Z, PS, P, and PB, which stand for Negative Big, Negative Medium, Negative Small, Zero, Positive Small, Positive Medium, and Positive Big, respectively, is shown in Figure 5.2.

Membership functions can be of a variety of shapes, the most usual being triangular, trapezoidal, or a bell shape. The triangular shape shown in Figure 5.2 and used for the controller described in this chapter can be expressed as

$$
F_i(x) = \begin{cases} \dfrac{x - \mu_i + \sigma_i}{\sigma_i} & \text{for } \mu_i - \sigma_i \leq x \leq \mu_i \\[2ex] -\dfrac{x - \mu_i - \sigma_i}{\sigma_i} & \text{for } \mu_i \leq x \leq \mu_i + \sigma \end{cases} \tag{5.1}
$$

where μ_i is the centroid of the ith membership function and σ_i is a constant that determines the spread of the ith membership function.

For simplicity, it is assumed that the membership functions are symmetrical and each one overlaps with the adjacent functions by 50%. In practice, the membership functions are normalized in the interval $[-L, L]$, which is symmetrical around zero. Thus, control signal amplitudes (fuzzy variables) are expressed in terms of controller parameters (gains), as shown in Figure 5.3. These parameters can be defined as

$$
K_j = \frac{2L}{X_{\text{range}_j}} \tag{5.2}
$$

where X_{range_j} defines the full range of the control variable X_j, that is,

$$
X_{\text{range}_j} = X_{\max_j} - X_{\min_j} \tag{5.3}
$$

and $X_{\max_j}, X_{\min_j}$ are the maximum and the minimum values of the control variable X_j.

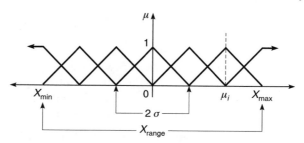

Figure 5.2 Seven triangular membership functions.

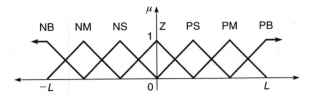

Figure 5.3 Normalized triangular membership functions.

Using the above definitions, the parameters of the membership function (σ_i, μ_i) can be calculated easily as

$$\sigma_i = \frac{X_{\text{range}_j}}{n-1} \tag{5.4}$$

$$\mu_i = X_{\text{min}_j} + (k-1) * \sigma_i \tag{5.5}$$

where $i = 1, \ldots, n$ and n is the number of linguistic variables.

The input and output gains K_j are referred to as the FLC parameters. The selection of these parameters is usually based on the previous knowledge of the controlled system.

5.3.3. Rule Creation and Inference

In general, fuzzy systems, as function estimators, map an input fuzzy set to an output fuzzy set $S: I^n \rightarrow I^p$. Fuzzy rules are the relations between the fuzzy sets. They usually are in the form "if A, then B," where A is the rule antecedent and B is the rule consequence. Each rule defines a fuzzy patch in the Cartesian product $A \times B$ (system state space). The antecedents of each fuzzy rule describe a fuzzy input region in the state space. This enables one to effectively quantize continuous state space so that it covers a finite number of such regions [15, 18]. In terms of associative memory definition (FAM), each rule represents an association $(A_i; B_i)$. The FAM structure of rules was introduced in 1987 [15]. The structure proposed firing all the rules at the same time (analogous to neural networks). This enables easier and faster very large scale integrated (VLSI) analogue and digital designs. A fuzzy system using two antecedents and one consequence $(A, B; C)$ is shown in Figure 5.4. The association $(A_i, B_i; C_i)$ or the rule of "A_i and B_i then C_i" maps inputs A, B to C', a partially activated version of C. The corresponding output fuzzy set C combines the partially activated sets $C'^1, \ldots, C'^m$, that is,

$$C = \sum_{i}^{m} C'^i \tag{5.6}$$

Consider a controller of two fuzzy variables, error (e) and derivative of output (e'), and one output fuzzy variable, control signal (μ), each quantized to seven fuzzy sets. This leads to a 7×7 FAM rule matrix, as shown in Figure 5.5. Every entity in the matrix represents a rule, for example,

If e is NB and e' is NM, then U is NB.

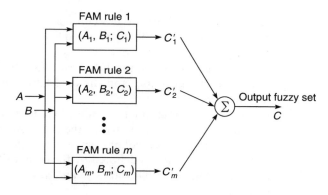

Figure 5.4 FAM system architecture for two fuzzy antecedent and one consequence.

Error derivative

Output error		NB	NM	NS	Z	PS	PM	PB
	NB	NB	NB	NB	NB	NM	NS	Z
	NM	NB	NB	NM	NM	NS	Z	PS
	NS	NB	NM	NM	NS	Z	PS	PM
	Z	NM	NM	NS	Z	PS	PM	PM
	PS	NM	NS	Z	PS	PM	PM	PB
	PM	NS	Z	PS	PM	PM	PB	PB
	PB	Z	PS	PM	PB	PB	PB	PB

Figure 5.5 Fuzzy control rules matrix.

The activation of the ith rule consequent is a scalar value w_i, which equals the minimum of the two antecedent conjuncts' values. For example, if e belongs to NB with a membership of 0.3 and e' belongs to NM with a membership of 0.7, then the rule consequence (w_i) will be 0.3. The rules' matrix can be extended to a multidimensional matrix based upon the number of control variables. The knowledge required to generate the fuzzy rules can be derived from an off-line simulation, an expert operator, and/or a design engineer. Some of the knowledge can be based on understanding of the behavior of the dynamic system being controlled. A lot of effort has been devoted to the creation of the fuzzy rules [15]. In some cases there is an upper hierarchical level of rules that generates the system rules, as in ref. 19. In other cases neural networks have been trained to generate the rules [20].

5.3.4. Fuzzy Inference

The well-known inference mechanisms in fuzzy logic are the correlation-minimum encoding and the correlation-product encoding. Consider fuzzy sets A and B to be fuzzy subsets of X, Y. The geometric set-as-points interpretation of finite fuzzy sets A and B as points in unit cubes allows the representation of the sets of vectors. Thus, A and B can be represented by numerical fit vectors $A = (a_1, \ldots, a_n)$ and $B = (b_1, \ldots, b_m)$, where $a_i = m_A(x_i)$ and $b_i = m_B(y_i)$. Let the relation between A and B be governed by the FAM rule (A, B). Using these definitions, the correlation-minimum encoding and the correlation-product encoding can be expressed as follows.

(i) The correlation minimum encoding is based on the fuzzy outer product notation. Expressed in matrices, the pairwise multiplication is replaced by the pairwise minima and the column sums with the column maxima [16]. This max–min composition relation is denoted by the composition operator ∘. Thus the fuzzy outer product of the fit row vectors A and B, which forms the FAM matrix M, can be given by

$$M = A^T \circ B \tag{5.7}$$

where

$$m_{ij} = \min(a_j, b_j) \quad \text{or} \quad a_j \hat{} b_j \tag{5.8}$$

and the cap operator $\hat{}$ indicates the pairwise minima.

(ii) The correlation-product encoding uses the standard mathematical outer product of the fit vectors A and B to form the FAM matrix M:

$$M = A^T B \tag{5.9}$$

where

$$m_{ij} = a_i b_j \tag{5.10}$$

Correlation-minimum encoding produces a matrix of clipped B sets, while correlation-product encoding produces a matrix of scaled B sets. In membership function representation, the scaled fuzzy sets $a_i B$ all have the same shape as B [16]. The clipped fuzzy sets $a_i \hat{} B$ are flat at or above a_i value. An example of two antecedent and one consequence rule that illustrates the difference between the two encoding techniques is shown in Figures 5.6 and 5.7. In this sense correlation-product encoding preserves more set information than the correlation-minimum encoding. The correlation-product encoding is used in the FLC application in this chapter.

5.3.5. Defuzzification Strategy

Defuzzification is a process of converting the FLC inferred control actions from fuzzy values to crisp values. This process depends on the output fuzzy set, which is generated from the fired rules [Eq. (5.6)]. The output fuzzy set is formed by either a correlation-minimum encoding or the correlation-product encoding as discussed in Section 5.3.4. Let the output fuzzy set be generated using the correlation-product encoding, defined by the function $m_o(y)$ shown in Figure 5.6, where

$$m_o(y) = \sum_{i=1}^{N} m_{o_i}(y) \tag{5.11}$$

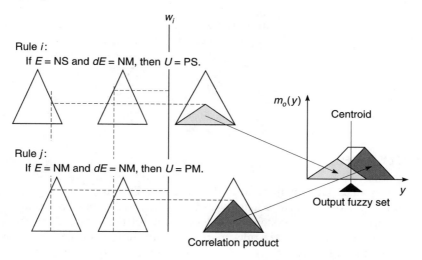

Figure 5.6 Generation of the output fuzzy set using correlation product encoding.

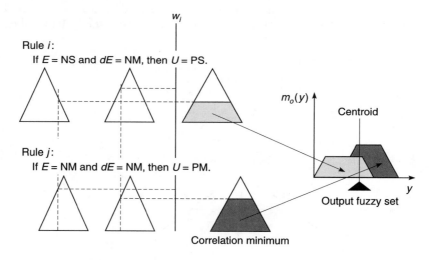

Figure 5.7 Generation of the output fuzzy set using correlation minimum encoding.

The control output u can then be expressed as

$$u_k = \frac{\displaystyle\int ym_o(y)\,dy}{\displaystyle\int m_o(y)\,dy}$$

(5.12)

For a discretized output universe of discourse $Y = \{m_1, \ldots, m_m\}$ that gives the discrete fuzzy centroid, Eq. (5.12) can be reduced to [15]

$$u_k = \sum_{i=1}^{m} y_i m_o(y_i) \left/ \sum_{i=1}^{m} m_o(y_i) \right. \tag{5.13}$$

5.4. FUZZY RULES

Fuzzy rules play a major role in the FLCs and have been investigated extensively [15, 16, 19]. However, rules usually can be generated using knowledge and operating experience with the system or understanding of the system dynamics.

In most cases, fuzzy rules map two input fuzzy variables, for example, the error e and the derivative of error $e\hat{}$, into one output fuzzy variable, the control signal u. In such a system each fuzzy variable can be easily quantized to a number of fuzzy sets, as mentioned in Section 5.3.2.

Consider a system with a highly oscillatory response with e and e' as given in Figure 5.8. Assume that both variables have three sets P, Z, and N. It can be seen from Figure 5.8 that if e is positive and e' is negative, the system will reduce the error

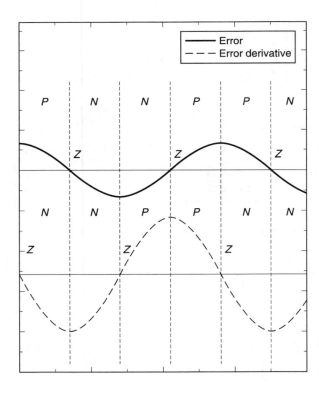

Figure 5.8 Variation of error and derivative of error for a highly oscillatory system.

itself, which means that the magnitude of the applied control should be minimum. This could be written in the form of a rule as follows:

> If e is positive and e' is negative, then the magnitude of the applied control should be zero.

Or:

> If $e \in P$ and $e' \in P$, then u is Z.

This rule corresponds to the rule region labeled A in both the rules generation of system dynamics (Figure 5.9) and the phase plane representation for the fuzzy variables (Figure 5.10).

In the case of negative e and negative e' the system tends to go to instability, which requires an opposite (positive) control action. This is interpreted as being in region B in both Figures 5.9 and 5.10. In region C the error is still negative while the

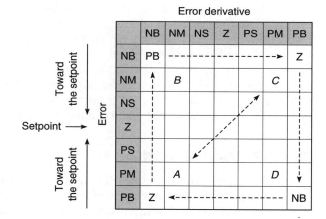

Figure 5.9 Rules generation by understanding the system dynamics.

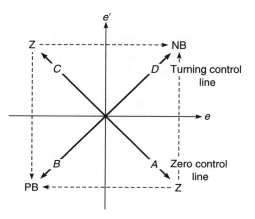

Figure 5.10 Phase plane for the controller fuzzy variables.

error derivative is positive. This implies that the system error is decreasing and thus the magnitude of the control action should be kept to a minimum (zero).

When the error becomes positive and the error derivative goes positive as well (region *D*), the system tends to go to instability again. This implies the necessity to apply opposite (negative) control action to compensate for this tendency toward instability.

When the fuzzy variable quantization is increased to seven fuzzy sets instead of three, the output control action will increase from zero (Z) control in rule region *A* to a minimum positive control (PB) in rule region *B*, as seen in Figure 5.9. Then the control action decreases from PB in rule region *B* to minimum control (Z) in rule region *C*. Also, when both *E* and *e'* approach their maximum positive, the control action goes to maximum negative (NB).

In the phase plane this can be seen as two lines separating different control actions, as seen in Figure 5.10. The first line separates the negative control from the positive control, while the second line represents the turning point of the control action.

A closer look at the rules shows that for this type of disturbance (e.g., positive step) the rules in region *B* represent the essential rules that control the rise time and the settling time whereas region *D* represents the main rules that deal with the system overshoot. The central rule in the central part, that is, *e* and *e'* around zero, is responsible for system steady state.

The preceding description yields a symmetrical set of rules that work in most control cases with some minor alterations. If the dynamic response of the controlled system is totally different from the one shown in Figure 5.8, then this requires a different understanding and a different set of rules.

5.5. FLC PARAMETER TUNING

Parameter tuning for the FLC plays an important role in achieving the controller goals. Previous experience with the controlled system is helpful in selecting the initial values of the FLC parameters. If sufficient information is not available about the controlled system, the selection of suitable FLC parameters can become a tedious trial-and-error process. Some efforts have been reported in the literature to automate the tuning of the FLC parameters at the design stage to get an optimal or near-optimal system performance [21–23].

Another algorithm to tune the FLC parameters off-line is proposed in this section. The objective of the proposed parameter tuning algorithm is to change the controller gains in an organized manner to achieve desired system response. The tuning algorithm tries to minimize three system performance indices (PIs). These indices are the system overshoot and the performance indices J_1, J_2, given as

$$OS = \frac{r - y}{r} \times 100\% \tag{5.14}$$

$$J_1 = \sum e^2 \tag{5.15}$$

$$J_2 = \sum te^2 \tag{5.16}$$

where r is the system reference, y is the system output, e is the system error, and t is the time.

An FLC with two inputs and a single output has two input parameters K_e, K_e' and one output parameter K_u. The three parameters K_e, K_e', and K_u are tuned using the guided search algorithm.

The algorithm changes the three parameters in overlap loops, simulates the system with the new parameters, and calculates the performance index. It also detects if one of the parameters degrades the performance indices or leads to instability. In this case it stops incrementing this parameter. If the desired performance indices are achieved, the search stops. Otherwise, it continues over the specified search range of the FLC parameters [24].

In case of a complete lack of information about the parameters, the search for the best parameters may require a large number of iterations in searching for a proper minimum. Using some practical information about signal levels, it is easy to set an operating range to the FLC parameters.

5.6. AUTOMATIC RULE GENERATION

In some cases the dynamic behavior of the controlled system is unpredictable and difficult to understand. This situation imposes the need to automate the rule generation process. A lot of effort has been devoted to achieve this goal. In some cases an upper hierarchical set of rules has been chosen to generate the controller rules [19]. This technique requires some understanding of the system dynamics to build this type of supervisory rules.

In some other cases, the rules were generated using an artificial neural network. The generation of rules is achieved by training the neural network using sampled data sets. Although effective, this method requires some neural network background and long training time [15].

In this section two effective automatic rule generation (ARG) methods are proposed. These methods are very similar to the neural network technique. They also use sampled data ensemble but employ a fuzzy system to generate the rules instead of a neural network. The rules can be obtained easily from the known desired performance (input-output data pairs) of the system when controlled by another well-designed controller, for example, an adaptive controller or a proportional-integral-derivative (PID).

In the case of a two-input, one-output FLC, the controller requires two fuzzy variables, and using a fuzzy mapping function (fuzzy rules), it generates one-output fuzzy control action. Instead, the ARG is a reverse process. It uses the three fuzzy variables (e, e', u) and generates the mapping function (fuzzy rules).

5.6.1. ARG Using Highest Match Method

The highest match (HM) algorithm properly fuzzifies the three input variables (e, e', u). Each fuzzy variable matches in two fuzzy subsets. The algorithm considers only the subsets with highest match (highest membership value). Thus if e matches in both X and PS with membership values 0.75 and 0.25, respectively, it will consider that e matches only in the Z subset. Similarly the e' and u subsets will be PS and NS, respectively. Thus for this sample of data, the rule having the antecedents $e = Z$ and $e' = PS$ will be assigned to a consequence $U = NS$; that is, the generated rule will be:

If $e = Z$ and $e' = PS$, then $u = NS$.

See Figure 5.11.

This process is repeated for the whole set of sampled data ensemble. At the end of the above-mentioned process, the same rule may be assigned to different consequences. To resolve this problem, the final rule consequence is generated by weighting all the consequences assigned to this rule by their number of occurrences. The generated rules are shown in Figure 5.12. Although there are some empty rules, the rule table tends to be symmetrical.

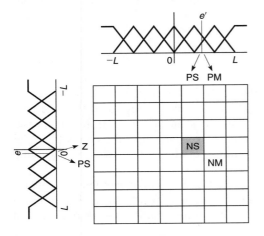

Figure 5.11 Filling the rules using the ARG highest match method.

Error derivative

	NB	NM	NS	Z	PS	PM	PB
NB	NB		NB	NB	NB	NM	PM
NM	NB						PM
NS	NB	NB	NM	NB	PS	PM	PB
Z	NM	NB	Z	Z	PS	PB	PB
PS	NB	NM	NS	Z			PB
PM	NB						PB
PB	NS	PB	Z	PB	PB	PB	PB

(System error — left vertical label)

Figure 5.12 Fuzzy rule matrix generated by the highest match method.

5.6.2. ARG Using Fuzzy Inference Engine Reverse Engineering

The algorithm mentioned in Section 5.6.1 focuses the information derived from the input-output data set on a single rule (the highest match) and ignores the effect of the other fired rules. This may lead to a considerable number of unassigned (empty) rules. The fuzzy inference reverse engineering (FIRE) technique avoids this drawback by considering the aggregated effect of all fired rules. Recall the defuzzification equation (5.13):

$$u_k = \sum_{i-0}^{n} w_i u_i \Big/ \sum_{i=0}^{n} w_i \qquad (5.17)$$

where w_i and u_i are the weight and the centroid of the ith rule, respectively.

   For a two-input, one-output FLC model using symmetrical membership function with 50% overlap, as seen in Figure 5.2, a maximum of four rules are fired at a time. Thus the above equation can be reduced to

$$u_k = \frac{w_1 u_1 + w_2 u_2 + w_3 u_3 + w_4 u_4}{w_1 + w_2 + w_3 + w_4} \qquad (5.18)$$

or

$$u = m_1 u_1 + m_2 u_2 + m_3 u_3 + m_4 u_4 \qquad (5.19)$$

where

$$m_i = w_i \bigg/ \sum_{i=0}^{4} w_i \qquad (5.20)$$

Using the input-output data set for e and e' with proper fuzzification, m_i can be calculated for the fired rules at each case of the input-output pairs. Assume that the fired rules are a, b, c, and d, as shown in Figure 5.13, and U is a vector of centroids of their consequences:

$$U = [u_a, u_b, u_c, u_d]$$

Thus Eq. (5.19) can be rewritten as

$$u_{\text{gen}} = m_a u_a + m_b u_b + m_c u_c + m_d u_d \qquad (5.21)$$

The output fuzzy control can be generated using Eq. (5.21) if the proper U vector is known. In other words, it corresponds to assigning the proper consequence to the proper rule.

   Fuzzifying the input data set for control signal will match in two fuzzy subsets, as seen in Figure 5.14. Assume that U_m is a vector containing the centroids of the matched subsets:

$$U_m = [u_1, u_2]$$

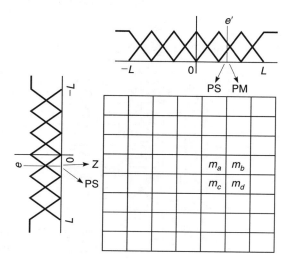

Figure 5.13 An example of rule firing.

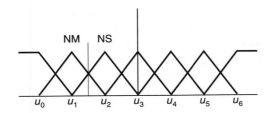

Figure 5.14 Fuzzification of the control signal.

The algorithm assigns different combinations of U_m elements (rule patterns) to the vector U. The selected rules pattern is the one that minimizes the difference between the u_{gen} and the actual u (read from the data set).

In addition, the algorithm also increases the matched subset vector U_m by two additional virtual subsets, one lower and one higher, to minimize the error. For this example U_m becomes

$$U_{me} = [u_0, u_1, u_2, u_3]$$

This process is repeated for the whole set of the sampled data. At the end, the same rule may be assigned to different consequences. The final rule consequence is generated by weighting all the consequences assigned to this rule by their number of occurrences. The following pseudocode summarizes the whole process:

```
while(not_end_of_file())
begin
read_sampled_data(e, e', u)
fuzzify (e, e', u)
find_fired_rules(e, e')
generate_rules_patterns(u)
select_appropriate pattern()
end
```

The generated rules are shown in Figure 5.15. The rule table tends to be symmetrical.

Error derivative

	NB	NM	NS	Z	PS	PM	PB
NB	NB	NB	NM	NS	NM	NS	Z
NM	NB	NB	NM	NS	NS	Z	PS
NS	NB	NM	NS	NS	Z	PS	PM
Z	NM	NB	NS	Z	PS	PB	PM
PS	NM	NS	Z	PS			PB
PM	NS	Z	PS	PS			PB
PB	Z	PS	PM	PS	PM	PB	PB

System error

Figure 5.15 Fuzzy rule matrix generated by the FIRE algorithm.

5.7. FUZZY-LOGIC-BASED POWER SYSTEM STABILIZER

A FLPSS with a set of FAM rules and continuous membership functions as discussed earlier has been designed. An organized method is followed to generate the rule base and to tune the parameters of the FLPSS.

The performance of the FLPSS is first studied using a simulation model of a synchronous machine connected to an infinite bus. The FLPSS has been tested on the power system under various fault and load disturbances to ensure its effectiveness, robustness, and reliability as a power system stabilizer.

5.7.1. Single-Machine Power System Model

A power system model consisting of a synchronous machine connected to a constant voltage bus through a double-circuit transmission line is used in the simulation studies. A schematic diagram of the model is shown in Figure 5.16.

The system is represented by a ninth-order nonlinear model including the governor and the automatic voltage regulator (AVR) exciter. The state equations representing the power system and the synchronous machine, governor, and AVR model parameters are given elsewhere [4]. The solution of the model differential equations was obtained using a fourth-order Runge-Kutta method with a simulation time step of 1 ms.

The control signal generated by the PSS is injected as a supplementary stabilizing signal to the AVR summing point. The system model has been designed to support both the CPSS and the FLPSS and can be easily extended to support other types of PSS.

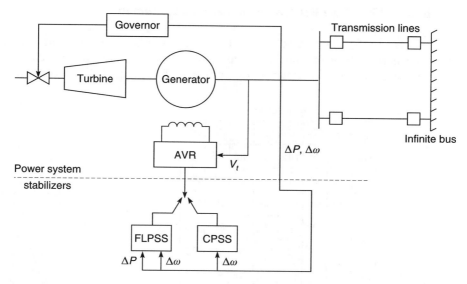

Figure 5.16 A schematic diagram of synchronous machine model with stabilizers.

5.7.2. Fuzzy Logic Power System Stabilizer

Using the description given in Section 5.3 for FLC design, a power system stabilizer based on the FLC algorithm has been developed. Since the goal of this application is to stabilize and improve the damping of the synchronous machine, speed deviation $\Delta\omega$ and active power deviation ΔP_e have been selected as the controller inputs. The controller output is then injected into the AVR summing point.

This configuration implies that the FLC has two input parameters, K_ω and K_p, and one output parameter, K_U, as seen in Figure 5.17. The selection of these parameters is usually subjective and requires previous knowledge of the fuzzy control variables (input and output signals). Also previous experience of the controlled system dynamics is commonly used in the creation of the fuzzy control rules. However, an organized approach as described in the next section has been adopted for the generation of rules and tuning of parameters for the FLPSS.

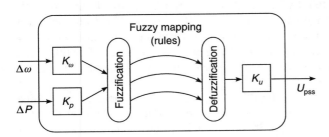

Figure 5.17 Schematic diagram of the FLPSS.

5.7.3. FLPSS Rule Generation and Parameter Tuning

Using the FIRE ARG method introduced in Section 5.6.2 and a sampled data set generated by using the CPSS, a proper set of rules was obtained. The rules used in all the following experiments are shown in Figure 5.18. The correlation-product

	Active power						
	NB	**NM**	**NS**	**Z**	**PS**	**PM**	**PB**
NB	NB	NB	NB	NB	NM	NS	Z
NM	NB	NB	NM	NM	NS	Z	PS
NS	NB	NM	NS	NS	Z	PS	PM
Z	NM	NM	NS	Z	PS	PM	PM
PS	NM	NS	Z	PS	PS	PM	PB
PM	NS	Z	PS	PM	PM	PB	PB
PB	Z	PS	PM	PB	PB	PB	PB

Speed deviation (row labels on left)

Figure 5.18 FLPSS rules generated by FIRE algorithm.

inference mechanism is used to generate the output fuzzy set for the FLPSS. The defuzzification process is based on the center-of-gravity method.

Once the proper rules are obtained, the proper parameter tuning should be done in order to achieve good performance. Tuning of FLPSS parameters can be a tedious trial-and-error process if not enough information is available about the range of the controller variables and how they change with different disturbances. The objective of the off-line tuning algorithm is to determine the controller parameters that will provide the desired system response [8].

As described in Section 5.5, the tuning algorithm tries to minimize three system PIs by varying the FLPSS parameters. In the present case, the output parameter was set to give the maximum allowable control action, while the other two parameters, K_ω, K_p, were tuned using the guided search algorithm.

The selected values of the FLPSS parameters are those that minimize the PIs using the guided blind search algorithm given in Section 5.5. Once the FLPSS had been tuned, the parameters were kept unchanged throughout subsequent studies. The tuned parameters of the FLPSS are

$$K_\omega = 100 \qquad K_p = 0.21 \qquad K_u = 0.1$$

5.7.4. Simulation Studies

A number of studies have been conducted to investigate the effectiveness of the proposed FLPSS. An illustrative set of tests is described here. The results are compared with the performance of a CPSS having a transfer function

$$G(s) = \frac{(1 + 0.3s)(1 + 0.3s)}{(1 + 0.055s)(1 + 0.055s)} \qquad (5.22)$$

The speed deviation and the output power deviation are sampled every 20 ms for control signal computation. The control output for both the FLPSS and CPSS is limited to 0.1 per unit (pu).

CPSS Parameters Selection. The CPSS parameters were tuned to give the optimal performance for the operating point of 0.95 pu generated power and a 0.95 power factor lag. With the selected parameters the response of the CPSS was very close to that of the FLPSS using the tuned parameters, as determined by the algorithm described in Section 5.7.3.

The parameters of both the FLPSS and the CPSS were then fixed at these values for all tests. The load angle response for the no-stabilizer case as well as with both controllers at the nominal operating point used in the selection of the CPSS parameters for a three-phase-to-ground fault at generator terminals is shown in Figure 5.19.

It is seen that the overshoot for both controllers is the same although the system settles slightly faster with the FLPSS.

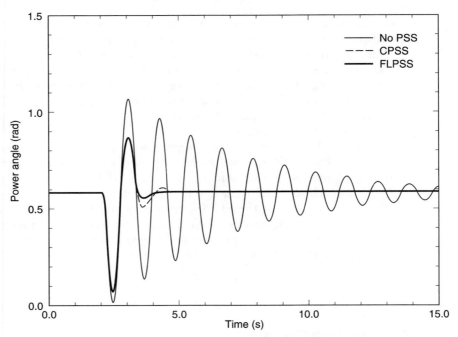

Figure 5.19 Response to three-phase-to-ground fault at power $= 0.95$ pu and pf $= 0.9$ lag.

Light-Load Test. With the generator operating at 0.2 pu power and a 0.9 power factor lag, a 0.1-pu step decrease in input torque reference was applied at 1 s and removed at 4 s so that the system returned to the original operating point. The response without stabilizer effects (Figure 5.20) shows large oscillations. The power angle response of the system with both FLPSS and CPSS stabilizers, also given in Figure 5.20, shows smaller overshoot and quick reaction with the FLPSS.

Leading Power Factor Operation Test. With the generator operating at 0.3 pu power and a 0.9 power factor lead, a 0.1-pu step decrease in input torque reference was applied at time 1 s and removed at time 4 s so that the system returned to the original operating point. The response of the system with both FLPSS and CPSS stabilizers is shown in Figure 5.21. The load angle response with the FLPSS shows very little overshoot and quick reaction.

Robustness. The fixed-parameter CPSS normally needs to be redesigned for each power system application and has to be retuned if the system configuration changes. However, the FLPSS has the capacity to accept imprecision and even vagueness in system parameters. Thus changes in the system configuration or parameters should have a minor effect on the performance of the FLPSS.

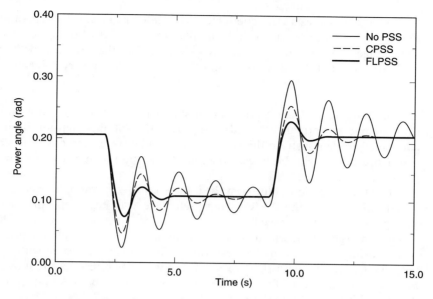

Figure 5.20 Response to ±0.1 input torque disturbance at power = 0.2 and
pf = 0.9 lag.

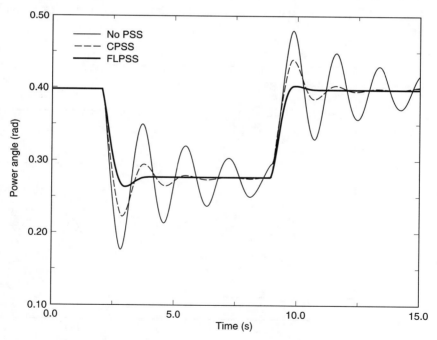

Figure 5.21 Response to ±0.1 input torque disturbance at power = 0.3 pu
and pf = 0.9 lead.

A number of tests were conducted with different machine inertia and transmission line impedances with which the oscillation frequency of the system varied over a wide range. The results of these tests showed that the response with the FLPSS deviated only slightly, illustrating that the FLPSS offers a very robust performance.

Stability Test. The introduction of stabilizing control signal not only improves the dynamic performance but also improves the stability margin. Two tests were conducted to demonstrate the effect of the stabilizers on the dynamic stability and transient stability. For both tests, the initial operating point is 0.9 pu power and 0.95 power factor lag.

For the dynamic stability test the input torque was gradually increased from the initial value. Since the AVR voltage reference was kept constant, the generator voltage remains at the specified value as long as the system is stable. The dynamic stability margin is described by the maximum power that the system can transfer before losing synchronism. The test was conducted for the system with no stabilizer, with the CPSS, and with the FLPSS. The power angle response is shown in Figure 5.22. The power the FLPSS was able to transfer was about 3.1 pu, compared with 2.35 pu with the CPSS and 1.6 pu with no stabilizer.

The transient stability was tested by applying a three-phase-to-ground fault near the sending end with different clearance times. The maximum clearance time for the power system with no PSS, with a CPSS, and with an FLPSS is given in Table 5.1. The FLPSS provides the largest clearance time, which indicates better transient stability.

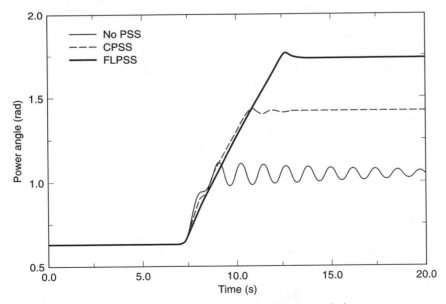

Figure 5.22 Power angle response to gradually increase in input torque.

TABLE 5.1 TRANSIENT STABILITY TEST RESULTS

	No PSS	CPSS	FLPSS
Clearance time, ms	300	310	315

5.8. FLC APPLICATION TO A MULTIMACHINE POWER SYSTEM

Multimode oscillations appear in an interconnected power system in which the generating units have quite different inertias and are weakly connected by transmission lines. The effectiveness of the FLPSS to damp multimode oscillations in a five-machine power system is illustrated in this section.

The fuzzy logic stabilizer applied to various generators is the one described in Section 5.7.2. The parameters of the FLCs were tuned by an extension of the techniques described in Sections 5.5 and 5.7.3.

The parameters of the FLPSS selected with the help of the above procedure are given in Table 5.2.

TABLE 5.2 FLPSS PARAMETERS USED IN MULTIMACHINE STUDY

Parameter	Generator 1	Generator 2	Generator 3
K_w	5	5	0.45
K_p	3.3	3.3	20
K_u	0.1	0.1	0.1

5.8.1. Multimachine System Configuration

A five-machine power system without an infinite bus, as shown in Figure 5.23, is used to test the FLPSS. This configuration represents two power generation areas physically far from each other. The interconnections between the two areas can be considered weak in comparison with the connection within the individual subsystems. Parameters for all generating units, transmission lines, load, and operating conditions are given in the literature [25].

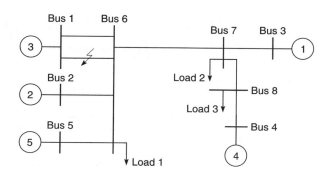

Figure 5.23 A five-machine power system configuration.

Different tests were applied to the multimachine model to study the response of the FLPSS. Results of an illustrative set of tests are given below.

5.8.2. Multimode Oscillations in the Multimachine System

When the multimachine power system is disturbed, multimode oscillations arise. A 0.25-pu step decrease in the mechanical input torque reference of generator 3 is applied at 1 s, and the system returns to the original condition at 10 s. Oscillations in Figure 5.24 show the local mode at about 1.3 Hz and the interarea mode at about

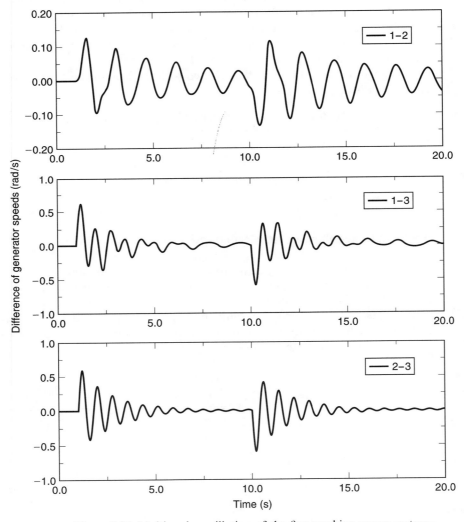

Figure 5.24 Multimode oscillation of the five-machine power system.

0.65 Hz under the above-mentioned disturbance without any PSSs installed. The speed deviation between generators 2 and 3 exhibits mainly local-mode oscillations. The speed deviation between generators 1 and 2 exhibits mainly the interarea-mode oscillations. Both the local- and interarea-mode oscillations exist in the speed deviation between generators 1 and 3.

5.8.3. Only One PSS Installed

The FLPSS is first installed on generator 3 only to study its effect on the stability and dynamic behavior of the system. None of the other generators had any PSS. The machine speed deviation and the accelerating power of generator 3 is sampled at the rate of 50 Hz and transferred to the FLPSS as inputs, and the control signal output

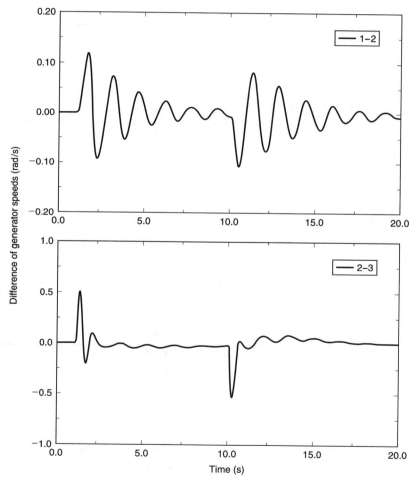

Figure 5.25 System response with FLPSS installed on generator 3 (operating point 1).

of the FLPSS, computed from the sampled data, is then transferred to the power system to control it. The iterative sample and control process is very similar to the physical process.

A 0.25-pu step decrease in the mechanical input torque reference of generator 3 is applied at 1 s, and the system returns to the original condition at 10 s. The FLPSS damps the local-mode oscillations very effectively, as shown in Figure 5.25. However, as expected, it has little influence on the interarea-mode oscillations. This is because the rated capacity of generator 3 is much less than generators 1 and 2, and the interarea-mode oscillations are introduced mainly by these large generators. Generator 3 does not have enough power to control the interarea-mode oscillations. To damp the interarea-mode oscillations, PSSs must be installed on generators 1 and 2, as shown in Section 5.8.4.

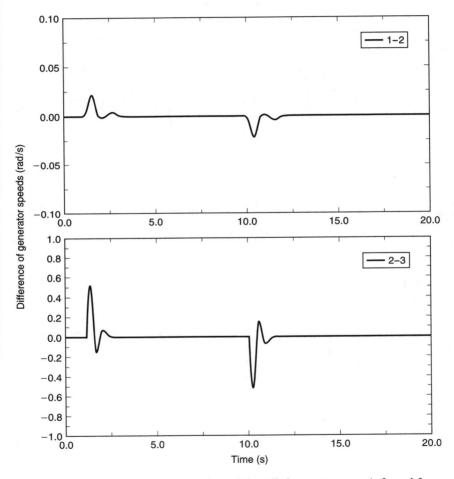

Figure 5.26 System response with FLC installed on generators 1, 2, and 3 (operating point 1).

5.8.4. Three PSSs Installed

To damp both the local and interarea modes of oscillation, two FLPSSs were additionally installed on generators 1 and 2. The system response with FLPSSs for a 0.25-pu step decrease in the mechanical input torque reference of generator 3 applied at 1 s and the system returning to the original condition at 10 s is given in Figure 5.26.

The system response shows that the FLPSS not only damps both modes of oscillations very effectively but also reduces significantly the overshoot in the speed difference between machines 1 and 2.

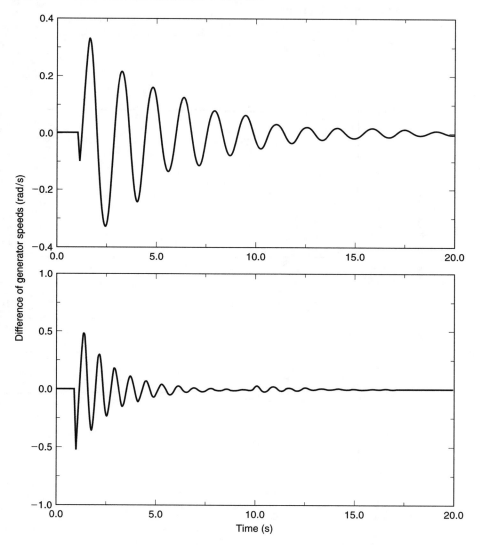

Figure 5.27 System response without stabilizer for three-phase-to-ground fault (operating point 1).

5.8.5. Three-Phase-to-Ground Fault Test

With the power system operating at the same operating conditions as Section 5.8.2, a three-phase-to-ground fault was applied at the middle of the transmission line between buses 1 and 6 at 1 s and cleared 100 ms later. At 10 s, the faulted transmission line was restored successfully. The open-loop response of the system under this disturbance is shown in Figure 5.27. When the proposed FLPSSs are installed on generators 1, 2, and 3, the response of the power system is improved, as shown in Figure 5.28. It can be seen that the FLPSS can improve the system performance very effectively.

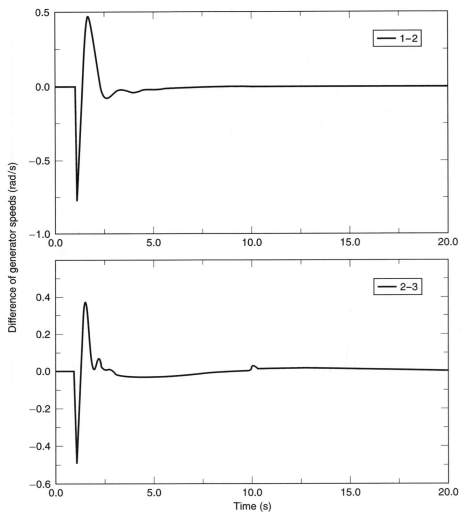

Figure 5.28 System response with FLPSS installed on generators 1, 2, and 3 for 3-phase-to-ground fault (operating point 1).

5.9. IMPLEMENTATION AND EXPERIMENTAL STUDIES

Simplicity and less intensive mathematical design requirements are the most important features of FLCs over most other control techniques. This feature allows the FLC to be easily implemented using inexpensive hardware technology. A low-cost microcontroller (the Intel 8051) has been used to implement the FLC-based PSS. Implementation of the PSS and experimental studies on a physical model of a power system are described in this section.

5.9.1. Hardware System Description

The microcontroller system used (the MCS8051) is a low-cost, reliable, general-purpose microcontroller available in the power research laboratory at the University of Calgary. The microcontroller used is the 8051FA, an Intel-designed 8-bit microcontroller. This 8051FA version incorporates an additional 128 bytes of internal random-access memory (RAM) and the programmable counter array (PCA). These features offer enhanced capabilities as a control and measurement tool. Additional components on the MCS8051 are 32 kbytes of flash read-only memory, 16 kbytes of static RAM, a keyboard liquid crystal display (LCD), and serial/printer port interfaces.

The microcontroller communicates with the outside environment through a data acquisition system (DAS). Output signals generated by the MCS8051 are written into a specific input-output (I/O) address and converted through digital-to-analog converters. A synchronous machine speed interface card has been designed and built to reshape the speed signal generated by the photo interrupter into a proper form.

A dedicated LCD and a small keyboard (KB) are used as a simple interface to communicate with users. Also a standard serial interface RS232 connects the MCS8051 with an Intel 486 based PC. A monitor program development software ICE51 is installed on the PC. The monitor program is used to develop and debug the application program on the MCS8051 using the serial interface link. After the completion of the design process of the application, the RS232 connection to the PC is removed. The MCS8051 is then a stand-alone unit with its LCD and KB user interface.

5.9.2. Implementation of the FLC Using 8051 Assembly Language

The 8051 8-bit microcontroller has limited mathematical capabilities. Unless there is a suitable compiler that supports floating-point manipulations, doing floating-point math is a difficult task. Despite the simplicity of the FLC algorithm, it does not mean that floating-point math is not required. It thus can become a burden to use an 8-bit processor with no floating-point math support. However, fuzzy logic implies that it can tolerate ambiguities. In other words it does not require high calculation accuracy.

Fuzzification Membership Function Realization. Most available hardware implementations of the fuzzifier tend to use a discretized form of the membership function
in the form of discrete tables. This does not give smooth values of membership unless
some of the values are processed using some interpolation. An easy way to implement the membership function is introduced here to overcome some of the short-
comings of the other methods.

Consider the triangular set membership functions, defined in Section 5.3.2 and
shown in Figure 5.29, that are symmetrical and have 50% overlap. For the example

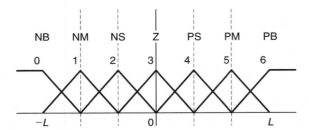

Figure 5.29 Membership functions for
seven fuzzy subsets.

of seven subsets, the input signal may lie in any of the six regions shown in Figure
5.29. Each region includes two different subsets. To simplify the problem, the membership functions are split into two sets of functions U_a, U_b, as shown in Figure
5.30(a) and (b), respectively. Each set of functions (U_a, U_b) has symmetric replicate

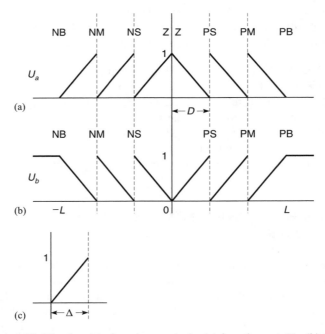

Figure 5.30 Membership function analysis: (a) function set U_a; (b) function
set U_b; (c) one triangular function.

functions and its negative side is a reflection of the positive. Thus, each new set can be characterized by the main function. For the set of functions U_a shown in Figure 5.30(a) the characteristic function is u_1, where

$$u_1(x) = 1 - \left(\frac{1}{\Delta}\right)x \quad \text{for} \quad 0 < x < \frac{1}{L} \tag{5.23}$$

while the characteristic function of U_b is u_2, where

$$u_2(x) = 1 - u_1(x) \tag{5.24}$$

In fact, from symmetry, it is possible to use only one triangular function to calculate the degree of fulfillment for all of the subsets (refer to Figure 5.30). This is clear because any triangle representing a subset is just a replicate shift of the one at the origin.

The correct calculation of the membership function can be verified by a simple experiment using a function generator and one D/A channel. If an applied signal of magnitude that covers the whole input range is used, it should be able to execute all the membership functions. A triangular wave with an amplitude varying between $+2L$ and $-2L$ is used as an input signal. The degree of fulfillment of the membership function is then output through a D/A and seen on an oscilloscope. It should show that the membership values vary smoothly and are correctly calculated.

Rules and Inference. The fuzzy subset linguistic tables defined in Section 5.3 were changed to integer numbers addressing the different subsets in order to suit assembly programming. Thus the fuzzy subsets NB, NM, MS, Z, PS, PM, and PB were assigned an integer value starting from 0 to 6, respectively. In this way, the fuzzy rules table becomes a table of integer values, as shown in Table 5.3.

The rule table is stored in the code segment of the MCS8051 memory map. Proper assembly routines are to fire and to extract the rules using code commands. The rule inference mechanism is built based on the min-max concept.

TABLE 5.3 FUZZY RULES REPRESENTED IN INTEGER NUMBERS

$\Delta\omega/P_e$	NB = 0	NM = 1	NS = 2	Z = 3	PS = 4	PM = 5	PB = 6
NB = 0	0	0	0	1	2	2	3
NM = 1	0	0	1	2	2	3	4
NS = 2	0	1	2	2	3	4	4
Z = 3	1	2	2	3	4	4	5
PS = 4	2	2	3	4	4	5	5
PM = 5	2	3	4	4	5	5	6
PB = 6	3	4	4	5	5	6	6

Defuzzification. The defuzzification process was done with the help of 32-bit math library using the center of gravity concept discussed in Section 5.3.5. Due to the symmetry of the membership functions, the simplified Eq. (5.13) was used in the defuzzification. The defuzzified value is then multiplied by the proper gain and converted to an analog signal representing the controller output.

Integrating FLC with Speed Deviation Measurement. The final form of controller includes the speed deviation measurement algorithm, a sampler, and the FLC algorithm. The FLC input, fuzzification, rule inference, and defuzzification modules were integrated together to form the final structure of the FLC, as shown in Figure 5.31. This was designed to run as a foreground task on the 8051 microcontroller.

The speed deviation measurement and the sampling process were designed to run in the background of the main application (control algorithm). Both are interrupt-driven tasks. The LCD and the KB were used as a simple user interface to display the FLC status (running/halted). The user can switch the controller (on/off) using the KB. After testing and verification of the FLC algorithm, the program was moved to a flash ROM. The program was installed such that it executes directly with power up. The total memory consumed by the integrated application (including the FLC algorithm, speed measurement, 32-bit math library, LCD/KB drivers, and A/Ds) is less than 2 kbytes of ROM and 54 bytes of internal 8051 RAM. The whole process also consumes less than 30 ms for a 12-MHz oscillator frequency. This shows that the FLC algorithm is very concise in size and time.

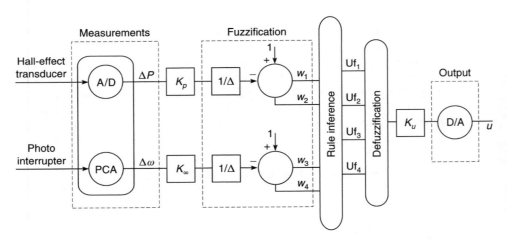

Figure 5.31 Building blocks of the FLC algorithm implementation.

5.9.3. The Micromachine Physical Power System Model

The micromachine system available in the University of Calgary Power Research Laboratory was used to test the FLC control algorithm. The system consists of a single machine (micromachine) connected to an infinite bus through a double-circuit transmission line. The overall physical model integrated with the MCS8051 basically represents the system shown in Figure 5.16.

A microalternator is a 3-kVA, 220-V three-phase synchronous machine. A 7.5-hp DC motor is used to drive the microalternator. A time constant regulator is used to change the effective field time constant of the microalternator in order to simulate a large generating unit. A simple AVR with the transfer function

$$G(s) = 100 \, \frac{1+s}{1+10.4s} \tag{5.25}$$

was used to simulate a high-gain, short-time-constant AVR. The stabilizing signal from the adaptive PSS is added to the summing junction after the AVR.

A lumped-element transmission line physical model is used to represent the transmission line. This model simulates the performance of a 500-kV, 300-km-long double-circuit transmission line connected to a constant-voltage bus. Consisting of six π sections, the transmission line gives a frequency response that is close to the actual transmission line response up to 500 Hz. The transmission line model could be configured according to a specific test.

Various disturbances, such as step changes in the generator terminal voltage and active power of the microalternator, symmetrical and unsymmetrical faults on the transmission line, and so on, can be applied. With these features, the dynamic performance of the power system and the effect of the PSS under various conditions can be investigated.

5.9.4. Experiments and Results

The behavior of the FLPSS was tested under a variety of disturbances at different operating conditions. Illustrative results for a few experiments are given next.

Voltage Reference Step Change. Setting the power system model to operate at the operating condition (where pf = power factor)

$$P = 0.73 \, \text{pu} \qquad \text{pf} = 0.95 \, \text{lag} \qquad V_t = 1.0 \, \text{pu}$$

a 4.5% step decrease in the reference voltage was applied at 2 s and removed at 12 s. Both the open-loop and FLPSS closed-loop responses for active power deviation are shown in Figure 5.32. The oscillation of the active power is effectively damped by the FLPSS within one cycle.

Input Torque Reference Step Change. With the microalternator operating at

$$P = 0.58 \, \text{pu} \qquad \text{pf} = 0.92 \, \text{lag} \qquad V_t = 1.0 \, \text{pu}$$

a 0.24-pu step decrease in the input torque reference was applied at 2 s and removed at 12 s. Test results with the FLPSS and a CPSS are shown in Figure 5.33. The CPSS was built using operational amplifiers to represent the transfer function given in Eq. (5.22) with the gain adjusted to provide best response at the system nominal operating point of 0.73 pu power and 0.95 power factor lag. The results show that the number of oscillations and the amplitude of the response with the FLPSS is reduced.

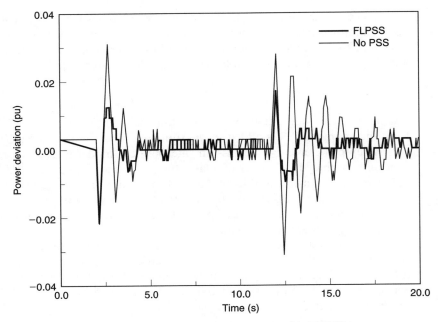

Figure 5.32 Response to ±4.5% change in V_{ref} with FLPSS at power = 0.74 pu, pf = 0.95 lag.

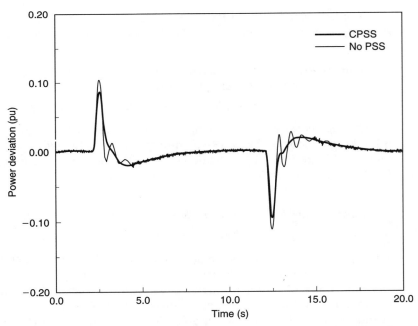

Figure 5.33 Response to −0.24-pu change in input power with FLC and CPSS at power = 0.58 pu, pf = 0.92.

Three-Phase-to-Ground Fault Test. With the system operating at

$$P = 0.81 \, \text{pu} \qquad \text{pf} = 0.87 \, \text{lag} \qquad V_t = 0.005 \, \text{pu}$$

a three-phase-to-ground fault was applied at the middle of one transmission line at 2 s. The transmission line was opened, by relay action, at both ends of the line 100 ms later. A successful reclosure was made after 600 ms and the line returned to normal operation.

The response of the active power for both the FLPSS and the CPSS is shown in Figure 5.34. Although both controllers have almost the same oscillation amplitude during the disturbance, the system settles much faster with the FLPSS than with the CPSS.

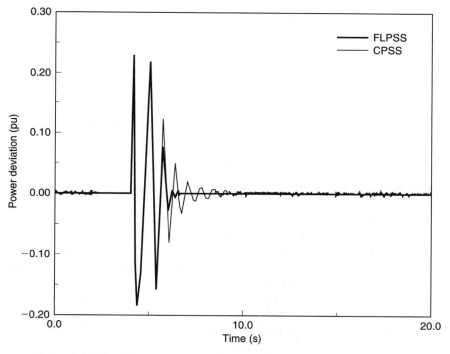

Figure 5.34 Response to three-phase-to-ground fault and successful reclosure at power = 0.81 pu, pf = 0.87 lag.

Unsymmetrical Short Circuit Tests. Under the same operating conditions as before, a two-phase-to-ground short circuit was applied at time 2 s at the middle of one transmission line. The line was opened for 100 ms where the fault was cleared and the system returned back to its initial conditions. The response given in Figure 5.35 shows that the FLPSS is able to significantly reduce the oscillation amplitude and the settling time.

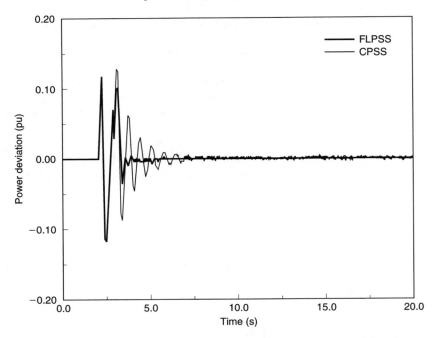

Figure 5.35 Response to two-phase-to-ground fault and successful reclosure (power $= 0.81$ pu, pf $= 0.87$ lag).

5.10. CONCLUDING REMARKS

Unlike the classical design approach, which requires a deep understanding of the system, exact mathematical models, and precise numerical values, a basic feature of the fuzzy logic controller is that a process can be controlled without the knowledge of its underlying dynamics. The control strategy learned through experience can be expressed by a set of rules that describe the behavior of the controller using linguistic terms. Proper control action can be inferred from this rule base that emulates the role of the human operator or a benchmark control action. Thus, fuzzy logic controllers are suitable for nonlinear, dynamic processes for which an exact mathematical model may not be available.

Using the principles of fuzzy logic control, a PSS has been designed to enhance the operation and stability of a power system. Results of simulation and experimental studies look promising.

References

[1] F. P. de Mello and T. F. Laskowski, "Concepts of Power System Dynamic Stability," *IEEE Transactions on Power Apparatus and Systems*, Vol. PAS-94(3), May/June 1975, pp. 827–833.

[2] E. V. Larsen and D. A. Swann, "Applying Power System Stabilizers, Part I to III," *IEEE Transactions on Power Apparatus and Systems*, Vol. PAS-100, No. 6, 1981, pp. 3017–3046.

[3] S.-J. Cheng, Y. S. Chow, O. P. Malik, and G. S. Hope, "An Adaptive Synchronous Machine Stabilizer," *IEEE Transactions on Power Systems*, Vol. PWRS-1, No. 3, 1986, pp. 101–109.

[4] G. P. Chen, O. P. Malik, G. S. Hope, Y. H. Qin, and G. Y. Xu, "An Adaptive Power System Stabilizer Based on Self-Optimizing Pole-Shifting Control Strategy," *IEEE Transactions on Energy Conversion*, Vol. 8, No. 4, 1993, pp. 639–645.

[5] Y. Zhang, G. P. Chen, O. P. Malik, and G. S. Hope, "An Artificial Neural Network Based Adaptive Power System Stabilizer," *IEEE Transactions on Energy Conversion*, Vol. 8, No. 1, 1993, pp. 71–77.

[6] B. Kosko, *Fuzzy Thinking*, Prentice-Hall, Englewood Cliffs, NJ, 1993.

[7] M. A. M. Hassan, O. P. Malik, and G. S. Hope, "A Fuzzy Logic Based Stabilizer for a Synchronous Machine," *IEEE Transactions on Energy Conversion*, Vol. 6, No. 3, 1991, pp. 407–413.

[8] K. A. El-Metwally and O. P. Malik, "Fuzzy Logic Power System Stabilizer," *IEE Proceedings on Generation, Transmission and Distribution*, Vol. 143, No. 3, 1996, pp. 263–268.

[9] Y. Y. Hsu and C. H. Cheng, "Design of Fuzzy Power System Stabilizers for Multi-machine Power Systems," *IEE Proceedings on Generation, Transmission and Distribution*, Vol. 137, Part C, No. 3, May 1990, pp. 233–238.

[10] C.-C. Su and Y.-Y. Hsu, "Fuzzy Dynamic Programming: An Application to Unit Commitment," *IEEE Transactions on Power Systems*, Vol. 6, No. 3, 1991, pp. 1231–1237.

[11] V. Miranda and J. T. Saraiva, "Fuzzy Modelling of Power Systems Optimal Load Flow," *IEEE Transactions on Power Systems*, Vol. 7, No. 2, 1992, pp. 843–849.

[12] A. R. Hasan, T. S. Martis, and A. H. M. Sadral Ula, "Design and Implementation of a Fuzzy Controller Based Automatic Voltage Regulator for a Synchronous Generator," *IEEE Transactions on Energy Conversion*, Vol. 9, No. 3, 1994, pp. 550–557.

[13] T. Hiyama, "Robustness of Fuzzy Logic Power System Stabilizers Applied to Multi Machine Power System," *IEEE Transactions on Energy Conversion*, Vol. 9, No. 3, 1994, pp. 451–459.

[14] L. A. Zadeh, "Fuzzy Sets," *Information and Control*, Vol. 8, 1965, pp. 338–353.

[15] B. Kosko, *Neural Networks and Fuzzy Systems: A Dynamic Approach to Machine Intelligence*, Prentice-Hall, Englewood Cliffs, NJ, 1992.

[16] C. C. Lee, "Fuzzy Logic in Control Systems: Fuzzy Logic Controller—Parts I and II," *IEEE Transactions on Systems, Man and Cybernetics*, Vol. 20, No. 2, 1990, pp. 404–435.

[17] M. Braae and D. A. Rutherford, "Selection of Parameters for a Fuzzy Logic Controller," *Fuzzy Sets and Systems*, Vol. 2, No. 3, 1979, pp. 185–199.

[18] J. E. H. Mamdani, "Twenty Years of Fuzzy Control: Experiences Gained and Lessons Learnt," *Proceedings of the IEEE Third International Conference on Fuzzy Logic*, San Francisco, 1993, pp. 339–344.

[19] N. Baaklini and E. H. Mamdani, "Prescriptive Methods for Deriving Control Policy in a Fuzzy-Logic Controller," *Electronics Letters*, Vol. 11, 1975, pp. 625–626.

[20] P. J. Antsaklis, "Neural Networks in Control Systems," *IEEE Control Systems Magazine*, Vol. 10, 1990, pp. 3–5.

[21] B. S. Zhang and J. M. Edmunds, "Self-Organizing Fuzzy Logic Controller," *IEE Proceedings D*, Vol. 139, No. 5, 1992, pp. 460–464.

[22] M. Maeda and S. Murakami, "A Self-Organizing Fuzzy Controller," *Fuzzy Sets and Systems*, Vol. 51, 1992, pp. 29–40.

[23] S. M. Smith and D. T. Comer, "Self-Tuning of a Fuzzy Logic Controller Using Cell State Space Algorithm," *Proceedings of the IEEE Second International Conference on Fuzzy Systems*, 1990, pp. 445–450.

[24] K. A. El-Metwally and O. P. Malik, "Parameter Tuning for a Fuzzy Logic Controller," *Proceedings of the IFAC Twelfth World Congress on Automatic Control*, Sydney, Australia, July 18–23, 1993, Vol. 2, pp. 581–584.

[25] G. P. Chen and O. P. Malik, "Tracking Constrained Adaptive Power System Stabilizer," *IEE Proceedings on Generation, Transmission and Distribution*, Vol. 142. No. 2, 1995, pp. 149–156.

T. Hiyama
Department of Electrical Engineering
and Computer Science
Kumamoto University
Kumamoto
860 Japan

Chapter 6

Fuzzy Logic Power System Stabilizer Using Polar Information

6.1. INTRODUCTION

Excitation control is well known as one of the effective means to enhance the overall stability of electric power systems [1, 2]. Conventional analog-type power system stabilizers have been implemented on actual power systems for this purpose and have provided enhancement of the overall stability to actual power systems. However, the parameters of a conventional power system stabilizer, such as gains and time constants, are fixed to ensure its optimal performance at a nominal operating point. Consequently, its performance is degraded whenever its operating point is shifted from nominal. Due to progress in computer technologies, microcomputer-based digital controllers, such as self-tuning and fuzzy logic power system stabilizers [3–13], are receiving increasing attention for the application to the generator excitation control in order to overcome the disadvantages of the widely used conventional power system stabilizers.

In this chapter, advanced fuzzy logic control rules are introduced based on using three-dimensional information of the generator acceleration, speed, and phase angle. All the three-dimensional information is derived through the digital filtering of measured real power output or speed of the generator [14, 15].

According to the information and simple fuzzy logic control rules, the stabilizing signal is revised at every sampling time and fed back to the excitation control loop of the study generator to provide it maximum damping. The rules are straightforward to minimize the computational burden, which is an advantage of the proposed fuzzy logic control scheme in real-time application.

To demonstrate the effectiveness of the advanced fuzzy logic control scheme and also to provide sufficient data for the actual installation of the proposed advanced fuzzy logic power system stabilizer (AFLPSS), simulation and experimental studies have been conducted. For the experimental studies, a personal computer (PC) based stabilizer was set on the 5-kVA laboratory system [12] at Kumamoto University and also on the analog network simulator (ANS) at the Research Laboratory of Kyushu Electric Power Co., Inc. (KEPCO). The fuzzy logic stabilizer (FLPSS), which requires only two-dimensional information, that is, the acceleration and speed of the study generator, was successfully tested at one of the hydropower stations in the KEPCO system in October 1992 when subjected to a small-sized reactance switching, faulty synchronization of the study generator, and so on [11]. On-site testing in the actual power system under operating conditions is severely restricted. The site tests did not involve large disturbances such as three-phase-to-ground faults. The ANS is capable of simulating various electromechanical transients for stability analysis subject to various types of small and even large disturbances without any risks associated with the site tests in actual power systems. Analog network simulators are increasingly being employed as an alternative to site tests in actual power systems [16]. The experimental results show the advantage of the advanced fuzzy logic control scheme to increase the controllable region through the comparison between the FLPSS and the AFLPSS.

Following the evaluation of the AFLPSS on the ANS, the prototypes have been set on two hydrounits in the KEPCO system between March and May 1994 for long-term evaluation of the proposed AFLPSS. This chapter also describes the actual installation, including the configuration of the prototype, and the results of the site tests.

6.2. ADVANCED FUZZY LOGIC POWER SYSTEM STABILIZER USING THREE-DIMENSIONAL POLAR INFORMATION [14]

6.2.1. Configuration of FLPSS and AFLPSS [8–13]

Figure 6.1 illustrates the basic configuration of the FLPSS and the AFLPSS for real-time excitation control. The input to the AFLPSS is the real power signal or the speed signal through the analog-to-digital (A/D) conversion interface from the study generator. Through digital filtering, the required information, that is, the acceleration, the speed, and the phase of the study generator, is derived directly from the measured signal on the study generator. According to the generator state and simple fuzzy logic control rules, the stabilizing signal is updated at every sampling time. The stabilizing signal is fed back to the excitation control loop of the study generator through the D/A conversion interface.

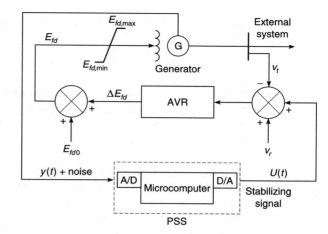

Figure 6.1 Microcomputer-based power system stabilizer.

6.2.2. Fuzzy Logic Control Scheme Using Two-Dimensional Information

In the proposed approach of defining the generator state, the speed/acceleration state of the study generator is utilized. The generator state is given by the point $p(k)$ in the speed/acceleration phase plane, as shown in Figure 6.2. The origin O is the desired equilibrium point. On the basis of the speed/acceleration state and a set of simple fuzzy logic control rules, the desired stabilizing signal is generated to shift

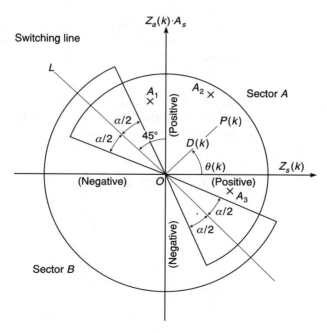

Figure 6.2 Phase plane.

the current state to the equilibrium point, that is, the origin O. The point $p(k)$ is given by

$$p(k) = [Z_a(k), A_s \cdot Z_s(k)] \tag{6.1}$$

where $Z_s(k)$ and $Z_a(k)$ are the speed deviation and the acceleration of the study generator. In addition, A_s is the scaling factor for the acceleration $Z_a(k)$.

The control rules are given in linguistic expressions as follows for the sample points A_1, A_2, and A_3 in sector A of the phase plane shown in Figure 6.2.

> *Rule 1*:   Slight deceleration control should be applied to the study generator at the sample point A_1 to prevent excessive shift of its speed deviation to the positive side, that is, to the first quadrant, because the speed deviation will be close to zero from the negative side, that is, from the second quadrant, with a relatively high positive acceleration. Therefore, the stabilizing signal should be positive small at this state to slightly increase the real power output from the study unit.
>
> *Rule 2*:   Strong deceleration control should be applied to the study unit at the sample point A_2 because both its speed deviation and acceleration are positive large. The stabilizing signal is positive large at this state to highly increase the real power output.
>
> *Rule 3*:   Slight deceleration control should be applied to the study unit at the sample point A_3 to quickly shift the speed deviation to zero because the unit is already in the region of slightly negative high acceleration but its speed deviation is still positive large. In this case, the stabilizing signal should be positive small to increase the real power output slightly.
>
> *Rule 4*:   In sector B, all the above situations are opposite to those in sector A. Therefore, the control strategies should be reversed to apply the acceleration control to the study unit. The stabilizing signal should be negative in sector B to decrease the real power output.

In this study, the polar variables angle $\theta(k)$ and radius $D(k)$ are derived in order to describe the stabilizer functions mathematically according to the above rules 1–4. The angle $\theta(k)$ and the radius $D(k)$ are given from the state variables $Z_s(k)$ and $Z_a(k)$ as follows:

$$D(k) = \sqrt{Z_s(k)^2 + [A_s \cdot Z_a(k)]^2} \tag{6.2}$$

$$\theta(k) = \tan^{-1}\left(\frac{A_s \cdot Z_a(k)}{Z_s(k)}\right) \tag{6.3}$$

As described in rules 1–4, the phase plane is divided into two sectors, A and B. These two sectors are described mathematically by using two angle membership functions. Two membership functions $N(\theta(k))$ and $P(\theta(k))$ are defined as shown in Figure 6.3 to represent sectors A and B, respectively. Namely, the values of $N(\theta(k))$ and $P(\theta(k))$ give the membership grades of deceleration and acceleration control, respectively,

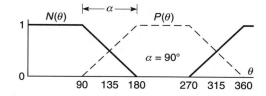

Figure 6.3 Angle membership functions $N(\theta(k))$ and $P(\theta(k))$.

for the study unit at the kth sampling time. In addition, the term α, shown in Figure 6.3, gives the overlap angle between sectors A and B.

Another radius membership function $G(k)$, which gives the gain factor, is also required as follows:

$$G(k) = \begin{cases} \dfrac{D(k)}{D_r} & \text{for } D(k) \leq D_r \\ 1.0 & \text{for } D(k) \geq D_r \end{cases} \tag{6.4}$$

where D_r is the radius member.

The stabilizing signal $U(k)$ is determined through the weighted averaging defuzzification algorithm given by

$$U(k) = \frac{N(\theta(k)) - P(\theta(k))}{N(\theta(k)) + P(\theta(k))} \cdot G(k) \cdot U_{\max}$$

$$= [1 - 2P(\theta(k))] \cdot G(k) \cdot U_{\max} \tag{6.5}$$

6.2.3. Advanced Fuzzy Logic Control Scheme [14]

The FLPSS, using only two-dimensional information, that is, the speed Z_s and the acceleration Z_a of the generator, is a robust stabilizer and has a wider stabilizable region compared with the analog-type conventional power system stabilizer (CPSS). However, stabilizing control up to the physically stabilizable limit is not possible by using the FLPSS with fixed parameters because the magnitude of the stabilizing signal U is kept small whenever the speed deviation and acceleration are close to zero even for an existing large phase deviation, as shown in Figure 6.4. The phase plane plot corresponding to Figure 6.4 is shown in Figure 6.5. As can be observed in these figures, the FLPSS shifts the speed/acceleration state of the study unit very close to the desired equilibrium point, the origin in the phase plane. The study unit, however, goes out of step because of the shortage of the deceleration control of the study unit. In these figures, additional deceleration control effort is required at the point where acceleration is zero and speed deviation is small. The FLPSS control signal is near zero at this point and does not force the speed error to zero.

In order to overcome this situation, additional integral information Z_p of the generator speed is taken into account to give the study unit enough damping by applying a larger positive signal additionally through the advanced fuzzy logic control scheme. For further improvements of the FLPSS, a switching surface is defined in the three-dimensional state space as shown in Figure 6.6, instead of the switching

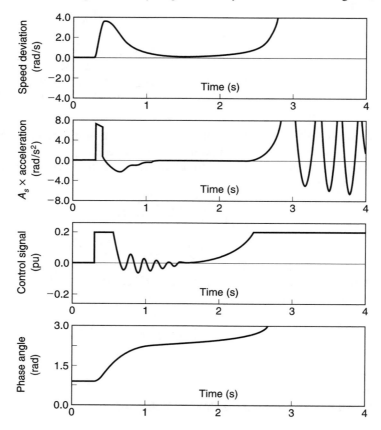

Figure 6.4 Typical time response indicating shortage of stabilizing effort.

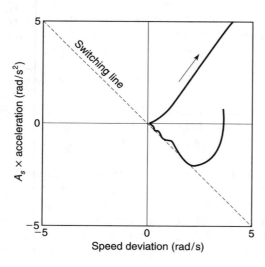

Figure 6.5 Phase plane plot of Figure 6.4.

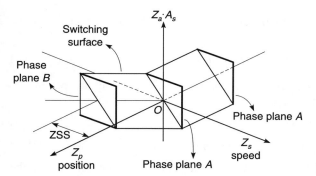

Figure 6.6 Switching surface for AFLPSS.

line used in the FLPSS. In this figure, the phase plane A is the one used in the FLPSS to determine the stabilizing signal according to the speed/acceleration state.

In the AFLPSS, this phase plane A is modified to the phase plane B when the integral information $Z_p(k)$ of the generator speed is positive, as shown in Figure 6.7. In Figure 6.7, the term A_s is the scaling factor for the acceleration signal $Z_a(k)$. In addition, sector A gives the region where deceleration control is required, and sector B is the region where acceleration control is required. These two sectors are defined by using two membership functions as shown in Figure 6.3. The stabilizing signal is determined by using these two membership functions together with the gain determined by the distance $D(k)$ of the generator state $p(k)$ from the origin O^*. By using this modification, the stabilizing signal is improved in such cases, as shown in Figures 6.4 and 6.5.

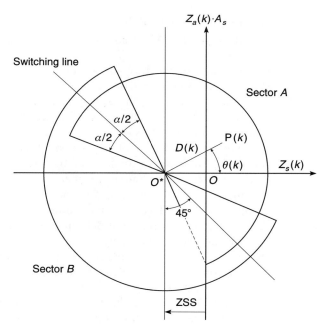

Figure 6.7 Phase plane for AFLPSS (phase plane B).

In the advanced fuzzy logic control scheme, the generator state is given by the point $p(k)$ in the phase plane B:

$$p(k) = [Z_p(k) + \text{ZSS}, A_s \cdot Z_a(k)] \tag{6.6}$$

where $\text{ZSS}(k)$ indicates the magnitude of shift of the origin O to the new origin O^*. The shift size $\text{ZSS}(k)$ is given by

$$\text{ZSS}(k) = \begin{cases} 0.0 & \text{for } Z_p(k) < Z_{p,\min} & (6.7) \\ S_g Z_p(k) & \text{for } Z_p(k) > Z_{p,\min} & (6.8) \end{cases}$$

where, in order to prevent excessive control action caused by the shifting of the origin, a dead band $Z_{p,\min}$ is also introduced.

In the advanced fuzzy logic control scheme, the polar variables, the angle $\theta(k)$ and the radius $D(k)$, are derived as follows:

$$D(k) = \sqrt{(Z_s(k) + \text{ZSS})^2 + [A_s \cdot Z_a(k)]^2} \tag{6.9}$$

$$\theta(k) = \tan^{-1}\left(\frac{A_s \cdot Z_a(k)}{Z_s(k) + \text{ZSS}}\right) \tag{6.10}$$

The stabilizing signal $U(k)$ is then determined through the same procedure given by Eq. (6.5) shown in Section 6.2.2. Here, it must be noted that the new origin O^* coincides with the origin O in steady state because the integral information $Z_p(k)$ becomes equal to zero in steady state when the study generator is stabilized.

The advanced stabilizing signal is obtained only through a slight modification of the quantities passed to the control routine as follows:

Required for FLPSS: $Z_s(k)$ and $Z_a(k)$
Required for AFLPSS: $Z_s(k) + \text{ZSS}(k)$ and $Z_a(k)$

On the basis of the three-dimensional information of the study unit and a set of simple fuzzy logic control rules, a desired stabilizing signal is generated at each sampling time to shift the generator state to the origin of the current phase plane shown in Figure 6.7, the origin of which is shifted to O^* by ZSS according to the phase information $Z_p(k)$.

The proposed control scheme has three basic parameters: the scaling factor A_s for $Z_a(k)$, the overlap angle α of the angle membership functions, and the fuzzy distance level for the radial member D_r. In addition, there are two supplementary parameters: the shift gain S_g and the dead band $Z_{p,\min}$. Two other factors are also involved in the AFLPSS: the maximum control effort $U_{\max}$ and the sampling interval ΔT. These factors are often determined by external criteria.

The adjustable parameters A_s, D_r, and α are tuned at a specific operating point subject to a specific disturbance shown later, and those parameters are fixed throughout the simulations and the experiments to demonstrate the robustness of the advanced fuzzy logic control scheme. The setting of the other parameters S_g and $Z_{p,\min}$ are also checked through the simulations and the experiments. It must be noted that the AFLPSS becomes equivalent to the FLPSS by setting the shift gain

S_g to zero. In addition, the gain of the AFLPSS is modified by the parameter D_r, and the phase compensation by the AFLPSS is modified through the parameter A_s.

6.2.4. Signal Conditioning [11–14]

In order to obtain the three-dimensional information of the study generator, signal conditioning of type 1, shown in Figure 6.8(a), is required, when using the speed signal as the input to the AFLPSS. When using the real power signal as the input, another signal conditioning of type 2, shown in Figure 6.8(b), is utilized. In Figure 6.8, R_1 and R_2 are reset filters, and the reset time constants T_{R1} and T_{R2} are set to 4.0 s and 0.5 s, respectively. The time constant of the second reset filter R_2 is set to be shorter in order to avoid excessive control action because of the delay of the resetting of the integrated speed signal $Z_p(k)$. In addition, I indicates an integrator. Through these signal conditioning actions, the acceleration $Z_a(k)$, the speed deviation $Z_s(k)$, and the signal related to the phase difference $Z_p(k)$ are determined at each sampling time. In steady state the quantities $Z_a(k)$, $Z_s(k)$, and $Z_p(k)$ become zero through the action of the reset filters.

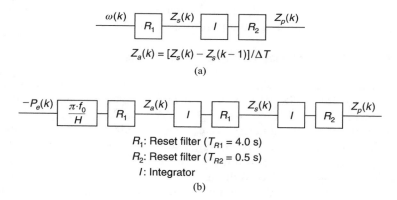

$$Z_a(k) = [Z_s(k) - Z_s(k-1)]/\Delta T$$

(a)

R_1: Reset filter ($T_{R1} = 4.0$ s)
R_2: Reset filter ($T_{R2} = 0.5$ s)
I: Integrator

(b)

Figure 6.8 Signal conditioning: (a) speed signal; (b) real power signal.

6.3. SIMULATION STUDIES USING ONE-MACHINE INFINITE-BUS SYSTEM [12,14]

6.3.1. One-Machine Infinite-Bus System

A one-machine infinite-bus system, shown in Figure 6.9, is used for numerical simulations. The generator constants are shown in Table 6.1, including the exciter and the transmission line constants. The simulations have been performed subject to the following disturbances:

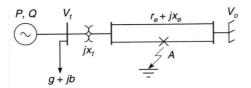

Figure 6.9 Model one-machine system.

TABLE 6.1 SYSTEM CONSTANTS

$H = 4.63\,\text{s}, D = 0.01, T'_{do} = 7.76\,\text{s}$
$X_d = 0.973, X'_d = 0.19, X_q = 0.55$
$K_a = 100.0, T_a = 0.05\,\text{s}$
$r_e = 0.03, x_e = 0.5, x_t = 0.1, g = 0.2, b = -0.1$

a. A three-phase-to-ground fault at point A for 0.1 s
b. A three-phase-to-ground fault at point A and the isolation of the faulted line after 0.1 s

In the simulations, a third-order machine model is utilized. For the fundamental transient stability analysis, the third-order model is sufficient.

6.3.2. Optimal Settings of Stabilizer Parameters

Through a sequential optimization technique, optimal parameters have been determined for CPSS, FLPSS, and AFLPSS subject to the above disturbance a at the operating point $P = 1.0\,\text{pu}$, $Q = 0.3\,\text{pu}$, and $V_t = 1.0\,\text{pu}$. In the sequential optimization, the parameters have been optimized on a one-by-one basis. The stabilizer performance has been evaluated by using the following discrete-time quadratic performance indices J_1 and J_2. For the optimal setting of stabilizer parameters, the index J_1 is utilized considering the control efforts by the stabilizers to reduce excessive control action. When investigating system stability, the second index J_2 is utilized to evaluate the damping characteristic of the study system. Throughout the simulations, the sampling interval ΔT is set to 1/120 s, and the maximum size of the stabilizing signal U_{max} is set to 0.2 pu. The block diagram of the CPSS is shown in Figure 6.10:

$$J_1 = \sum_k \{[\Delta\omega(k) \cdot k\,\Delta T]^2 + [U(k) \cdot k\,\Delta T]^2\} \tag{6.11}$$

$$J_2 = \sum_k [\Delta\omega(k) \cdot k\,\Delta T]^2 \tag{6.12}$$

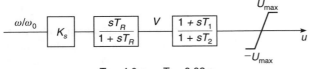

$T_R = 4.0\,\text{s}$ $T_2 = 0.03\,\text{s}$

Figure 6.10 Block diagram of CPSS.

TABLE 6.2 OPTIMAL SETTINGS OF STABILIZER
PARAMETERS

CPSS	$K_s = 58.0, T_1 = 0.29\,\text{s}$
FLPSS	$A_s = 0.2, D_r = 0.85, \alpha = 30.0$
AFLPSS(1)	$A_s = 0.2, D_r = 0.85, \alpha = 30.0$
	$S_g = 2.4, Z_{p,\min} = 0.3$
AFLPSS(2)	$A_s = 0.2, D_r = 0.85, \alpha = 30.0$
	$S_g = 5.1, Z_{p,\min} = 0.3$

Note: Input signal = speed signal.

The newly introduced parameter, that is, the shift gain S_g, is optimized subject to disturbance b at the same operating point described above and also at another operating point, $P = 1.11\,\text{pu}$, $Q = 0.3\,\text{pu}$, and $V_t = 1.0\,\text{pu}$, where the system becomes unstable when applying the AFLPSS with the shift gain determined at the former operating point. In Table 6.2, AFLPSS(1) indicates the parameters obtained at the former operating point, and AFLPSS(2) indicates the parameters obtained at the latter operating point. Furthermore, the dead band $Z_{p,\min}$ is optimized at the former operating point, where the system becomes stable when applying the FLPSS, in order to avoid the excessive control action caused by the shift of the origin in the phase plane.

In the table, D_r is the distance parameter, and the gain is determined by $D(k)/D_r$. The term α is the overlap angle between sectors A and B.

Figure 6.11 shows the variation of the performance index J_1 according to the changes of the shift gain S_g at the operating point $P = 1.11\,\text{pu}$ and $Q = 0.3\,\text{pu}$. As shown in this figure, the study system cannot be stabilized at this operating point using the lower setting of the shift gain S_g.

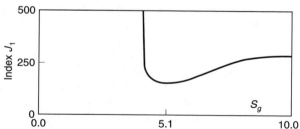

Figure 6.11 Effect of shift gain S_g at operating point of $P = 1.11\,\text{pu}$ and $Q = 0.3\,\text{pu}$ ($A_s = 0.2$, $D_r = 0.85$, $\alpha = 30.0$).

6.3.3. Stabilizer Performance

Table 6.3 indicates the stable regions subject to disturbance b achieved by CPSS, FLPSS, AFLPSS(1), and AFLPSS(2). Here, SMP gives the physically stabilizable maximum power output subject to disturbance b and is determined from first swing stability when applying the positive maximum stabilizing signal plus $U_{\max}$ to the study unit immediately after applying disturbance b to the system. As shown in the

TABLE 6.3   CRITICAL REAL POWER OUTPUT

Fault Duration (steps)	Critical Power Output (pu)				
	CPSS	FLPSS	AFLPSS(1)	AFLPSS(2)	SMP
0	1.13	1.16	1.17	1.18	1.20
3	1.12	1.15	1.17	1.18	1.20
6	1.09	1.12	1.17	1.17	1.20
9	1.05	1.09	1.14	1.16	1.16
12	1.01	1.05	1.10	1.11	1.11
15	0.97	1.00	1.05	1.06	1.06
18	0.93	0.96	1.01	1.01	1.01

Note: 1 step = 1/120 s.

table, a wider stable region is obtained by applying the AFLPSS. In particular, the AFLPSS(2) is able to stabilize the study unit very close to the physically stabilizable limit.

Figure 6.12 shows a comparison between CPSS, FLPSS, and AFLPSS(1) subject to disturbance a, where the operating point of the study unit is set to $P = 1.03$ pu and $Q = 0.3$ pu. As shown in the figure, the CPSS is not able to stabilize the study unit. In this case, there is no significant difference between the FLPSS and the AFLPSS(1). Figure 6.13 shows the frequency response characteristics obtained through the describing function method for the CPSS and the FLPSS. Both of them have almost the same characteristics; therefore, the CPSS and the FLPSS have the same performance under small signal disturbances. However, as shown in Figure 6.12, significant stabilizing effects are achieved by the FLPSS when under large disturbance conditions.

Figure 6.14 shows a comparison between AFLPSS(1) and AFLPSS(2) subject to disturbance b under the critical operating point $P = 1.11$ pu and $Q = 0.3$ pu. The shift gain S_g of 2.4 is not sufficient to stabilize the study unit. The tuning of the shift gain S_g should be performed at a critical operating point to enlarge the stabilizable region, as shown in the case of AFLPSS(2) with the shift gain $S_g = 5.1$. Figure 6.15 shows the speed/acceleration phase plane plot corresponding to Figure 6.14.

Figure 6.16 shows the variation of the performance index J_2 according to the change of the fault duration time subject to disturbance b, where the operating point of the study unit is set to $P = 1.0$ pu and $Q = 0.3$ pu. Again a wider stabilizable region is achieved by the AFLPSS(2).

6.3.4. AFLPSS Using Power Signal

Without changing the parameter setting AFLPSS(2) for the AFLPSS using the speed signal, the AFLPSS using the real power output signal is realized. Figure 6.17 shows a comparison between the AFLPSS using the speed signal and the AFLPSS using the real power signal. Almost the same control effect is observed for both

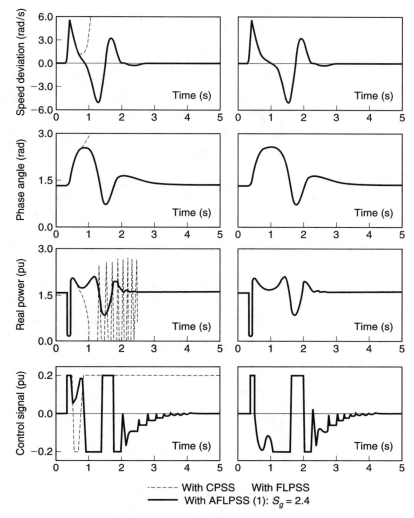

Figure 6.12 Comparison studies at operating point $P = 1.03$ pu and $Q = 0.3$ pu.

stabilizers with the same parameter settings. In this case, disturbance b is applied to the system under the operating point $P = 1.08$ pu and $Q = 0.3$ pu.

6.3.5. Experimental Evaluation on 5-kVA Laboratory System [12]

The FLPSS is set up by using a personal computer as shown in Figure 6.18. The central processing unit (CPU) and the floating-point unit (FPU) on the PSS are the Intel 80386SX and the Intel 80387SX. The clock frequency is 10 MHz. A 12-bit A/D

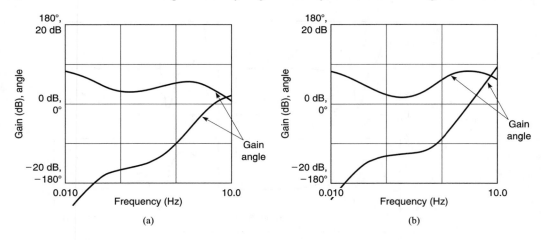

Figure 6.13 Frequency response characteristics of (a) CPSS and (b) FLPSS.

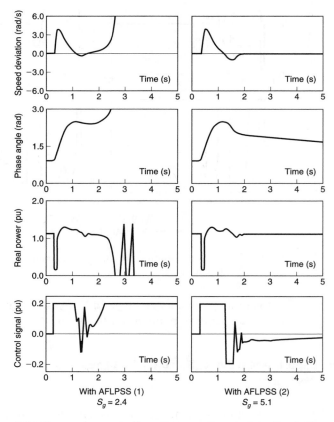

Figure 6.14 Comparison studies at operating point $P = 1.11$ pu and $Q = 0.3$ pu.

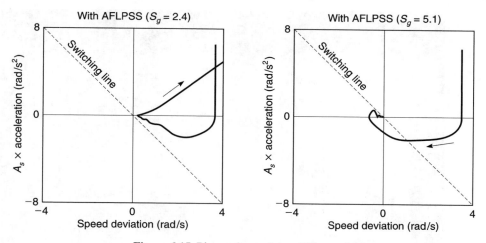

Figure 6.15 Phase plane plots of Figure 6.14.

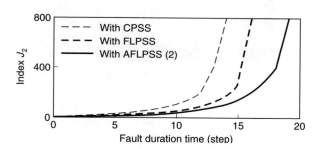

Figure 6.16 Performance index J_2 and fault duration time (1 step = 1/120 s).

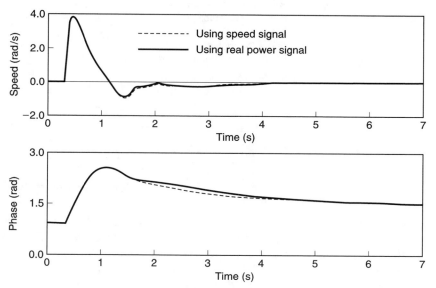

Figure 6.17 AFLPSS using real power signal.

Figure 6.18 PC-based FLPSS and AFLPSS.

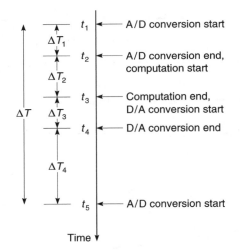

Figure 6.19 Control procedure.

conversion interface and a 16-bit D/A conversion interface are installed on the PSS to obtain the real power output signal and also to feed the stabilizing signal back to the PSS terminal of the automatic voltage regulator (AVR). The control procedure is shown in Figure 6.19. The sampling interval ΔT is set to 20 ms throughout the experiments, and the required computation time is less than 2 ms to generate a stabilizing signal at each sampling time.

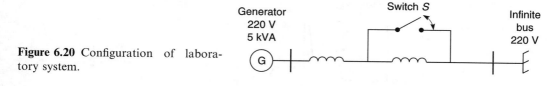

Figure 6.20 Configuration of laboratory system.

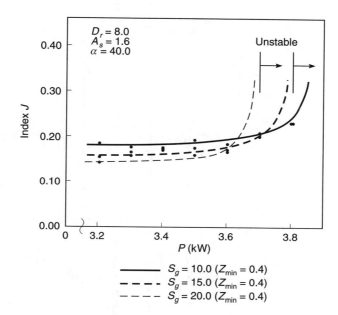

Figure 6.21 Stable regions.

To demonstrate the effectiveness of the proposed AFLPSS, experiments are performed by using a laboratory system rated at 5 kVA, 220 Vac, and 60 Hz. Figure 6.20 illustrates the basic configuration of the laboratory system. Disturbances are added to the laboratory system by line switching to change the length of the short transmission line connecting the generator to a commercial power source. In the experiments, the speed signal is used as the input to the FLPSS and also to the AFLPSS. Figure 6.21 shows the stable region of the laboratory system. The parameter settings are also shown in the figure. A larger stable region is achieved by the AFLPSS. When applying the FLPSS to the system, the critical power output is 3.3 kW. Figure 6.22 shows the system response at the operating point $P = 3.4$ kW. The system is stabilized by applying the AFLPSS with the shift gain $S_g = 20.0$.

In the experiments, the sampling interval ΔT is set to 20 ms, and the maximum size of the stabilizing signal U_{max} is set to 0.2 pu. In Figure 6.22, the speed deviation of 0.1 V is equivalent to the angular velocity of 5.236 rad/s, and the stabilizing signal of 1.0 V is equivalent to 0.1 pu in the per-unit system.

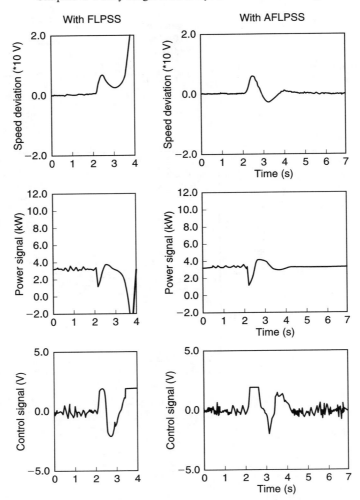

Figure 6.22 Comparison between FLPSS and AFLPSS.

6.4. EVALUATION ON ANALOG NETWORK SIMULATOR

6.4.1. Analog Network Simulator

Figure 6.23 shows the analog network simulator (ANS) at the Research Laboratory of KEPCO. The ANS was manufactured at Toshiba Corp. The simulator consists of the following modules:

- Generator modules (12 units)
- Transformer modules for generators (12 units)
- Main transformers for substations (10 units)

Figure 6.23 Overview of ANS.

- Transmission line modules (50 units)
- Load modules (10 units)
- Circuit breaker modules (16 units)
- Static condensers, shunt reactors, and lightning arrestors
- DC transmission module (1 unit) and static var compensator (SVC) module (1 unit)

All the modules are operated and monitored by a host computer and two terminals. The simulator is capable of handling up to a 12-machine system.

6.4.2. Study System

Figure 6.24 shows the configuration of a power system used for the experiments on the ANS. This system represents one of the local areas of KEPCO. In the system, unit 1 is a 20.25-MVA hydroplant that was one of the candidates to which the actual installation of the proposed AFLPSS was being planned for long-term evaluation. The PC-based AFLPSS was set on unit 1. The input signal to the FLPSS and also to the AFLPSS is specified to the real power signal from the study generator. The signal conditioning of type 2, shown in Figure 6.8(b), is utilized to obtain the three-dimensional information. The first block is not shown for simplicity.

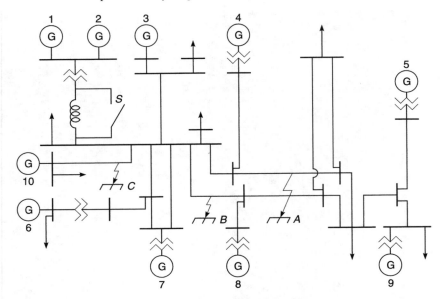

Figure 6.24 Configuration of study system.

The effectiveness of the AFLPSS was tested subject to the following disturbances:

1. Three-phase-to-ground fault at locations of A, B, and C in the study system for four cycles
2. Two-phase-to-ground fault for four cycles at the same locations described earlier
3. Single-phase-to-ground fault for four cycles at the same locations
4. Reactance switching by opening the switch S

For disturbances (a)–(c), the faults were removed after four cycles.

6.4.3. Parameter Tuning

The basic parameters A_s, D_r, and α are tuned and also evaluated through the experiments by using the following discrete-time quadratic performance index subject to disturbance 1:

$$J = \sum_k \Delta\omega(k)^2 \tag{6.13}$$

The tuned parameters, which are common for both the FLPSS and the AFLPSS, at the site are as follows:

$$A_s = 9.0 \qquad D_r = 1.8 \qquad \alpha = 65.0$$

where the maximum size of stabilizing signal U_{max} is set to 0.2 pu. Throughout the experiments, all these parameters were fixed to the values shown above. Here, it must be noted that the shift gain S_g was set to zero during the tuning.

6.4.4. Test Results

Throughout the experiments, three quantities (i.e., the real power output, the speed deviation, and the stabilizing signal) were monitored on the PC-based data acquisition system. All the results shown in this section are the ones monitored on the data acquisition system instead of using the monitoring system on the ANS.

Figures 6.25 and 6.26 show the time responses of unit 1 subject to disturbances 1 and 3, respectively, at location C, where the time responses of the same unit without any stabilizer are also illustrated for comparison studies. Almost the same stabilizing effects are observed for disturbance 2 and also at locations A and B. Here, the shift

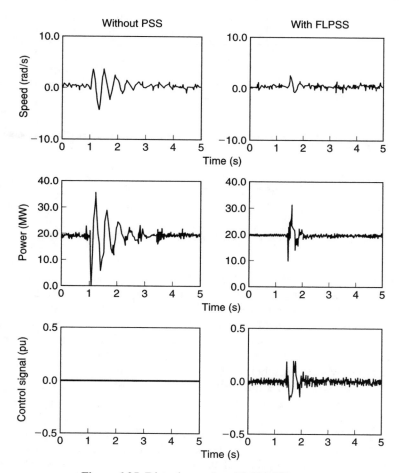

Figure 6.25 Disturbance 1 at 20.25 MW output.

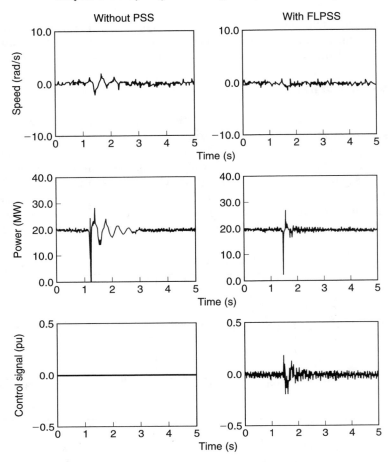

Figure 6.26 Disturbance 3 at 20.25 MW output.

gain S_g is set to zero. Therefore, only the performance of the FLPSS is evaluated with satisfactory results showing the improvement of the damping of unit 1.

In order to investigate the effectiveness of the AFLPSS, the switching disturbance 4 is considered. According to the restriction of the ANS operation, the steady-state maximum power setting from generator modules cannot exceed 1.25 pu of the machine base. Therefore, a large reactance is set for disturbance 4 to make unit 1 unstable by switching below the rated power output.

Figures 6.27–6.29 indicate the advantage of the AFLPSS compared with the FLPSS. The stabilizable region is enlarged by the application of the AFLPSS to unit 1 subject to disturbance 4. Here, the shift gain S_g was increased up to the values 15.0 and 20.0 for which the study system was stabilized at the respective operating points. It must be noted that there is no significant difference in the control functions achieved by the AFLPSS and the FLPSS at a more stable operating point of unit 1. In these experiments, the dead band $Z_{p,\min}$ is set to zero. During the experiments,

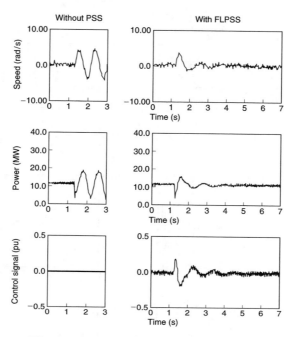

Figure 6.27 Disturbance 4 at 12 MW output.

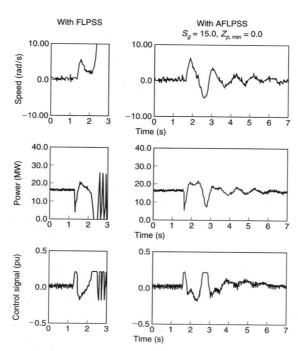

Figure 6.28 Disturbance 4 at 17 MW output.

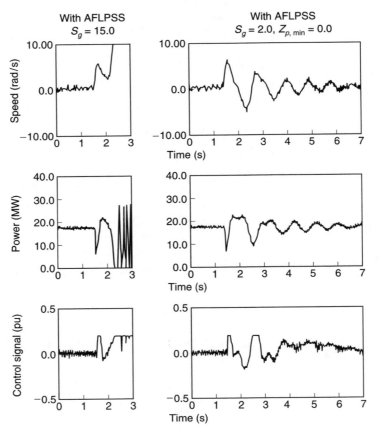

Figure 6.29 Disturbance 4 at 18 MW output.

the size of the dead band $Z_{p,\min}$ was changed up to 0.08 without any significant effects.

As shown in Figures 6.25–6.29, the observed mode of oscillation is only the local mode on unit 1 and there is no interarea oscillations. According to the former studies, it has been shown that the damping of oscillations achieved by the FLPSS or by the AFLPSS is more significant compared with that obtained using conventional analog-type PSS, where there is no significant interarea oscillations.

In order to investigate the effective damping of the low-frequency interarea modes by the FLPSS and also by the AFLPSS, experiments being carried out using the ANS and a four-machine laboratory system composed of machines with ratings of 60–90 kVA. The comparisons with conventional analog-type PSSs are also considered. It has also been shown that both the local and interarea modes of oscillations are effectively damped by using the FLPSS or the AFLPSS, when comparing with the damping achieved by the conventional analog-type PSSs.

6.5. ACTUAL INSTALLATION ON HYDROUNITS

6.5.1. Basic Configuration of Prototype

The block diagram of the PC-based prototype of the AFLPSS is shown in Figure 6.30, including the monitoring unit active power (P), reactive power (Q), voltage and frequency (PQVF), the protection unit, and the uninterruptible power system (UPS). The overview of the prototype is also shown in Figure 6.31. The PQVF monitors the real and the reactive power output, the terminal voltage, and the system frequency.

The protection unit has the following functions:

- When a continuous maximum level of stabilizing signal is detected for the specified period, it opens the switch S to disconnect the AFLPSS from the unit. The time setting can be changed in the range of 1.0–10.0 s.
- When interruption of the power supply from the UPS is detected, it also opens the switch S to disconnect the AFLPSS from the unit.

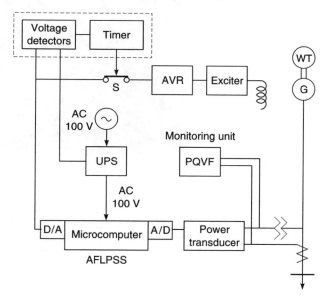

Figure 6.30 Basic configuration of prototype.

6.5.2. Site Tests

The first prototype was installed on a hydrounit (30.2 MVA, 11 kV, 600 rpm) at the Kurokawa No. 1 Hydro Power Station on March 14, 1994. This unit has a brushless AC exciter (160 kW, 260 V). The block diagram of the excitation system is shown in Figure 6.32 along with the constants.

Figure 6.31 Overview of prototype.

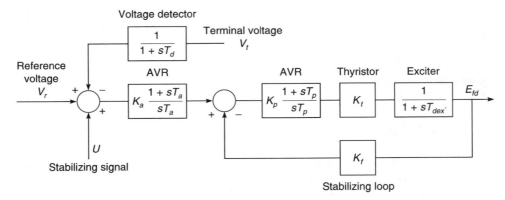

Figure 6.32 Block diagram of excitation system.

According to the ANS experimental results and the digital simulations, the adjustable parameters are set to the following values:

$$A_s = 10.0 \qquad D_r = 2.0 \qquad \alpha = 55.0 \qquad S_g = 10.0 \qquad Z_{P,\min} = 0.0$$

This unit was also one of the candidates for actual installation. The types of AVR and exciter are almost the same as the candidate unit shown in Section 6.4.

This station was selected because of the number of faults around this area reported in a year. The maximum size of the stabilizing signal $U_{\max}$ is now set to 0.05 pu because of the regulation for the unit. During long-term evaluation, this setting will be changed up to 0.1 pu to investigate the effect of the maximum stabilizing effort. The following are the parameter values for Figure 6.32:

$$K_a = 1.5 \qquad T_a = 1.6\,s \qquad K_p = 2.0 \qquad T_p = 0.28 \qquad K_t = 4.0$$

$$K_f = 0.17 \qquad T''_{dex} = 0.28\,s \qquad T_d = 0.026\,s$$

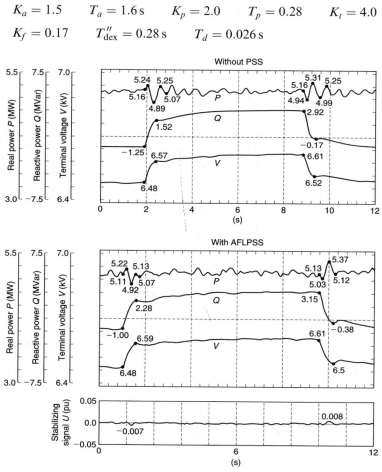

Figure 6.33 Typical result of site test (3% step change of reference voltage at 5 MW output).

The second prototype was installed on the unit (23.4 MVA, 6.6 kV, 200 rpm) at the Kawabaru Hydro Power Station on May 26, 1994, after the site tests. This unit also has a brushless AC exciter (155 kW, 160 V). The block diagram of the excitation system is almost the same as the one shown in Figure 6.31, where a digital AVR is included in the unit; therefore, the tuning of the AFLPSS parameters was performed at the site using step changes of the reference voltage under the unit operation of 5 MW output. Typical results are shown in Figure 6.33. In this case the tuned parameters are as follows:

$$A_s = 3.0 \qquad D_r = 0.2 \qquad \alpha = 50.0$$

The shift gain S_g and the dead band $Z_{p,\min}$ are now set to zero. During long-term evaluation, these settings will be modified to investigate the efficiency of the AFLPSS. The maximum size of the stabilizing signal is now set to 0.05 pu; however, the size will be also changed up to 0.1 pu during the evaluation.

6.6. CONCLUSION

This chapter has presented the evaluation of the AFLPSS on the ANS and also its actual installation on the hydrounits. The AFLPSS is set up by using a personal computer for the evaluation using the ANS. The results show the advantages of the proposed AFLPSS compared with the former FLPSS. The stable region is further enlarged by the application of the AFLPSS. To implement the advanced fuzzy logic control scheme in the former conventional one, only minor modifications are required for the control algorithm in the control software. The installation of the two AFLPSS prototypes has been successfully performed on the hydrounits in the KEPCO system for long-term evaluation of the proposed AFLPSS. Further experiments are also ongoing to aim the installation of the AFLPSS to large thermal plants considering multimode oscillations.

References

[1] E. V. Larsen and D. A. Swann, "Applying Power System Stabilizers: Parts I, II, and III," *IEEE Transactions on Power Apparatus and Systems*, Vol. PAS-100, No. 6, 1981, pp. 3017–3041.

[2] P. Kundur, M. Kleine, G. J. Rogers, and M. Zwyno, "Application of Power System Stabilizers for Enhancement of Overall System Stability," *IEEE Transactions on Power Systems*, Vol. PWRS-4, 1989, pp. 614–621.

[3] D. Xia and G. T. Heydt, "Self-Tuning Controller for Generator Excitation Control," *IEEE Transactions on Power Apparatus and Systems*, Vol. PAS-102, 1983, pp. 1877–1885.

[4] A. Ghosh, G. Ledwich, O. P. Malik, and G. S. Hope, "Power System Stabilizer Based on Adaptive Control Techniques," *IEEE Transactions on Power Apparatus and Systems*, Vol. PAS-103, No. 8, 1984, pp. 1983–1989.

[5] S. J. Cheng, O. P. Malik, and G. S. Hope, "Self-Tuning Stabilizer for a Multi-machine Power System," *IEE Proceedings*, Part C, Vol. 133, No. 4, 1986, pp. 176–186.

[6] Y. Y. Hsu and K. L. Liou, "Design of Self-Tuning PID Power System Stabilizers for Synchronous Generators," *IEEE Transactions on Energy Conversion*, Vol. EC-2, No. 3. 1987, pp. 343–348.

[7] O. P. Malik, G. S. Hope, S. J. Cheng, and G. Hancock, "A Multi-Micro-computer Based Dual-Rate Self-Tuning Power System Stabilizer," *IEEE Transactions on Energy Conversion*, Vol. EC-2, No. 3, 1987, pp. 355–360.

[8] T. Hiyama, "Application of Rule-Based Stabilizing Controller to Electrical Power System," *IEE Proceedings*, Part C, Vol. 136, No. 3, 1989, pp. 175–181.

[9] T. Hiyama, "Rule-Based Stabilizer for Multi-machine Power System," *IEEE Transactions on Power Systems*, Vol. PWRS-5, No. 2, 1990, pp. 403–411.

[10] T. Hiyama and T. Sameshima, "Fuzzy Logic Control Scheme for On-line Stabilization of Multi-machine Power System," *Fuzzy Sets and Systems*. Vol. 39, 1991, pp. 181–194.

[11] T. Hiyama, S. Oniki, and H. Nagashima, "Experimental Studies on Micro-computer Based Fuzzy Logic Stabilizer," *Proceedings of the Second International Forum on Application of Neural Network to Power Systems*, Yokohama, Japan, 1993, pp. 212–217.

[12] T. Hiyama, "Real Time Control of Micro-machine System Using Micro-computer Based Fuzzy Logic Power System Stabilizer," *IEEE Transactions on Energy Conversion*, Vol. EC-9, No. 4, 1994, pp. 724–731.

[13] T. Hiyama, "Robustness of Fuzzy Logic Power System Stabilizers Applied to Multimachine Power System," *IEEE Transactions on Energy Conversion*, Vol. EC-9, No. 3, 1994, pp. 451–459.

[14] T. Hiyama, M. Kugimiya, and H. Satoh, "Advanced PID Type Fuzzy Logic Power System Stabilizer," *IEEE Transactions on Energy Conversion*, Vol. EC-9, No. 3, 1994, pp. 514–520.

[15] T. H. Ortmeyer and T. Hiyama, "Frequency Response Characteristics of the Fuzzy Polar Power System Stabilizer," *IEEE Transactions on Energy Conversion*, Vol. 10, No. 2, 1995, pp. 333–338.

[16] H. Doi, M. Goto, T. Kawai, S. Yokokawa, and T. Suzuki, "Advanced Power System Analog Simulator," *Transactions on Power Systems*, Vol. 5, No. 3, 1990, pp. 962–968.

Chapter 7

T. Hiyama
Department of Electrical Engineering
and Computer Science
Kumamoto University
Kumamoto
860 Japan

Fuzzy Logic Switching of FACTS Devices

7.1. INTRODUCTION

Power electronic devices have been receiving increasing attention in applications to improve power system stability and also to increase the maximum power transmission through existing AC transmission lines. Such transmission systems are designated as flexible AC transmission systems (FACTS) [1, 2].

In this chapter, we present a fuzzy logic control switching control scheme for FACTS devices such as thyristor-controlled series capacitor (TCSC) modules [3], static var compensators (SVCs) [4–8], and braking resistors (BRs) [9, 10] to enhance the overall stability of electric power systems. Only real power flow measurement is required at the location of the FACTS device to generate the switching control signal at every sampling time according to the phase/speed state of the study system. In order to obtain the phase/speed state of the electric power system, signal conditioning is required for the measured real power flow through reset filters and integrators. According to the phase/speed state and simple fuzzy logic control rules, the switching control signal is determined via on-line computation. The rules are straightforward and do not require a heavy computational burden [10–14, 16].

In order to investigate the effectiveness of the proposed fuzzy logic switching control scheme, nonlinear simulations have been performed using a five-machine infinite-bus system as the model. In addition, a coordination with power system stabilizers (PSSs) or a different type of FACTS device has also been considered in this chapter. The study shows that the stable region would be highly enlarged by the application of FACTS devices using the fuzzy logic switching control scheme. The coordination is also effective to increase the stable region.

7.2. FUZZY LOGIC SWITCHING CONTROL SCHEME [10, 16–19]

The proposed switching controller is set up using a microcomputer and analog-to-digital (A/D) and D/A conversion interfaces for real-time control. Figure 7.1 illustrates the basic configuration of the proposed switching controller for FACTS devices. All filtering is achieved on the microcomputer by solving difference equations for the reset filters and the integrators. The required input signal to the controller is only the real power signal P at the location of the FACTS device. Through digital filtering, the required two-dimensional information [i.e., the speed information $Z_s(k)$ and the phase information $Z_p(k)$] is derived directly from the measured real power signal P.

The output of the first reset filter, $Z_a(k)$, is a local measure of the acceleration at the site of the FACTS device. The signal $Z_s(k)$ is obtained by integrating $Z_a(k)$ and then applying a reset filter to remove any offset. This signal is a measure of the speed or frequency deviation at the site. The signal $Z_p(k)$ is similarly obtained from $Z_s(k)$ and is the measure of phase deviation at the site. The reset filter at the last stage is also necessary to remove any offset from $Z_p(k)$. Here, it must be noted that all these signals become zero through the resetting by the reset filters in the steady state when the study system is stabilized.

The digital reset filters and the integrators, shown in Figure 7.1, are given by the following discrete-time transfer functions in the z-plane. From the discrete-time transfer functions, a set of difference equations is derived. These difference equations are solved on the microcomputer to get the signals $Z_p(k)$ and $Z_s(k)$:

$$H_{DR_i}(z^{-1}) = \frac{(2T_{R_i}/\Delta T)(1 - z^{-1})}{(1 + 2T_{R_i}/\Delta T) + (1 - 2T_{R_i}/\Delta T)z^{-1}} \tag{7.1}$$

$$H_{DI}(z^{-1}) = \frac{\Delta T}{2}\frac{1 + z^{-1}}{1 - z^{-1}} \tag{7.2}$$

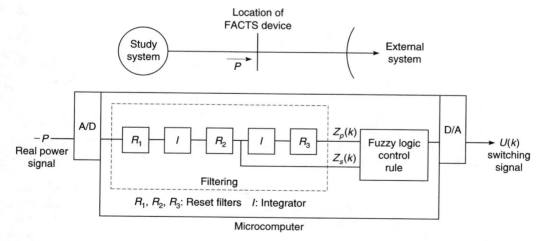

Figure 7.1 Basic configuration of switching controller.

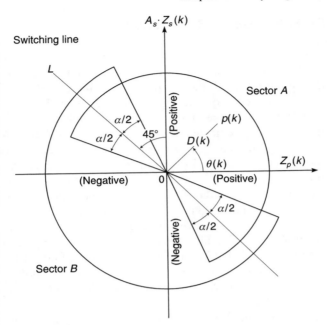

Figure 7.2 Phase plane.

The state of the study power system is given by the point $p(k)$ in the two-dimensional state space as shown in Figure 7.2:

$$p(k) = [Z_p(k), A_s Z_s(k)] \tag{7.3}$$

In the above equation, $Z_p(k)$ and $Z_s(k)$ give the phase/speed state and A_s is the scaling factor for $Z_s(k)$. From these signals $Z_p(k)$ and $Z_s(k)$, the polar variables, the radius $D(k)$, and the angle $\theta(k)$ are derived as follows:

$$D(k) = \sqrt{Z_p(k)^2 + [A_s Z_s(k)]^2} \tag{7.4}$$

$$\theta(k) = \tan^{-1}[A_s Z_s(k)/Z_p(k)] \tag{7.5}$$

As shown in Figure 7.2, the phase plane is divided into two sectors, A and B. In the first quadrant, which is a part of sector A, both $Z_p(k)$ and $Z_s(k)$ are positive; therefore, deceleration control is required for the study system. In this case, the real power flow should be increased at the location of the FACTS device through the switching control of the FACTS device. On the contrary, in the third quadrant, which is a part of sector B, both $Z_p(k)$ and $Z_s(k)$ are negative, and acceleration control is required for the system. In this case, the real power flow should be decreased at the location of the FACTS device.

These two sectors are expressed mathematically by using two polar membership functions as shown in Figure 7.3. The membership functions are best defined in terms of the polar coordinates on the phase plane. The membership function $N(\theta(k))$ gives the grade of the deceleration control, and the function $P(\theta(k))$ gives that of the acceleration control, where $N(\theta(k)) + P(\theta(k)) = 1.0$, as shown in Figure

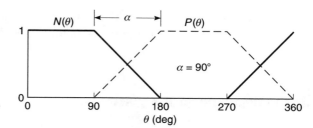

Figure 7.3 Two polar membership functions.

7.3. The switching between the deceleration and the acceleration control is taking place in the second and in the fourth quadrants gradually.

By using these two membership functions, the weighted averaging defuzzification algorithm is expressed by

$$U(k) = \frac{N(\theta(k)) - P(\theta(k))}{N(\theta(k)) + P(\theta(k))} G(k)$$

$$= [1 - 2P(\theta(k))]G(k) \tag{7.6}$$

$$G(k) = \begin{cases} D(k)/D_r & \text{for } D(k) \leq D_r \\ 1.0 & \text{for } D(k) \geq D_r \end{cases} \tag{7.7}$$

In the above equations, $G(k)$ indicates the gain factor, which is determined from the radius $D(k)$. The proposed control scheme has three adjustable parameters: the scaling factor A_s for $Z_s(k)$, the overlap angle α of the angle membership functions shown in Figure 7.3, and the fuzzy distance level for the radial member D_r. These parameters are optimized to give the maximum damping to the study system. The sampling interval ΔT is another factor involved in the switching controller. This factor is often determined by external criteria. The stabilizing signal $U(k)$ is in the range between -1.0 and 1.0. According to the value of the stabilizing signal $U(k)$, the number of energized modules for the TCSC, or the firing angle of the thyristor switch for the SVC or the BR, is determined as shown later.

7.3. EVALUATION OF CONTROL SCHEME

To tune the adjustable controller parameters, such as α, D_r, and A_s, the discrete-time quadratic performance index J_T is defined using the speed deviation $\Delta\omega(k)$ of each generator as follows:

$$J_T = \sum_{i=1}^{N} J_i \qquad J_i = \sum_{k=0}^{} \Delta\omega_i(k)^2 \cdot T_k^2 \tag{7.8}$$

$$T_k = k \cdot \Delta T \tag{7.9}$$

Here, the origin of the time T_k is set to the instant when the disturbance is applied to the system, and N denotes the total number of generators in the study system.

Through a sequential optimization technique, these adjustable parameters are optimized at a specific operating point and are subject to a specific disturbance.

7.4. MULTIMACHINE STUDY SYSTEM [14]

In order to investigate the effectiveness of the proposed fuzzy logic switching control scheme, nonlinear simulations are performed for the five-machine infinite-bus study system shown in Figure 7.4. The system represents the East Kyushu Subsystem of the Kyushu Electric Power Co. All the generators are thermal ones. Both units 4 and 5 have fast-acting thyristor excitation systems. In addition, the units are equipped with conventional power system stabilizers (CPSSs). Unit 4 is identical to unit 5, including the automatic voltage regulator (AVR), the governor, and the CPSS. The block diagrams of the excitation system and the CPSS are shown in Figure 7.5 and in Figure 7.6(a), respectively. When considering the coordinated control with the fuzzy

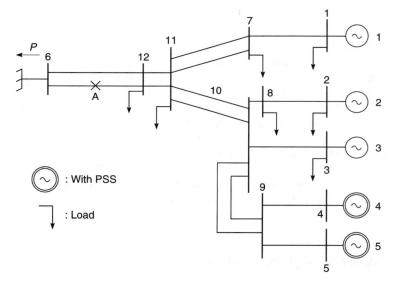

Figure 7.4 Model multimachine system.

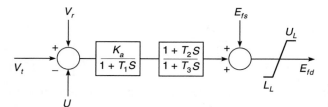

Figure 7.5 Block diagram of excitation system.

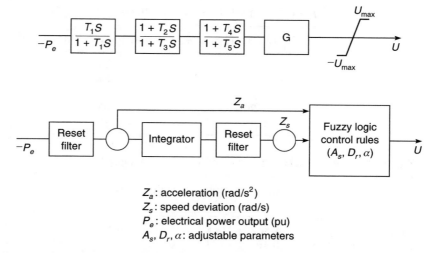

Z_a: acceleration (rad/s^2)
Z_s: speed deviation (rad/s)
P_e: electrical power output (pu)
A_s, D_r, α: adjustable parameters

Figure 7.6 Block diagram of (a) CPSS and (b) FLPSS.

logic power system stabilizer (FLPSS), all the CPSSs on units 4 and 5 are replaced by
the FLPSSs shown in Figure 7.6(b):

$$K_a = 150.0 \qquad T_1 = 0.04 \text{ s} \qquad T_2 = 0.02 \text{ s}$$
$$T_3 = 0.6 \text{ s} \qquad U_L = 4.4 \text{ pu} \qquad L_L = -3.5 \text{ pu}$$

where E_{fs} = excitation in the steady state
E_{fd} = excitation
V_r = reference terminal voltage
V_t = terminal voltage
U = supplementary stabilizing signal
U_L = upper limit of excitation
L_L = lower limit of excitation

Also,

$$T_1 = 4.0 \text{ s} \qquad T_2 = 0.2 \text{ s} \qquad T_3 = 0.5 \text{ s} \qquad T_4 = 0.1 \text{ s}$$
$$T_5 = 0.05 \text{ s} \qquad G = 0.55 \qquad U_{\max} = 0.1 \text{ pu}$$

and

$$A_s = 0.30 \qquad D_r = 0.06 \qquad \alpha = 30.0° \qquad U_{\max} = 0.1 \text{ pu}$$

Table 7.1 shows the relation between the load size and the power flow to the
infinite bus. The power generation from units 1–5 is fixed in steady state, and the
power transmission P_I to the infinite bus is increased by reducing the load size. The
operating condition at 100% load size is shown in Table 7.2.

TABLE 7.1   TYPICAL LOADING CONDITIONS

East Kyushu Load Size %	Power Flow P_I (MW)
100	743
90	848
80	953
70	1055
60	1155

TABLE 7.2   OPERATING CONDITION AT 100% LOAD (100 MVA BASE)

	Unit 1	Unit 2	Unit 3	Unit 4	Unit 5
Real power (pu)	2.50	1.68	1.68	6.90	6.90
Reactive power (pu)	0.47	0.23	0.11	1.67	1.67

	Load 1	Load 2	Load 3	Load 7	Load 8	Load 11	Load 12
Real power (pu)	0.10	0.08	0.08	2.85	1.47	3.46	3.70
Reactive power (pu)	0.06	0.0	0.0	1.00	0.21	−0.10	0.45

7.5. SIMULATION STUDIES

7.5.1. Thyristor-Controlled Series Capacitor Modules [16]

Figure 7.7 shows the basic configuration of the TCSC modules. From the value of the switching signal $U(k)$, the number of energized modules N_{om} is determined, as shown in Table 7.3, where $2M$ indicates the total number of modules and it is assumed, for simplicity, that half of them are energized in steady state. For deceleration control of the study system, the number of energized modules is increased. On the contrary, the number of energized modules is reduced for the acceleration control of the study system.

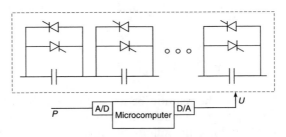

Figure 7.7 Configuration of TCSC modules.

TABLE 7.3   NUMBER OF ENERGIZED MODULES

Size of Control Signal $U(k)$	N_{om}
$1 = U(k)$	$2M$
$1 - 1/M \leq U(k) < 1$	$2M - 1$
$1 - 2/M \leq U(k) < 1 - 1/M$	$2M - 2$
$\vdots$	$\vdots$
$1/M \leq U(k) < 2/M$	$M + 1$
$-1/M < U(k) < 1/M$	$M*$
$-2/M < U(k) \leq -1/M$	$M - 1$
$\vdots$	$\vdots$
$-1 + 1/M < U(k) \leq -1 + 2/M$	2
$-1 < U(k) \leq -1 + 1/M$	1
$-1 = U(k)$	0

* M modules are energized at the steady state.

In the simulations, the switched series capacitors are installed on bus 6. Preliminary studies showed this location to be more effective than other locations closer to the individual generators. The total number of switched capacitor modules $2M$ is specified as 12, and 6 modules are energized in steady state. In the transient state, the number of modules energized during each sampling interval is changed in real time according to the proposed fuzzy logic switching control scheme. The reactance of each module is set to -0.001 pu in this study. The compensation rate is low enough so as not to cause subsynchronous resonance.

In the simulation studies, the following types of disturbances are considered:

(a) A three-phase-to-ground fault at point A in the middle of one of the parallel transmission lines between 12 and 13. The fault duration time is four cycles.

(b) A three-phase-to-ground fault at the same point A. The faulted line is isolated after four cycles.

The simulations are performed under various loading conditions and also considering the coordination with the power system stabilizers, where the following assumptions are made:

1. The electrical transients caused by the switching of the capacitor modules are neglected.

2. The thyristor switching delay is neglected.

3. The metal oxide varistors have negligible effect in the postfault period.

Optimal Setting of Adjustable Parameters. All the parameters of the fuzzy logic switching controller for the capacitor modules are optimized subject to disturbance b and under the load size of 100%. A sequential optimization technique is applied to get the optimal adjustable parameter values shown in Table 7.4. The optimal parameters are shown in Table 7.5. Here, it must be noted that the controller

TABLE 7.4   SEQUENTIAL OPTIMIZATION

Step	A_s	D_r	α	Index J_T
0	0.10	0.10	90.0	510.3
1	0.50	0.10	90.0	100.0
2	0.50	0.02	90.0	60.8
3	0.50	0.02	30.0	53.6
4	0.30	0.02	30.0	37.8
5	0.30	0.02	30.0	37.8

TABLE 7.5   OPTIMAL PARAMETER SETTING

$$A_s = 0.30, \quad D_r = 0.02, \quad \alpha = 30.0$$

Reset filter time constant: $T_{R1} = 4.0$ s, $T_{R2} = 0.4$ s

parameters are fixed to the same values shown in Table 7.5 throughout all the simulations under various loading conditions in order to investigate the robustness of the proposed control scheme.

In addition, the sampling interval ΔT is set to $1/120$ s; therefore, the number of energized capacitor modules is changed every half cycle.

Coordination with Power System Stabilizer. In order to investigate the efficiency of the coordinated stabilizing control using the switched series capacitors and the generator excitation control, the FLPSS is also simultaneously considered for comparison with the coordination between the TCSC and the CPSS. The configurations of the CPSS and the FLPSS are shown in Figure 7.6. The sampling interval ΔT is also set to the same value of $1/120$ s.

Typical Simulation Results. Figures 7.8 and 7.9 show typical system responses subject to both disturbances a and b under the loading condition with the load size of 100%. In these figures, the simulation results are also indicated when considering the coordination between the series capacitor switching control and the generator excitation control with PSSs. The PSS signals are shown at the bottom. As shown in these figures, the system damping is highly improved by applying the proposed fuzzy logic switching of series capacitors. Moreover, the shorter settling time is achieved by coordination with the PSSs, especially with the FLPSSs. Both the FLPSS and the CPSS are tuned separately for the system without any switched capacitor modules. Therefore, rigorously speaking, simultaneous control is applied to the study system through the switched series capacitors and the power system stabilizers.

Table 7.6 indicates the stable region obtained using various combinations of stabilizing control through the switching of capacitor modules and the generator excitation control. Table 7.6 also includes the level of the performance index J_T given by Eq. (7.8) for each stable case. This index provides an indication of the

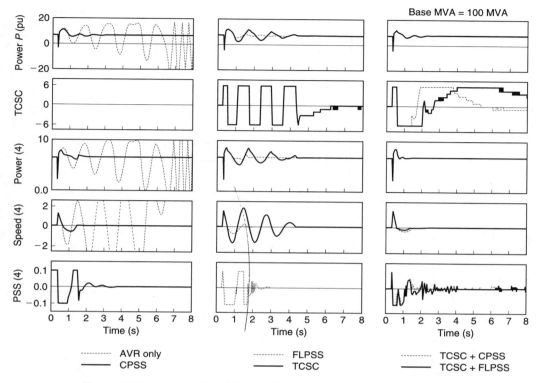

Figure 7.8 Typical results subject to disturbance a under loading condition of 100% load.

magnitude of the transient effect for the various stable cases. The widest stable region is achieved by the coordination with the FLPSSs. Here, it must be noted that the stable maximum power transmission to the infinite bus is 300 MW when considering only the AVR control without PSS on units 4 and 5 subject to disturbance b.

Figure 7.10 shows comparisons of the time response of the coordinated control between the switched capacitor modules and the FLPSSs subject to disturbance b at critical loading conditions.

7.5.2. Static Var Compensator [17, 18]

In this section, the same fuzzy logic control scheme is applied to the thyristor-controlled reactor (TCR) type SVC. Switching control of the SVC is well known as an effective means for voltage control. The modulation of the real power flow is also possible, and it provides stability enhancement of electric power systems.

The firing angle of the TCR is determined from the switching control signal $U(k)$ in on-line computation from the measured real power flow at the location of the SVC. The firing angle is updated every sampling instant to give the system maximum damping. Figure 7.11 gives the basic configuration of the SVC. The

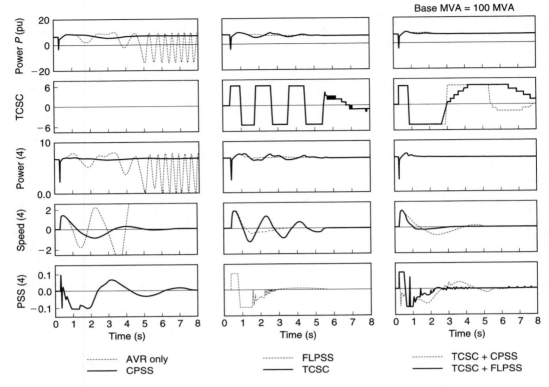

Figure 7.9 Typical results subject to disturbance b under loading condition of 100% load.

TABLE 7.6 PERFORMANCE INDEX J_T AND STABLE REGIONS SUBJECT TO DISTURBANCE b

Power Flow P (MW)	Load Size (%)	AVR Only	CPSS	FLPSS	TCSC CPSS	TCSC FLPSS	TCSC
748	100	Unstable	22.1	2.2	37.8	10.1	1.7
848	90		49.0	3.5	25.1	16.6	2.3
900	85		Unstable	5.6	26.6	26.2	2.6
952	80			10.8	41.7	28.3	3.1
962	79			13.0	51.3	27.7	3.3
973	78			Unstable	66.3	29.3	3.5
983	77				87.1	38.1	3.8
994	76				147.6	68.1	4.0
1003	75				Unstable	Unstable	4.3
1055	70						10.0
1075	68						16.8
1085	67						Unstable

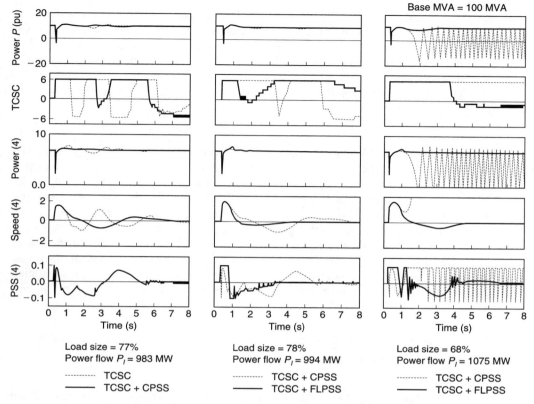

Figure 7.10 Effects of coordinates control through TCSC and PSS.

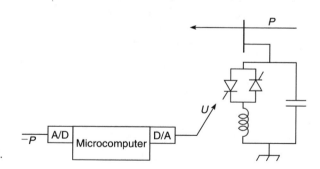

Figure 7.11 An SVC with a TCR.

controller generates firing control signals to the thyristor switching unit to modify the reactive power absorbed by or injected from the SVC in the transient condition.

For deceleration control of the study system, the injected reactive power from the SVC should be increased in order to increase the real power flow at the location of the SVC; then the SVC susceptance is capacitive. The amount of the reactive power injected from the SVC is modified by changing the setting of the firing angle of the thyristor switch. In contrast, for the acceleration control of the study system, the reactive power absorbed by the SVC should be increased in order to reduce the real power flow at the location of the SVC; then the SVC susceptance is inductive. The amount of reactive power absorbed by the SVC is also modified by changing the thyristor firing angle.

The SVC susceptance $B_{SVC}(k)$ is determined from the switching control signal $U(k)$ as follows:

$$B_{SVC}(k) = B_{SVC,MAX} \cdot U(k) + B_{SVC0} \tag{7.10}$$

where B_{SVC0} is the SVC susceptance in the steady state and $B_{SVC,MAX}$ gives the maximum size of the susceptance change. In this study, B_{SVC0} is set to zero. Therefore, the firing angle of the thyristor switch is set to $111°$ in the steady state. Namely, it is assumed that the reactive power absorbed by the SVC is zero in the steady state. The SVC susceptance varies from $-B_{SVC,MAX}$ to $B_{SVC,MAX}$ in this study. Therefore, $B_{SVC,MAX}$ gives the capacity of the SVC. Figure 7.12 shows the relation between the size of the stabilizing signal $U(k)$ and the thyristor firing angle.

The SVC is set on one of the bus bars in the transmission system shown in Figure 7.4. The simulations are performed under various loading conditions and also considering the coordination with the PSSs, where the following assumptions are made:

1. The electrical transients caused by the thyristor switching are neglected.
2. The thyristor switching delay is neglected.

Optimal Setting of Adjustable Parameters. All the parameters of the fuzzy logic switching controller for the SVC are optimized subject to disturbance b and under the load size of 100%, as described earlier in this chapter. The size of the SVC, that is, $B_{SVC,MAX}$, is set to 2.0 pu, which is 200 MVAR (megavolt ampere reactive). A

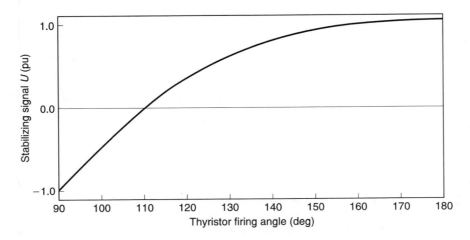

Figure 7.12 Size of stabilizing signal vs. firing angle.

TABLE 7.7 OPTIMAL PARAMETERS OF FUZZY LOGIC SWITCHING CONTROLLER FOR SVC

Location	A_s	D_r	a (deg)
Bus 10	0.01	0.04	10.0
Bus 11	0.01	0.05	10.0
Bus 12	0.01	0.07	10.0

Time constants of reset filters:
$T_{R1} = 3.5$ s, $T_{R2} = 0.6$ s, $T_{R3} = 0.45$ s

sequential optimization technique is applied to get the optimal adjustable parameter values. The optimal parameters are shown in Table 7.7 for the SVC located at bus 10, 11, or 12. In addition, the sampling interval ΔT is set to 1/120; therefore, the firing angle of the thyristor switching circuit is changed at every half cycle.

Table 7.8 shows the coordinated parameter settings for the FLPSS or the CPSS with the installation of the SVC at bus 12. Through the simulations with the settings shown in Figure 7.6, simultaneous control of the SVC with the PSSs is investigated. On the other hand, through the simulations with the settings shown in Table 7.8, a coordinated control of the SVC with the PSSs is evaluated. The parameter settings, given by Table 7.8, are determined at the 100% loading condition. These settings are not modified for the simulations under different loading conditions to investigate the robustness of the proposed control schemes. The sampling interval ΔT is set to 1/120 s for the FLPSS in the simulations.

TABLE 7.8 OPTIMAL PARAMETERS FOR PSS ON UNITS 4 AND 5 TUNED WITH SVC ON BUS 12

FLPSS*:	$A_s = 0.1$, $D_r = 0.17$, $\alpha = 90°$
	Reset filter time constant $T_R = 4.0$ s
CPSS*:	$G = 0.10$, $T_1 = 4.0$ s, $T_2 = 0.2$ s, $T_3 = 0.5$ s, $T_4 = 0.1$ s, $T_5 = 0.05$ s

Typical Simulation Results. Figure 7.13 shows the damping of oscillations achieved by the SVC on one of the buses 10, 11, or 12 subject to disturbance b. Figure 7.14 shows the critical power flow to the infinite bus subject to disturbance b. As shown in Figures 7.13 and 7.14, suitable locations for the SVC are at bus 11 or 12. In this study, the location of the SVC is fixed to bus 12 for the following simulations. Throughout these simulations, the parameter settings shown in Table 7.7 are fixed even for different loading conditions and also for different settings of SVC size to evaluate the robustness of the proposed control schemes.

Table 7.9 indicates the values of performance index J_T and the stable regions achieved by various combinations of stabilizing equipment. The widest stable region is achieved by the coordinated fuzzy logic stabilizing control between the SVC and FLPSS*s (the coordinated FLPSS) with the fixed parameters shown in Table 7.8. A wider stable region is also obtained by the simultaneous fuzzy logic stabilizing

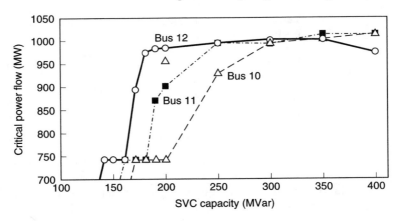

Figure 7.13 Damping of oscillation achieved by SVC.

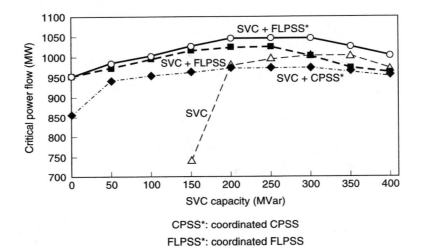

CPSS*: coordinated CPSS

FLPSS*: coordinated FLPSS

Figure 7.14 Critical power flow P_I achieved by SVC.

control between the SVC and the FLPSSs with the fixed parameters shown in Figure 7.6(b). These results indicate the robustness of the proposed control schemes. Furthermore, the coordination between the CPSSs and the SVC is also effective to enlarge the stable region. Here, it must be noted that the FLPSS* indicates the coordinated FLPSS with the parameters shown in Table 7.8, and the CPSS* denotes the coordinated CPSS.

Figure 7.15 shows the effects of the size of SVC for stability enhancement of the study system. The robustness is also indicated for the coordination between the fuzzy logic switching controller of the SVC and the PSSs. In addition, the simultaneous stabilizing control using the fuzzy logic switching controller of the SVC and the FLPSSs is also robust.

TABLE 7.9 VALUES OF PERFORMANCE INDEX J_T FOR TIME DURATION OF 16 s AND STABLE REGIONS

Power Flow P (MW)	Load Size (%)	AVR Only	SVC	FLPSS	SVC CPSS*	SVC FLPSS	SVC FLPSS*
748	100	Unstable	20.6	2.2	7.2	4.7	1.1
848	90		43.5	3.7	11.7	4.7	1.2
859	89		47.8	4.0	12.4	4.7	1.3
869	88		52.9	4.5	13.2	4.7	1.3
880	87		58.7	5.1	14.0	4.7	1.3
890	86		65.2	5.7	15.0	4.6	1.4
952	80		129.0	14.9	26.1	4.6	1.8
962	79		131.0	Unstable	30.6	4.7	1.9
973	78		126.7		52.2	4.8	2.1
983	77		101.1		Unstable	5.0	2.3
994	76		Unstable			5.3	2.5
1024	73					7.4	3.4
1035	72					Unstable	3.4
1045	71						4.9
1055	70						Unstable

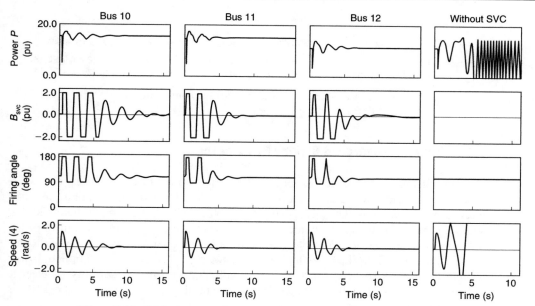

Figure 7.15 Critical power flow P_l achieved by various combinations of SVC and PSS.

Figure 7.16 illustrates the time responses of the study system simulated for various combinations of the stabilizing control equipment. The damping of the study system is highly improved by the coordinated fuzzy logic stabilizing control between the SVC and the FLPSS*s.

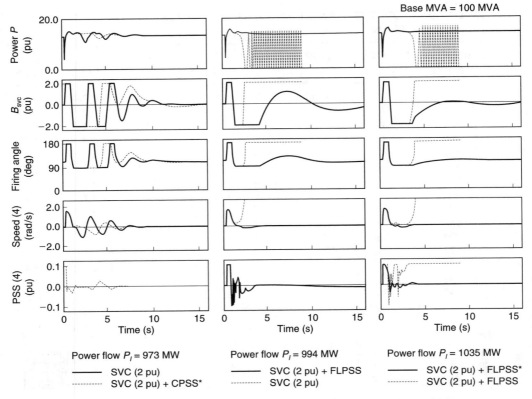

Figure 7.16 Stabilizing effects achieved by various combinations of SVC and PSS.

7.5.3. Thyristor-Controlled Braking Resistor [19]

Stabilizing control of a BR is one of the effective methods to absorb the excessive energy caused by system disturbances and provides a stability enhancement of electric power systems. Usually a bang-bang-type switching control scheme is considered for the BR using mechanical switches. However, the optimal location of the switching line moves according to the changes of the operating condition. To overcome such a situation, the fuzzy logic control switching scheme is also applied for the switching of a thyristor-controlled BR to enhance the overall stability of electric power systems. In addition, the coordinated stabilizing control with the SVC is also considered.

The basic configuration of the BR is shown in Figure 7.17. In the case of the BR, its location is the generator terminals; therefore, the required measurement is that of the real power output on which the BR is installed. For deceleration control of the study generator, the BR is switched on the generator terminals to absorb excess energy from the generator. The amount of energy absorbed by the BR is modified by changing the firing angle of the thyristor switch. Conversely, when

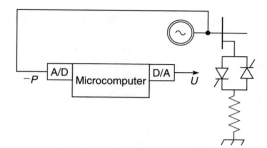

Figure 7.17 Basic configuration of braking resistor.

acceleration control is required for the system, the BR cannot supply any real power to the system, and it is removed from the generator terminal so as not to absorb energy from the generator.

The BR conductance $g_{BR}(k)$ is determined from the switching control signal $U(k)$ as follows:

$$g_{BR}(k) = \begin{cases} g_{BR,MAX} \cdot U(k) & \text{for } U(k) \geq 0.0 \\ 0.0 & \text{for } U(k) < 0.0 \end{cases} \qquad (7.11)$$

where $g_{BR,MAX}$ gives the capacity of the BR. In the proposed fuzzy logic switching control scheme, a bang-bang (i.e., non-fuzzy) type control scheme is introduced by setting the parameter α, which gives the overlap angle between sectors A and B, to zero and also by setting the gain factor $G(k)$ to 1.0.

Coordination with SVC. The coordinated stabilizing control with an SVC unit on the transmission system is also investigated. When considering the switching control of the SVC in the transient period, the SVC susceptance is determined by Eq. (7.10). The switching signal $U(k)$ is derived from the real power flow signal measured at the location of the SVC. Here, it is also assumed that the steady-state susceptance B_{SVC0} is zero. Under critical conditions, the appropriate amount of reactive power is supplied to the transmission bus bar, where the SVC unit is set on, in order to increase the bus bar voltage and also to get a smaller phase difference for the transmission of the same amount of the real power flow through the transmission lines. For this emergency control, the phase signal $Z_p(k)$ at the location of the SVC is utilized to change the firing angle setting B_{SVC0} by ΔB_{SVC0}:

$$\Delta B_{SVC0} = \beta \quad \text{after } Z_P(k) \geq Z_{P,MAX} \qquad (7.12)$$

where ΔB_{SVC0} is a step change of the SVC susceptance after the emergency condition is detected. The values of β and $Z_{P,MAX}$ are determined under critical conditions.

The thyristor-controlled BR is set on one of the generator terminals. Furthermore, when considering the coordination with the SVC, the SVC unit is set on bus 11. The simulations are performed under various loading conditions, where the following assumptions are taken into account:

1. The electrical transients caused by the thyristor-controlled BR are neglected.
2. The thyristor switching delay is neglected.

Optimal Setting of Adjustable Parameters. All the parameters of fuzzy logic switching controller for the BR are optimized subject to disturbance b and under the load size of 100%. The size of the BR is set to 2.0 pu, which is 200 MW. The optimal parameters are shown in Table 7.10 for the BR location at each generator terminal separately. In this case, the performance index J_T is calculated for a period of 8 s. In the other cases shown later, the performance index J_T is obtained for a period of 16 s. Here, it must be noted that the controller parameters are fixed at the same values shown in Table 7.10 throughout all the simulations under various loading conditions in order to investigate the robustness of the proposed switching control scheme. In addition, the sampling interval ΔT is set to 1/120 s; therefore, the firing angle of the thyristor switching circuit is performed every half cycle.

The adjustable parameters for the SVC controller located on bus 11 are shown in Table 7.11. In this case the BR with the size of 2.0 pu is set on bus 4. The performance index J_T is reduced to 3.1 from 3.7 by the additional control from the SVC.

TABLE 7.10   OPTIMAL PARAMETER SETTING FOR BR ($g_{BR, MAX} = 2.0$ pu)

Location of BR	A_s	D_r	α	Index J_T
Bus 1	0.3	0.02	90.0	5.4
Bus 2	0.6	0.02	40.0	5.4
Bus 3	0.7	0.02	70.0	5.1
Bus 4	0.2	0.02	30.0	3.7
Bus 5	0.2	0.02	30.0	3.7

TABLE 7.11   OPTIMAL PARAMETER SETTING FOR SVC WITH BR ON UNIT 4 ($B_{SVC, MAX} = 2.0$ pu)

Location of SVC	A_s	D_r	α	Index J_T
Bus 11	0.1	0.3	30.0	3.1

Typical Simulation Results. Throughout the simulations, the adjustable control parameters are fixed to the same values shown in Tables 7.10 and 7.11 to investigate the robustness of the proposed switching control scheme.

Table 7.12 shows the stabilizing effect from the BR, which is set on the generator bus bars 1–4 separately. The BR capacity is specified to 2.0 pu in this case. Also shown in the table is the stabilizing effect from the CPSSs actually equipped on units 4 and 5. Much wider stable regions are achieved by the BR at each location. In all the simulations shown later, the CPSSs are removed from units 4 and 5 to get the exact effects from the BR and also from the SVC, which is set on bus 11 in the transmission system. From Table 7.12, the optimal location of the BR is bus 4 because of the smaller values of the performance index J_T and the larger stable region.

TABLE 7.12 PERFORMANCE INDEX J_T UNDER DIFFERENT LOADING CONDITIONS ($g_{BR,MAX} = 2.0$ pu)

Power Flow P (MW)	CPSS, Units 4 and 5	Location of BR			
		Unit 1	Unit 2	Unit 3	Unit 4
743	24.8	5.4	5.4	5.1	3.7
848	62.7	6.8	6.7	5.5	3.7
900	Unstable	6.8	6.7	5.9	3.7
952		8.2	8.1	7.4	3.6
1003		11.4	10.0	12.3	3.5
1055		15.5	11.8	23.8	3.3
1105		19.8	22.9	Unstable	4.2
1115		Unstable	Unstable		Unstable

Table 7.13 also indicates the stabilizing effects of the BR with the rating of 1.0 pu. In this case, the BR at bus 4 has significant effects for the stabilization of the model system. The stable region is much larger than that obtained by the CPSSs. This table also shows that bus 4 is the optimal location of the BR in the model system.

Table 7.14 shows the effect of the coordination with the SVC. The BR size is set to 2.0 pu. In this case, only the BR has a significant effect for the stabilization of the study system, and the SVC has very little effect. Therefore, the coordination is only considered under critical conditions, and the setting of the firing angle is changed at the time when the condition, given by Eq. (7.12) is satisfied. As shown in this table, the stable region is enlarged by the coordination under the critical condition.

Table 7.15 also illustrates the effect of the coordination with the SVC. In this case, with the smaller size BR, the SVC has a significant effect on the system performance. The reason for this can be seen in the time-domain performance shown in Figures 7.18 and 7.19.

The switching control parameters shown in Table 7.11 are obtained with the BR placed at bus 4 in order to consider the coordinated control of the BR with the SVC. When considering the simultaneous control using the separately tuned BR and SVC,

TABLE 7.13 PERFORMANCE INDEX J_T UNDER DIFFERENT LOADING CONDITIONS ($g_{BR,MAX} = 1.0$ pu)

Power Flow P (MW)	CPSSS, Units 4 and 5	Location of BR			
		Unit 1	Unit 2	Unit 3	Unit 4
743	24.8	445.2	654.4	926.5	54.8
848	62.7	Unstable	Unstable	Unstable	91.0
900	Unstable				132.2
952					187.7
1003					209.6
1055					Unstable

TABLE 7.14 PERFORMANCE INDEX J_T BY COORDINATION WITH SVC ($g_{BR, MAX} = 2.0$ pu)

Power Flow P (MW)	Location of BR: Unit 4	
	Without SVC	SVC at Bus 11 $\Delta B_{SVC0} = 0.5$ pu
743	3.7	3.7
848	3.7	3.7
900	3.7	3.7
952	3.6	3.6
1003	3.5	3.5
1055	3.3	3.3
1105	4.2	4.8
1115	Unstable	5.0
1125		Unstable

Note: $B_{SVC0} = 0.0$, $Z_{P, MAX} = 0.2$.

TABLE 7.15 PERFORMANCE INDEX J_T BY COORDINATION WITH SVC ($g_{BR, MAX} = 1.0$ pu)

Power Flow P (MW)	Without SVC	Location of BR: Unit 4	
		With SVC at Bus 11	
		$B_{SVC, MAX} = 1.0$ pu	$B_{SVC, MAX} = 2.0$ pu
743	54.8	14.2	6.8
848	91.0	17.5	7.4
900	132.2	19.7	7.6
952	187.7	22.0	7.8
1003	209.6	26.4	9.0
1055	Unstable	23.0	12.4
1065		Unstable	12.9
1075			13.1
1085			Unstable

Note: $B_{SVC0} = 0.0$, $\Delta B_{SVC0} = 0.0$.

the maximum stabilizable power flow to the infinite bus is 1065 MW, which is only 10 MW less than the maximum obtained by the coordination of the BR with the SVC. This fact indicates the robustness of the proposed control scheme. The separately tuned control parameters for the SVC are as follows: $A_s = 0.01$, $D_r = 0.05$, and $\alpha = 10.0$.

Typical time responses are shown in Figures 7.18 and 7.19, where the BR is placed at bus 4. As shown in Figure 7.18, the SVC has only secondary effects on the stabilization in the case when the BR rating is large enough. However, the SVC has significant effects when the BR rating is small and the coordination gives larger

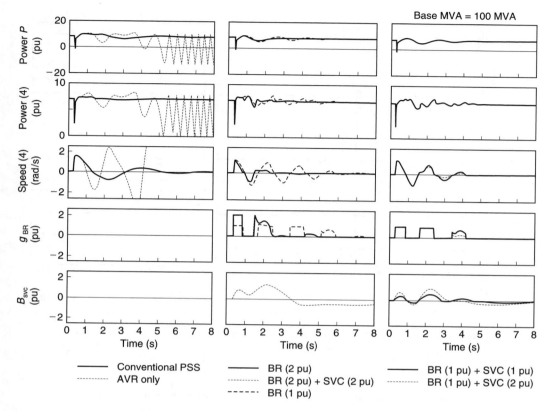

Base MVA = 100 MVA

Conventional PSS
AVR only

BR (2 pu)
BR (2 pu) + SVC (2 pu)
BR (1 pu)

BR (1 pu) + SVC (1 pu)
BR (1 pu) + SVC (2 pu)

Figure 7.18 Typical simulation results (power flow = 743 MW).

stable regions. Figure 7.20 shows the speed deviations of the other units, 1, 2, and 5. As shown in the figure, each unit has mainly its local mode together with a small level of the global mode. However, the power flow to the infinite bus has mainly the global mode, that is, the interarea mode. This indicates the effectiveness of the proposed control scheme for the damping of the multimode oscillations.

Figure 7.19 shows the stabilizing effects of various combinations of the BR and the SVC. It also indicates the effect of the emergency control given by Eq. (7.12). These figures suggest that the BR has the greatest effect on the early part of the transient, while the SVC effect is seen later in the transient period.

Through all the simulations, the controller parameters are fixed to the same values obtained under the load size of 100% and subject to disturbance b. The simulation results indicate the robustness of the proposed switching control scheme because of the wider stable region achieved by the switching controllers with the fixed parameters. The thyristor switch for the BR might be replaced by the mechanical switch; however, the switching control scheme becomes a bang-bang type, and the fine stabilization might not be possible.

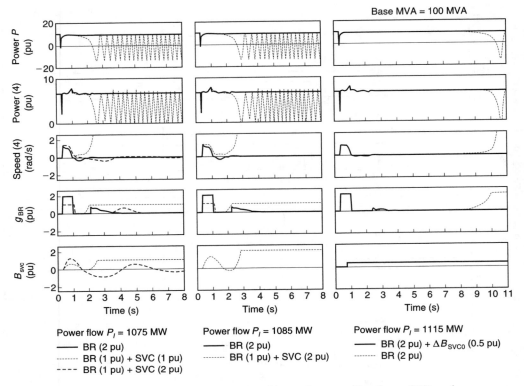

Base MVA = 100 MVA

Power flow P_l = 1075 MW
— BR (2 pu)
······ BR (1 pu) + SVC (1 pu)
---- BR (1 pu) + SVC (2 pu)

Power flow P_l = 1085 MW
— BR (2 pu)
······ BR (1 pu) + SVC (2 pu)

Power flow P_l = 1115 MW
— BR (2 pu) + ΔB_{SVC0} (0.5 pu)
······ BR (2 pu)

Figure 7.19 Stabilizing effects achieved by various combinations of BR and SVC.

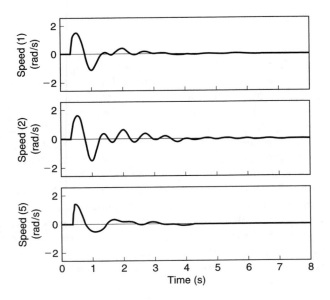

Figure 7.20 Speed deviations of units 1, 2, and 5.

7.6. CONCLUSIONS

In this chapter, the fuzzy logic switching control scheme has been proposed for FACTS devices such as thyristor-controlled series capacitor modules, static var compensators, and braking resistors. The stable region is highly enlarged by the fuzzy logic switching of these FACTS devices. Coordination with the power system stabilizers is also effective in enlarging the stable region. The coordination between the BR and the SVC is also effective when the rating of the BR is small. Through the simulations, the robustness of the proposed control scheme is verified. The proposed control scheme is simple and does not require heavy computation on the micro-computer-based switching controller. Therefore, real-time application is readily available.

References

[1] "Flexible AC Transmission Systems (FACTS); Scoping Study, Vol. 1, Part 1; Analytical Studies," EPRI Report EL-6943, Electric Power Research Institute, Palo Alto, CA.

[2] "Flexible AC Transmission Systems (FACTS); Scoping Study, Vol. 2, Part 1; Analytical Studies," EPRI Report EL-6943.

[3] B. Pilvelait, T. H. Ortmeyer, and D. Maratukulam, "Advanced Series Compensation for Transmission Systems Using a Switched Capacitor Module," *IEEE Transactions on Power Delivery*, Vol. 6, No. 2, 1993, pp. 584–592.

[4] C. J. Wu and Y. S. Lee, "Damping of Synchronous Generator by Static Reactive Power Compensator with Digital Controller," *IEE Proceedings*, Part C, Vol. 138, No. 5, 1991, pp. 427–433.

[5] E. Lerch, D. Povh, and L. Xu, "Advanced SVC Control for Damping Power System Oscillations", *IEEE Transactions on Power Systems*, Vol. 6, No. 2, 1991, pp. 458–465.

[6] S. Paul, S. K. Basu, and R. Mondal, "A Microcomputer Controlled Static Var Compensator for Power Systems Laboratory Experiments," *IEEE Transactions on Power Systems*, Vol. 7, No. 1, 1992, pp. 370–376.

[7] E. Z. Zhou, "Application of Static Var Compensators to Improve Power System Damping," *IEEE Transactions on Power Systems*, Vol. 8, No. 2, May 1993, pp. 655–661.

[8] L. Avgquist, B. Lundin, and J. Samuelsson, "Power Oscillation Damping Using Controlled Reactive Power Compensation—A Comparison Between Series and Shunt Approaches," *IEEE Transactions on Power Systems*, Vol. 8, No. 2, 1993, pp. 687–700.

[9] A. H. M. A. Rahim and D. A. H. Alamgir, "A Closed-Loop Quasi-optimal Dynamic Braking Resistor and Shunt Reactor Control Strategy for Transient Stability," *IEEE Transactions on Power Systems*, Vol. 3, No. 3, 1988, pp. 879–886.

[10] T. Hiyama, T. Sameshima, and C. M. Lim, "Fuzzy Logic Switching Control of Braking Resistor and Shunt Reactor for Stability Enhancement of Multimachine Power System," *Proceedings of International Conference on Automation, Robotics, and Computer Vision*, 1990, pp. 695–699.

[11] T. Hiyama and T. Sameshima, "Fuzzy Logic Control Scheme for On-line Stabilization of Multi-machine Power System," *Fuzzy Sets and Systems*, Vol. 39, 1991, pp. 181–194.

[12] M. A. M. Hassan and O. P. Malik, "Implementation and Laboratory Test Results for a Fuzzy Logic Based Self-Tuned Power System Stabilizer," *IEEE Transactions on Energy Conversion*, Vol. 8, No. 2, 1993, pp. 221–228.

[13] T. Hiyama, "Real Time Control of Micro-machine System Using Micro Computer-Based Fuzzy Logic Stabilizer," *IEEE Transactions on Energy Conversion*, Vol. 9, 1994, pp. 724–731.

[14] T. Hiyama, "Robustness of Fuzzy Logic Power System Stabilizers Applied to Multimachine Power System," *IEEE Transactions on Energy Conversion*, Vol. 9, 1994, pp. 451–459.

[15] J. W. Ballance and S. Goldberg, "Subsynchronous Resonance in Series Compensated Transmission Lines," *IEEE Transactions on Power Apparatus and Systems*, Vol. 92, 1973, pp. 1649–1658.

[16] T. Hiyama, M. Mishiro, H. Kihara, and T. H. Ortmeyer, "Coordinated Fuzzy Logic Control for Series Capacitor Modules and PSS to Enhance Stability of Power System," *IEEE Transactions on Power Delivery*, Vol. 10, 1995, pp. 1098–1104.

[17] T. Hiyama and H. Kihara, "Fuzzy Logic Switching of SVC for Stability Enhancement of Power System Considering Coordination with PSS," *Proceedings of Intelligent Systems Applications in Power*, Vol. 1, 1994, pp. 321–327.

[18] H. Kihara, T. Hiyama, H. Miyauchi, and T. H. Ortmeyer, "Coordinated Fuzzy Logic Control between SVC and PSS to Enhance Stability of Power Systems," *Transactions of Institute of Electrical Engineers of Japan*, in press.

[19] T. Hiyama, M. Mishiro, H. Kihara, and T. H. Ortmeyer, "Fuzzy Logic Switching of Thyristor Controlled Braking Resistor Considering Coordination with SVC," *IEEE Transactions on Power Delivery*, Vol. 10, 1995, pp. 2020–2026.

C. N. Lu
R. C. Leou
Department of Electrical Engineering
National Sun Yat-Sen University
Kaohsiung, Taiwan
80424 Republic of China

Chapter 8

Effects of Uncertain Load on Power Network Modeling

8.1. INTRODUCTION

Problems in power system analysis, such as load flow, optimal power flow, fault current calculation, contingency evaluation, and penalty factor calculations, rely on using a model of the system. Network data and operating point data are required for network modeling. The network data are defined as the network configuration and branch admittances that specify the bus admittance matrix, and the operating point data are defined as a set of data that includes P and Q injections at a PQ bus and P injection and voltage at a PV bus that determine the operating point. For on-line analysis, the model must reflect real-time conditions of the system. For off-line studies, this model is specified by the user through the input data.

The power system model needed for on-line analysis is a solved power network. The model is divided into two parts, one representing the internal network from which the control center receives telemetered data and the other representing the external network, which consists of the rest of the interconnected system [1]. The state estimator (SE) uses a set of noisy real-time measurements to construct real-time models of the internal network. Since an SE is used as a filter for the real-time measurements, the resulting model is useful to the operator as a check for missing or suspicious data and is used as a basis for other application functions such as contingency evaluation and optimal power flow. Consequently, it is important to reduce the uncertainties in the measurements to a level that allows satisfactory control.

The response of the external system to a contingency makes the flows from the external system to the boundary buses different from their precontingency values.

For on-line contingency evaluation, an external network model that accounts for the response of the external system must be added to the internal system representation. The external network may be represented by an equivalent model, an unreduced model, or a combination of the two. Without interutility data exchange, the external network is, in general, unobservable. Complete network and load data of the external system even for the base case are usually not available in a real-time environment. Most, if not all, of the external system load and generation data required to complete the external load-flow model may have to be produced. In load-flow-based external-network modeling methods, the errors in the assumptions of network parameters, topology, and bus load/generation introduce the boundary bus injection errors known as boundary mismatch [2–7].

Power flow analysis is indispensable for power system off-line studies. The power flow problem is, in essence, the calculation of line power flow given the load/generation schedule and network data. A power flow program is a necessary tool in the complete study of voltage, loading, and losses in a system. For planning purposes, off-line models considering plausible changes in the future are used in load flow study. In planning a new or reinforced system and in predicting the near-future state of an existing system, various nodal loads must be estimated. It is impossible to estimate these precisely, but they could be predicted subject to certain variations. In order to assess system states under various conditions, it is necessary to calculate the power flows for all combinations of possible generation and loads. The computational time for this study would be very high and the meaning of the results would be hard to determine.

In system planning and operation activities it is important to have a detailed knowledge of individual bus loads. Much attention has been given to modeling generation and transmission/distribution equipment. Load representation has received less attention and continues to be an area of greater uncertainty. Many studies have shown that load representation can have a significant impact on analysis results. Even with a long history of research efforts on this subject, accurate modeling of loads continues to be a difficult task due to several factors, including [8]:

- The large number of diverse load components and lack of precise information on the composition of the load
- Uncertainties regarding the characteristics of many load components
- Information and location of load devices not directly accessible to the electric utility
- Changing load composition through time

Due to the uncertain nature of bus load, load representations based on probability, possibility, and fuzzy set theories have been proposed in the literature. During the past two decades the relatively new theories of fuzzy sets and possibility have developed rapidly, and several applications relating to the electric power systems planning have been reported [9–22].

In this chapter, the effects of uncertain loads and network parameters are discussed. Load modeling approaches using probability and possibility concepts are described. An example of fuzzy set theory application to the external-network modeling problem that involves a great deal of uncertain data is presented. In this example, information about the external network is imprecise. Based on the system dispatchers' knowledge about the external-network load, generator, and system behavior, scheduled bus injections at the external area are adjusted in order to find an acceptable external-network model for real-time system operations and analysis.

8.2. EFFECTS OF DATA UNCERTAINTY ON POWER NETWORK MODELING

Since there always exist errors in the branch length and calculated or measured parameters of lines and transformers in the system, the network parameters used in system studies contain various degrees of uncertainties. Transmission line data are usually estimated at their 50°C values, and self and mutual impedances are calculated based on certain circuit geometries. The ambient temperature is random and line sag changes the geometry of the line. Therefore, the actual line parameters are uncertain, and this uncertainty could significantly influence the analysis results. In addition to network parameter uncertainties, network topology could also be uncertain due to unscheduled outages and erroneous circuit breaker status data collected from a supervisory control and data acquisition (SCADA) system. In general, uncertainties involved in power network modeling can be divided into three categories: topological, temporal, and numerical. In this chapter, only the temporal uncertainties pertaining to bus load are discussed. The effects of uncertain load on real-time and off-line power network modeling are described in the following section.

8.2.1. Measurement Uncertainty and Its Effect on Real-Time Network Modeling

In a SCADA system each analog measurement is the final product of a chain of instruments beginning with the instrument transformers, going through transducers and converters, scaling, and ending in the data base at the control center. Real-time measurement plays an important role in on-line network modeling. The uncertainty in each analog measurement is due to a combination of random and systematic errors. Random errors are caused by the degree of precision of various instruments in the measurement streams. Systematic errors are introduced by the gains, zero offsets, and nonlinearity of the instruments as well as the linkages between the instruments and the scaling procedures [23]. Measurement nonsimultaneity occurs because the data from the various measurement devices are scanned at different times. Thus, two measurements used in an SE might occur, say, as much as 10 s apart. Changes may occur in the system over this time span due to changes in demand, generation, and network configuration. Measurements are used in the SE and various real-time control applications wherein the uncertainties propagate. In order to obtain accurate network analysis results, it is important to estimate the uncertainties in the analog measurements. This would increase the level of confidence in the simulation results, allowing closer and more effective control.

For real-time security analysis, the unreduced model of the external network is represented by the topology of the network, branch parameters, and scheduled data at various buses in the network. In current practice, in order to fill in plausible data for the external network, the unobservable region is assumed to be in its normal condition; that is, all devices that are normally in service are assumed operational, all breakers are assumed in their normal state, and bus injections are estimated by extrapolation based on load distribution factors and/or economic dispatch [2]. Currently, load flow and state-estimation-based methods are used for external-network modeling. It has been pointed out that many of the currently used external-network modeling techniques require data that far exceed the quantity and quality of the data that most utilities have available [8]. Therefore, one must remember that most data used in the external-network modeling are approximate. The main problem of external-network modeling is that the assumptions made can frequently be wrong. If there is a mismatch between the actual and assumed values, in load-flow-based methods, the errors will be reflected at the boundary of internal and external networks as boundary mismatch. The primary reasons for the boundary mismatch are (1) outages of facilities (topology errors) in the external system are not reflected in the model and (2) load and generation distribution in the external system is out of line. Due to the deficiency in handling constraints on pertinent data, many of the currently used external-network modeling techniques fail to provide acceptable solutions. When there is a major discrepancy between the base-case power flow and the measurements or expected values, the network application loses its credibility with the users, and the subsequent contingency analysis will be inaccurate [7].

8.2.2. Effects of Uncertain Load on Off-line Network Modeling

Off-line models used in power flow analysis are only approximations; uncertainties are involved in the parameter values and magnitudes of the demands and generations assumed for the system load and generation buses. Forecast loads contain uncertainty; the further into the future the forecast, the larger the data variance. If loads are uncertain, then the power generations, branch flows, and bus voltage are also uncertain. For a given set of parameter and bus data, a power flow solution would be nothing but a snapshot of the system at a given instant. Solutions obtained would be valid only for a single specific system configuration and operating condition. Since the system evolves through time, in planning activities, it seems reasonable to ask not what the system looks like at a given instant, but rather to ask for the range of all plausible system conditions that might be encountered as a result of expected uncertainties in demands and other system parameters [16]. Having loads as uncertain defines the problem within a domain of possible future load scenarios; decisions on setting or reinforcing system components may then be evaluated.

8.2.3. Uncertainties Involved in Distribution Network Modeling

Distribution networks contain large numbers of components, and much of the required data for network modeling are not available or uncertain. Presently, in the process of creating a distribution system model, engineers need to fill in the missing information, such as the power factor of the circuit, the phase of the load, coincident factors, and load distribution along the feeder. Thus, a good deal of uncertainties are involved in the distribution network analysis. Clearly, the simulated results will be more accurate if there is no missing information. However, it would be very costly and it is not always feasible for the operator to collect detailed information for all of their electrical networks. In planning activities, distribution system load flow techniques are used. Since the loads on feeder sections and laterals are not readily available, it is therefore required to devise an approach to estimate and represent loads at various feeder points.

In many distribution automation projects, meters are now installed along the feeders, providing the possibility of building a quasi-real-time distribution system model [25]. However, for economic reasons, the number of remote meters in a distribution system is limited and is not sufficient for state estimation. Thus estimated feeder loads must be included in the SE as pseudomeasurements in order to obtain a complete estimate of the system states. The estimation and representation of the feeder loads would affect the distribution system SE results and, therefore, the real-time model of the distribution system.

8.3. MODELING UNCERTAIN LOAD IN NETWORK ANALYSIS

For networks with constant configuration, line parameters, and a set of measured or estimated values of node loads, the network modeling problem is to find the set of corresponding values of system states and branch flows. Traditional methods of load flow solution require specific values for loads, and any variation of values will require a new solution. In this case, a large number of conventional deterministic load flow cases need to be carried out. Once all these studies have been done, the range of values for a specified quantity can be identified. In practice, it is not feasible to carry out individual load flows for every change in load due to the prohibitively large amount of calculations and the difficulty in analyzing and synthesizing the results of so many load flows. A practical way to overcome the difficulties is by selecting a limited number of variations of loads. Often this is done arbitrarily, and the results are based on partial information, rendering them inaccurate. The answer may be under- or overestimated and lead to wrong decisions. Two useful alternatives to modeling uncertainty are as follows:

1. *Probabilistic approach*: In this approach, data are given by their mean values (mathematical expectations) and their extreme limits of deviations (errors)

from mean values; that is, the probability distribution of all uncertainties are assumed.

2. *Unknown-but-bounded (possibility) approach*: In this approach, only limits on the uncertainties are assumed, with no assumptions about probability distributions. These limits may be upper and lower bounds or individual uncertainties.

In many cases probabilistic models provide helpful tools for dealing with complex situations; however, these models could be misleading if the uncertainty represents an event that will occur rarely or no measurement data that are available. Under these situations, due to the lack of actual samples, the assumption will not be closely related to the theoretical *probability* of occurrence. For this type of uncertainty, unknown-but-bounded models could be more appropriate.

8.3.1. Probabilistic Approach

The probabilistic approach is widely used in the SE for real-time network modeling. Each bus load measurement is modeled as a sum of the true value and its associated random error. The random error serves to model the uncertainty in the measurement. If the measurement error is unbiased, the probability density function of the random error is usually chosen as a normal distribution with zero mean. The maximum-likelihood concept or, equivalently, the weighted least-squares estimation can be used to estimate the system states. Field experiences in using the SE for real-time system operations have shown that the probabilistic approach can provide fairly accurate real-time models.

In many previous studies it has been shown that the power flow problems could be analyzed probabilistically instead of deterministically [9–13]. In these methods consumer demands and power generations are modeled as random variables and described by probability density functions that, when convolved, produce power flows that are also described by probability density functions. What the probabilistic load flow does is obtain ranges of output variables based on the ranges of input data in one direct calculation. Probabilistic methods are proposed in many power system studies due to the prohibitively large computer requirements for handling a huge number of individual deterministic problems resulting from many different combinations and magnitudes of the input variables.

There are many analytical models for defining input variable density functions. Some of these examples are normal distributions to represent nodal load forecasts, binomial distributions to represent a set of identical generator units, discrete variables when neither of the preceding distributions are appropriate, and single or one-point values when a power has a unity probability of occurrence. In the case of normal distributions, the expected value of a load or a generation must be predicted and characterized by a standard deviation obtained from experience with forecast errors [9–13]. The errors in the specified data can be included in a probabilistic analysis, and the analysis therefore takes these into account in the computed results. To handle the uncertainties in future load, one may have an imprecise definition of load, covering a range of values by a probabilistic approach, but in the planning exercises it may not make sense to adjust any kind of probability distribution to load values [15], and in many situations, it can be argued that insufficient data are available for this type of analysis or that the data that are available are inaccurate.

8.3.2. Possibilistic Approach

Considering a situation in which a value is uncertain, it is a practical and logical process for treating uncertainty with whatever information that is available. This information can be objective (we are certain that the possible distribution is between the two data sets) or subjective (the information is obtained from experience or from the opinion of experts). Suppose that information available is such that we can accept that the uncertain value belongs to the referential set R (a set of real numbers). In many situations encountered, it is possible to locate the value inside a closed interval of R, that is, an interval of confidence of R: $[a_1, a_2]$. An interval of confidence is one way of describing the uncertainty by using lower and upper bounds [24].

Compared to the probabilistic approach, it is appropriate to say that a model in which input data assigned with error bounds can represent data in a more practical way since in many situations the distribution of errors is often unknown.

A fuzzy number is an extension of the concept of interval of confidence. Instead of considering the interval of confidence at one unique level, it is considered at several levels and more generally at all levels from 0 to 1. The fuzzy sets are defined by a membership function $\mu_F(u)$ that tells us to what degree the element u belongs to the fuzzy set F. The membership function can have any value between 0 and 1, and the shape of the membership function can be determined according to the domain expert's judgments. Fuzzy methods do not necessarily need data from the past. However, historical data may be used as a basis for human judgment and subjective estimation. It makes sense to use probability theory and its operational setting in order to derive membership functions. But membership functions need not be related to probability. The term $\mu_F(u)$ can be the degree of proximity between u and an ideal prototype of F, that is, based on the idea of distance rather than probability [25].

Fuzzy modeling of power systems can take into account the qualitative aspects and vagueness or uncertainty that cannot be modeled by a probabilistic approach. New fuzzy load flow analysis tools enable one to incorporate in power system modeling information about loads and generated power that is neither deterministic nor probabilistic and obtain results under the form of fuzzy information about voltages, active and reactive power flows, and losses.

In load flow and optimal power flow studies, loads may be specified within a conservative range with additional possible values of load represented as fuzzy values. In the literature, the load possibilities are translated into trapezoidal possibility distribution by assigning a degree of membership to each possible value of the load. Figure 8.1 presents the possibility distribution for real load at bus 1, which is expected to be between $P_1^{(2)}$ and $P_1^{(3)}$. In a fuzzy load flow environment, it is important that linguistic declarations about loads can be translated into fuzzy numbers. When one is concerned with one fuzzy value for some bus bar, the adoption of a

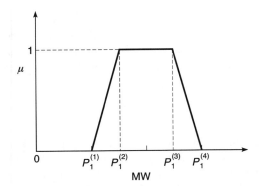

Figure 8.1 A membership function for the megawatt load.

suitable membership function that can be understood as a possibility distribution is a problem. Building fuzzy load diagrams is surely one subject open to further study.

The external-network modeling procedure begins by estimating the plausible scheduled injections for each external bus. External bus load schedules are often estimated using load group parameters in conjunction with real-time data to determine all zone, load group, and individual load MW and MVAR. Due to lack of accurate information about scheduled loads and generations, it seems adequate to characterize the bus injections and other quantities not by unique numerical values but by data intervals. Consequently, for an external-network solution to appear reasonable, it must lie within the range of the plausible system conditions that might be encountered. In the external-network modeling approach proposed in ref. 6, for each scheduled data item [megawatts, megavolt ampere reactive (MVAR), voltage] two sets of confidence intervals are established. The first set indicates the maximum likelihood of the scheduled values and the second indicates their strict limits. This is then translated into a trapezoidal possibility distribution similar to that shown in Figure 8.1. The intervals are defined based on planning data provided by the neighboring system or from past experience. Generally, the confidence interval is narrower for those quantities about which more information is available (e.g., external jointly owned generator output).

8.3.3. Modeling Distribution Feeder Loads

Deterministic techniques have been used in planning distribution networks. Probabilistic methods have also been proposed for distribution system analysis. In the probabilistic approach, it is assumed that we have some statistical data from the past or we can assign some subjective probability distributions for the data concerned. However, due to the quick-changing phenomena of distribution systems, it is not always justified to assume that the future will be similar to the past. Furthermore, in many cases there is no historical or empirical data available.

In distribution system analysis, a deterministic approach to estimate loads on feeder nodes is based on the assumption of load conformity where the individual loads change in the same pattern [27]. From the telemetered feeder current I_{FD}, the load current of branching point I, I_i, can be estimated as

$$I_i = I_{FD} \frac{C_i}{\sum_{i=1}^{N} C_i} \tag{8.1}$$

where C_i = sum of rated capacities of transformers at branching point I
N = number of branching points supplied by the feeder

The loads estimated by Eq. (8.1) are not accurate since the loads in a distribution system are nonconforming. The peak load of industrial customers may occur in the afternoon while that of the commercial sector may take place in the evening.

Much of the available data relating to feeder loads are neither deterministic nor probabilistic. For instance, statements such as "load at bus A is approximately 2 MW" or "load at bus B is mainly industrial type" are clearly neither deterministic nor probabilistic. To handle linguistic variables of knowledge representation used by domain experts, the fuzzy set approach provides a means to deal with this kind of imprecision or uncertainty. Linguistic terms can be used to describe the uncertain hourly load pattern. In ref. 26, each load type is divided into five load levels: very small (VS), small (S), medium (M), large (L), and very large (VL). The ranges of each load level are separately defined for each load type based on the load curves obtained from a survey. Examples of hourly load patterns of each load type that use these five linguistic variables are shown in Table 8.1. The load pattern for any branching point can be approximated based on load patterns shown in Table 8.1 according to the estimated load composition of that branching point.

Probabilistic and possibilistic approaches are fundamentally different in representing uncertainties. Because knowledge about the data that one has are different for these two approaches—one is quantitative while the other is qualitative—a

TABLE 8.1 EXAMPLES OF HOURLY LOAD LEVELS FOR DIFFERENT LOAD TYPES

Hour	Commercial	Industrial	Residential	Hour	Commercial	Industrial	Residential
1	M	S	VS	13	M	S	VS
2	S	S	VS	14	M	S	VS
3	VS	S	VS	15	M	S	VS
4	VS	S	VS	16	M	S	VS
5	VS	S	VS	17	L	S	VS
6	VS	S	S	18	L	S	S
7	VS	S	M	19	VL	S	M
8	VS	S	S	20	VL	S	S
9	S	M	S	21	VL	M	S
10	S	L	S	22	VL	L	S
11	S	VL	S	23	L	VL	S
12	M	L	M	24	M	L	M

Source: Ref. 26.

comparison between the two approaches is not easy and should be avoided. However, in handling uncertain data, both approaches are cooperative rather than competing. In some cases the probabilistic point of view is more appropriate, while in other situations the possibilistic point of view could lead to a more natural interpretation of the results. What is really important, in practice, and what fuzzy sets are good for, is to correctly represent the pieces of knowledge provided by a person and to model the inexactly expressed information.

8.4. FUZZY SET THEORY APPLICATION TO EXTERNAL-NETWORK MODELING

Fuzzy set theory has been applied to many power system problems to take into account the effects of uncertain loads [14–27]. In this section, a fuzzy linear programming (LP) technique applied to the problem of external-network modeling is presented. As described in Sections 8.1 and 8.2, when the external system is represented by an unreduced load flow model, the operating point data (loads, generations, and specified voltages) are required. Some external-system data may be available in real time, for example, due to partial data exchange between control centers. But in general these data are not all available to the internal system in real time, and so extrapolated values or off-line data have to be augmented. Errors made in the assumptions for network topology, load level, and distribution and generation commitment and dispatch can produce external models that are inaccurate. In order to provide a more acceptable external-network model for on-line analysis, when there is a large mismatch between the measured data and the network solution, an adjustment in the solution is required.

Since the external power system is not of direct interest, the criterion for the adequacy of the external model is not the exact replication of the system. Instead, the external model is considered quite adequate if it can accurately reflect the effects of the external system when subsequent analysis is done on the internal system; that is, the goal is not to compute an exact model of the external network (although this is desirable) but to produce a model that is acceptable to the power system dispatcher and accurate enough to represent the effects of the external network on the predictive studies of the internal network.

8.4.1. Fuzzy LP-Based External-Network Modeling [26]

The fuzzy representation of external-network bus loads was described in Section 8.3. Figure 8.1 shows an example of fuzzy load representation. In the proposed method, two confidence intervals are defined for each analog data. Figure 8.2 shows a simplified flow diagram of the external-network modeling approach. In step 1, by assigning all boundary buses adjacent to observable and unobservable areas as slack buses and using estimated one-point values of injections that are usually the mean values of confidence intervals, a deterministic load flow solution is found for the combination network of the external and boundary regions. The

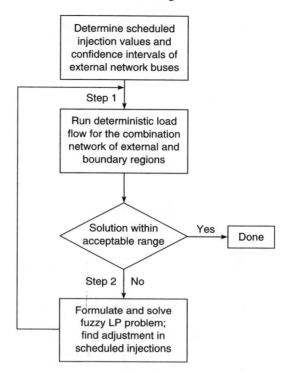

Figure 8.2 Flow diagram of fuzzy LP-based external-network-modeling approach.

quality of the solution is then evaluated by checking whether the injections, line flows, and voltage are within their confidence intervals. If the first confidence interval is not violated, then the current solution is acceptable and the external-network modeling process is complete. Otherwise, it goes to step 2.

In step 2, since some of the confidence intervals are exceeded in the current solution, in order to obtain a more acceptable solution, a modification to the current solution is performed by formulating an LP problem to reduce the deviations. The LP problem is based on the following linear model, similar to that used in the fast decoupled load flow formulation:

$$B' \Delta\theta = \Delta P \tag{8.2}$$

$$B'' \Delta V = \Delta Q \tag{8.3}$$

In the LP problem, changes of scheduled bus injections (loads or generations) of external-network buses are chosen as unknown variables. The objective of the LP problem is the minimization of changes in scheduled bus injections such that the solution will be within the acceptable ranges. Since each analog data has its own acceptable range, the allowable changes of bus injections, tie line flows, or scheduled voltages from their current values are enforced in the constraint set. Data that cannot be expressed directly by the increment of active and reactive power injections, such as scheduled tie line flows and voltage, can be transformed to functions using Eqs. (8.2) and (8.3). The LP problem is formulated as minimizing

$$\sum_i \Delta P_i + \Delta Q_i \qquad (8.4)$$

subject to

$$\sum_i \Delta P_i = 0 \qquad \sum_i \Delta Q_i = 0$$

$$L_{p_i}^{\text{Low Limit}} \leq f(\Delta P) \leq L_{p_i}^{\text{High Limit}} \qquad (8.5)$$

$$L_{q_i}^{\text{Low Limit}} \leq f(\Delta Q) \leq L_{q_i}^{\text{High Limit}}$$

The lower and higher limits on the allowable changes are computed based on current values and limits of the first confidence intervals. The above formulation can be simplified to minimizing

$$C^T X \qquad C, X \in R^n \quad \text{such that } AX \leq b, \ b \in R^m A_{m,n}, \ X \geq 0 \qquad (8.6)$$

8.4.2. Fuzzy Linear Programming [28]

The LP formulation of (8.4) and (8.5) can be further modified to include possible uncertainties in high and low limits of the scheduled data (the second confidence intervals). The LP formulation in Eq. (8.5) is transformed into the following form requiring minimization of

$$C^T X \quad \text{such that } a_i^T X \preceq b_i' \qquad i = 1, 2, \ldots, m_1$$

$$d^{T_j} X \leq b_j \qquad j = m_1 + 1, \ldots, m_1 + m_2 \qquad (8.7)$$

$$X \geq 0 \qquad m = m_1 + m_2$$

The roles of the constraints in Eq. (8.7) are different from those in the classical LP equation (8.6), where the violation of any single constraint by any amount renders the solution infeasible. The system dispatcher would accept a solution within the range between the first and second confidence intervals but with less degree of satisfaction. Fuzzy LP offers a way to allow for this vagueness. Here we use $\preceq$ to denote the fuzzy version of $\leq$ and it has the linguistic interpretation "essentially smaller than or equal to." An interpretation of Eq. (8.7) is to find a solution such that the predetermined level b_0 of the value of objective function $C^T X$ is least exceeded, and the first set of restrictions are satisfied as well as possible while the second set of restrictions are strictly enforced. For the first m_1 constraints, the solution is prepared to tolerate violations t_i up to $b_i + p_i$. In our problem, the b_i and $b_i + p_i$ represent the first and second confidence interval limits, respectively. All solutions $X \geq 0$ are feasible if they have the following property:

$$a_i^T X \leq b_i + t_i \qquad t_i \leq p_i, \ i = 1, \ldots, m_1$$

$$d_i^T \leq b_j \qquad j = m_1 + 1, \ldots, m_1 + m_2 \qquad (8.8)$$

Even though the violations t_i up to p_i are tolerable, the larger are the t_i, the smaller the degree of satisfaction will be. This degree of satisfaction can be defined by $\mu_i(X)$:

$$\mu_i(X) = \begin{cases} 1 - \dfrac{t_i}{P_i} & a_i^T X = b_i + t_i \\ 1 & a_i^T X \le b_i \end{cases} \qquad (8.9)$$

The membership function $\mu_i(X)$ of the fuzzy set "feasible solutions" with respect to the ith constraint can be understood as a measure of the truth of the statement "X satisfies $a_i^T \le b_i$ as much as possible." The membership function of the objective function can also be defined as [29]

$$\mu_0(X) = \begin{cases} 1 - \dfrac{t_0}{P_0} & C^T X = b_0 + t_0 \\ 1 & C^T X \le b_0 \end{cases} \qquad (8.10)$$

Here, P_0 is the maximum acceptable violation of the level b_0.

Based on the membership functions selected and the definition given by Bellman and Zadeh [29], the membership function of the fuzzy set solution of problem (8.7) is

$$\mu(X) = \min_i \{\mu_i(X)\} \qquad (8.11)$$

Assuming that the dispatcher is interested in a crisp evaluation of the optimal solution, we would suggest the maximizing solution to (8.11), which solves the following problem:

$$\max \min\{\mu_i(X)\} = \max \mu(X) \qquad (8.12)$$

With these membership functions, we formulate the following mathematical model with normalized μ's to find the most satisfactory decision on the scheduled injection modification [28]:

Maximize $\mu(x)$

Subject to $\begin{cases} AX - t \le b, \ A = (a_i^T), \ b = (b_i) \\ t \le P, \ t = (t_i), \ P = (P_i) \end{cases} \quad i = 0, \ldots, m_1$

$DX \le b' \qquad D = (d_j) \qquad b' = (b_j) \qquad j = m_1 + 1, \ldots, m_1 + m_2$

$$X, t \ge 0 \qquad (8.13)$$

Let $\mu(X) = \lambda$ and, from Eq. (8.9),

$$\mu_i = 1 - \frac{t_i}{P_i} \ge 1 - \frac{(AX)_i - b_i}{P_i} \ge \lambda \qquad (8.14)$$

we obtain

$$\lambda P_i + (AX)_i \le P_i + b_i \qquad (8.15)$$

$$t_i = (AX)_i - b_i \qquad (8.16)$$

From (8.15) and (8.16), we obtain

$$\lambda P + t \le P \qquad AX - t \le b \qquad t \le P \qquad (8.17)$$

The crisp substitute of the formulation (8.13) is as follows:

$$\max \lambda \qquad \lambda P + t \le P$$
$$AX - t \le b \qquad t \le P \qquad DX \le b' \tag{8.18}$$

After the fuzzy LP problem is set up, either the revised simplex method or the dual simplex method can be used to solve (8.18) and find the changes in scheduled injections. When an adjustment in the scheduled injections is obtained, the load flow is executed again, and the process continues until the solution becomes acceptable; that is, the solution is within the specified confidence intervals.

In the above formulation, λ indicates the degree to which the constraints are satisfied. The equality $\lambda = 1$ means that solution satisfies all first confidence intervals. If $0 < \lambda < 1$, the fuzzy LP formulation provides a solution in which some of the first confidence intervals are exceeded but the second confidence intervals are satisfied. If $\lambda = 0$, it tells us that, even when relaxed to the second confidence intervals, the constraints cannot all be satisfied. In this case the external-network parameters and their confidence intervals must be checked and modified in order to obtain an acceptable solution. The violations of the confidence intervals provide us starting points of detection and identification of assumption errors in the external-network model [31, 32].

8.4.3. Test Results

The proposed method was tested on the IEEE 30 and 118 bus test systems. A load flow solution was used as a base-case solution for all tests [33]. External-network modeling errors were introduced by adding errors to the base-case external network bus injections as well as by removing branches from the system. Recall that the base-case load flow solution is used only for simulation purposes. In the real world, the external-network state is unknown.

Tables 8.2 and 8.3 show the base-case load flow solution, the first and final estimated external-network solutions at the boundary area. Table 8.2 shows the results when some scheduled injections of external-network bus are incorrect and Table 8.3 shows the results when both injection and topological errors are present. In these two cases the objective function values of the fuzzy LP problems are equal to 1, meaning that solutions are all within acceptable ranges. The differences in injections at the first-, second-, and third-tier external buses under the columns of base case and first load flow solutions are the assumption errors purposely introduced in the bus injections. It can be seen from Tables 8.2 and 8.3 that, in the first load flow solution, there is a significant boundary mismatch. Intuitively, the mismatch grows higher in the presence of a topological error.

In the final load flow solution the boundary bus mismatch is reduced by adjusting the first and second tier external-network bus injections. The differences between the base-case solution and the final load flow solution at the first-, second-, and third-tier buses result from a lack of sufficient information regarding the external bus injections. As can be seen from Tables 8.2 and 8.3, the boundary bus mismatch is redistributed to the remote buses based on the information available for the buses.

TABLE 8.2 TEST RESULTS ON 118 BUS SYSTEM WITH SCHEDULED
INJECTION ERRORS

Bus		Base-Case Load Flow Solution		First Load Flow Solution with Assumption Errors		Final External-Network Solution	
Type	Number	P	Q	P	Q	P	Q
Boundary	1	4.446	1.523	4.184	1.634	4.441	1.530
bus	24	−1.470	0.085	−0.739	0.443	−1.352	0.332
	33	−0.147	−0.039	0.188	−0.139	−0.205	−0.019
	34	−0.012	0.112	0.299	0.186	0.016	0.185
	38	−0.030	−0.451	0.562	−0.483	−0.086	−0.449
	68	−1.763	−1.005	−2.878	−0.816	−1.794	−0.998
	70	−1.527	−0.472	−1.479	−0.464	−1.568	−0.509
First-tier	15	−1.350	0.624	−0.450	1.388	−0.450	0.889
external bus	19	−0.675	−0.375	−1.725	−1.125	−0.725	−0.621
	23	−0.105	−0.045	−1.305	−0.495	−0.823	−0.077
	74	−1.02	−0.405	−0.270	−0.455	−0.270	−0.455
	75	−0.705	−0.165	−1.755	0.285	−1.234	0.285
	77	−0.915	3.462	−0.965	2.237	−0.965	2.134
Second-tier	13	−0.510	−0.240	−0.460	−0.190	−0.460	−0.190
external bus	18	−0.900	0.484	−0.150	1.148	−0.150	0.731
	20	−0.270	−0.045	−1.17	−0.495	−1.170	−0.495
	78	−1.065	−0.390	−0.165	0.510	−0.165	0.510
	80	7.282	1.278	8.332	1.121	7.332	1.005
	82	−0.810	−0.405	−0.210	0.345	−0.961	0.345
	118	−1.495	−0.225	−2.245	−0.675	−2.245	−0.675
Third-tier	2	−0.300	−0.135	−1.200	−1.035	−1.200	−1.035
external bus	4	−0.585	−0.180	−1.335	0.270	−1.335	0.270
	10	7.2	0.215	8.700	0.832	8.700	0.846
	27	−1.065	2.125	−0.165	1.986	−0.165	1.856
	117	−0.300	−0.120	−1.050	0.330	−1.050	0.330

A bus with higher confidence bus injection data will have lower bus mismatch in the final solution, while buses with less accurate information (wider confidence interval) will absorb higher bus mismatch. Since acceptable ranges are defined and enforced, the solution is acceptable and meaningless results such as negative load or negative generation will not exist in the solution.

A contingency analysis was performed on the obtained external-network models to evaluate their accuracy. Branch outages were introduced in the internal network near the boundary. Prior to the outage, the branch carried 41 MW and 16.3 MVAR. Tables 8.4 and 8.5 show some of the internal line flows and bus voltages near the boundary. It can be seen that the model obtained by the proposed method provides accurate results in contingency analysis. It can be seen by comparing the results in Tables 8.4 and 8.5 that the accuracy is lower when there is a topological error in the

TABLE 8.3   TEST RESULTS ON 118 BUS SYSTEM WITH SCHEDULED
INJECTION AND TOPOLOGICAL ERRORS

Bus		Base-Case Load Flow Solution		First Load Flow Solution with Assumption Errors		Final External-Network Solution	
Type	Number	P	Q	P	Q	P	Q
Boundary	1	4.446	1.523	4.184	1.634	4.441	1.530
bus	24	−1.470	0.085	−1.545	0.371	−1.571	0.271
	33	−0.147	−0.039	0.482	−0.212	−0.243	−0.007
	34	−0.012	0.112	0.535	0.163	0.002	0.152
	38	−0.030	−0.451	0.922	−0.477	−0.182	−0.449
	68	−1.763	−1.005	−2.878	−0.816	−1.794	−0.998
	70	−1.527	−0.472	−1.480	−0.464	−1.568	−0.509
First-tier	15	−1.350	0.624	−0.450	1.612	−0.450	0.599
external bus	19	−0.675	−0.375	−1.725	−1.125	−0.725	−0.664
	23	−0.105	−0.045	−1.305	−0.495	−1.694	−0.495
	74	−1.02	−0.405	−0.270	−0.455	−0.270	−0.455
	75	−0.705	−0.165	−1.755	0.285	−1.234	0.285
	77	−0.915	3.462	−0.965	2.237	−0.965	2.134
Second-tier	13	−0.510	−0.240	−0.460	−0.190	−0.460	−0.090
external bus	18	−0.900	0.484	−0.150	1.335	−0.270	0.491
	20	−0.270	−0.045	−1.170	−0.495	−0.170	−0.495
	78	−1.065	−0.390	−0.165	0.510	−0.165	0.510
	80	7.282	1.278	8.332	1.121	7.332	1.005
	82	−0.810	−0.405	−0.210	0.345	−0.961	0.345
	118	−1.495	−0.225	−2.245	−0.675	−2.245	−0.675
Third-tier	2	−0.300	−0.135	−1.200	−1.035	−1.200	−1.035
external bus	4	−0.585	−0.180	−1.335	0.270	−1.335	0.270
	10	7.2	0.215	8.700	0.831	8.700	0.832
	27	−1.065	2.125	−0.165	2.055	−0.165	1.824
	117	−0.300	−0.120	−1.050	0.330	−1.050	0.330

nearby external model. This result agrees with the previous study, which indicates that topological errors have greater effects on the accuracy of the external-network model.

The major difference between the proposed method and previous SE-based external-network modeling methods is that instead of adjusting measurement weighting and rerunning the SE to obtain an acceptable solution, the proposed method uses load flow and fuzzy LP formulations to adjust the external-network solution. By taking into account the load and generation uncertainties, this approach provides a flexible means for obtaining more acceptable external solutions. Experience has shown that the stability of solution procedure and overall solution quality are improved. The computational requirements associated with the method are acceptable.

TABLE 8.4   EFFECTS OF INJECTION ERRORS ON CONTINGENCY ANALYSIS

Internal Area Line Flow (MW)		Contingency Analysis on Base-Case Solution		Contingency Analysis on Obtained External-Network Model	
From	To	P	Q	P	Q
23	24	164.1	2.9	169.6	7.3
24	72	89.5	−12.3	95.0	−7.5
70	71	−5.70	31.5	−11.1	27.4
72	71	15.6	−32.6	20.9	−28.8
70	74	12.3	−0.2	24.2	1.4
70	75	7.5	−12.9	0.3	−10.5
Internal Area Bus Voltage (pu)		$|V|$		$|V|$	
Bus 24		0.820475		0.847891	
Bus 70		0.895784		0.896367	
Bus 71		0.883728		0.886484	
Bus 72		0.822611		0.836926	

TABLE 8.5   EFFECTS OF TOPOLOGICAL ERRORS ON CONTINGENCY ANALYSIS

Internal Area Line Flow (MW)		Contingency Analysis on Base-Case Solution		Contingency Analysis on Obtained External-Network Model	
From	To	P	Q	P	Q
23	24	164.1	2.9	176.5	8.9
24	72	89.5	−12.3	101.3	−8.4
70	71	−5.70	31.5	−15.7	34.9
72	71	15.6	−32.6	26.0	−34.4
70	74	12.3	−0.2	14.1	−0.6
70	75	7.5	−12.9	9.8	−13.5
Internal Area Bus Voltage (pu)		$|V|$		$|V|$	
Bus 24		0.820475		0.829942	
Bus 70		0.895784		0.896392	
Bus 71		0.883728		0.884012	
Bus 72		0.822611		0.823000	

8.5. CONCLUDING REMARKS

In this chapter the effects of uncertain load in real-time and off-line power network modeling are discussed. Different approaches for representing uncertain load in power systems are described. In deterministic methods, the formulation of network modeling problems assumes that the data provided are absolutely precise and provide results totally compatible with the given data apart from round-off errors. However, in practice, load data can only be known within some finite precision, this being more the case as the study represents conditions that are more distant into the future. In order to avoid a large amount of repeated calculations and to take into account load uncertainties, it seems beneficial to represent the uncertain load using the two approaches discussed in this chapter.

In building network models for power system analysis, if one has complete knowledge of the network parameters, load and generation data, then the numeric representation of each of the data is a certain or deterministic number. If knowledge is not complete but one knows that there is a frame of repetition of events with fixed laws governing it, one could represent the information by random members, with associated probability distributions. However, if the knowledge available is not sufficient for building the density function of the distribution but one still has qualitative information on the data to be represented, a possibility representation or use of fuzzy numbers could be useful. Probability theory and possibility theory are two complementary views of uncertainty. In some situations, the probability point of view is more appropriate, while in other cases the possibility point of view could be more natural in interpreting the situation. There are situations where both have to be used conjointly (e.g., fusion of probabilistic information and fuzzy set information, computation with ill-defined probabilities, etc.). Uncertainties in loads or generations that do not have definite probability distribution can be incorporated in power system models by using the possibility approach to give a better representation of system behavior. As a consequence of dealing with fuzzy events, a whole set of load scenarios is analyzed at one time, discarding the need for expensive simulation studies.

References

[1] A. Bose and K. A. Clements, "Real-Time Modelling of Power Networks," *Proceedings of the IEEE*, Vol. 75, No. 12, 1987, pp. 1607–1622.

[2] F. F. Wu and A. Monticelli, "A Critical Review of External Network Modelling for Online Security Analysis," *Electrical Power and Energy Systems*, Vol. 5, 1983, pp. 222–235.

[3] A. Monticelli and F. F. Wu, "A Method That Combines Internal State Estimation and External Network Modelling," *IEEE Transactions on Power Apparatus, and Systems*, Vol. PAS-104, 1985, pp. 91–103.

[4] C. N. Lu, K. C. Liu, and S. Vemuri, "An External Network Modelling Approach for On-line Security Analysis," *IEEE Transactions on Power Systems*, Vol. 5, No. 2, 1990, pp. 565–573.

[5] P. Sanderson, R. Curtis, D. Athow, C. N. Lu, K. C. Liu, and C. Letter, "Real Time Complete Model Estimation for Contingency Study: Field Experience," *IEEE Transactions on Power Systems*, Vol. 6, No. 4, 1991, pp. 1480–1484.

[6] C. N. Lu, R. C. Leou, K. C. Liu, and M. Unum, "A Load Flow and Fuzzy Linear Programming Based External Network Modelling Approach," *IEEE Transactions on Power Systems*, Vol. 3, No. 3, 1994, pp. 1293–1301.

[7] External Network Modelling Task Force, "External Network Modelling—Recent Practical Experience," *IEEE Transactions on Power Systems*, Vol. 9, No. 1, 1994, pp. 216–225.

[8] IEEE Task Force Report, "Load Representation for Dynamic Performance Analysis," *IEEE Transactions on Power Systems*, Vol. 8, No. 2, 1993, pp. 472–482.

[9] B. Borkowska, "Probabilistic Load Flow," *IEEE Transactions on Power Apparatus and Systems*, Vol. PAS-93, No. 3, 1974, pp. 752–759.

[10] R. N. Allan and M. R. G. Al-Shakarchi, "Probabilistic AC Load Flow," *Proceedings of the IEE*, Vol. 123, No. 6, 1976, pp. 531–536.

[11] J. F. Dopazo, O. A. Klitin, and A. M. Sasson, "Stochastic Load Flow," *IEEE Transactions on Power Apparatus and Systems*, Vol. PAS-94, 1975, pp. 299–309.

[12] R. N. Allan, A. M. Leite da Silva, and R. C. Burchett, "Evaluation Methods and Accuracy in Probabilistic Load Flow Solutions," *IEEE Transactions on Power Apparatus and Systems*, Vol. PAS-100, No. 5, 1981, pp. 2539–2546.

[13] R. N. Allan, B. Borkowska, and C. H. Grigg, "Probabilistic Analysis of Power Flows," *Proceedings of the IEE*, Vol. 121, No. 12, 1974, pp. 1551–1556.

[14] V. Miranda, M. A. C. C. Matos, and J. T. Saraiva, "Fuzzy Load Flow—New Algorithms Incorporating Uncertain Generation and Load Representation," paper presented at the Power System Computation Conference, Gratz, Austria, Aug. 1990, pp. 621–627.

[15] J. T. Saraiva, V. Miranda, and L. M. V. G. Pinto, "Impact on Some Planning Decisions from a Fuzzy Modelling," *IEEE Transactions on Power Systems*, Vol. 9, No. 2, 1994, pp. 819–825.

[16] Z. Wang and F.-L. Alvarado, "Interval Arithmetic in Power Flow Analysis," *IEEE Transactions on Power Systems*, Vol. 7, No. 3, 1992, pp. 1341–1349.

[17] K. Tomsovic, "A Fuzzy Linear Programming Approach to the Reactive Power/Voltage Control Problem," *IEEE Transactions on Power Systems*, Vol. 7, No. 1, 1992, pp. 287–293.

[18] N. Kagan and R. N. Adams, "Electrical Power Distribution Systems Planning Using Fuzzy Mathematic Programming," *Electrical Power and Energy Systems*, Vol. 16, No. 3, 1994, pp. 191–196.

[19] V. Miranda and J. T. Saraiva, "Fuzzy Modelling of Power System Optimal Load Flow," *IEEE Transactions on Power Systems*, Vol. 7, No. 2, 1992, pp. 843–849.

[20] J. T. Saraiva and V. Miranda, "Impact in Power System Modelling Form Including Fuzzy Concepts in Models," paper presented at the IEEE/NTUA Athens Power Technical Conference, Athens, Greece, Sept. 5–8, 1993.

[21] K. H. Abdul-Rahman and S. M. Shahidehpour, "Static Security in Power System Operation with Fuzzy Real Load Conditions," *IEEE Transactions on Power Systems*, Vol. PWRS-10, pp. 77–87, February 1995.

[22] R. P. Broadwater, H. E. Shaalan, W. J. Fabrycky, and R. E. Lee, "Decision Evaluation with Interval Mathematics: A Power Distribution System Case Study," *IEEE Transactions on Power Delivery*, Vol. 9, No. 1, 1994, 59–67.

[23] M. M. Adibi and J. P. Stovall, "On Estimation of Uncertainties in Analog Measurements," *IEEE Transactions on Power Systems*, Vol. 5, No. 4, 1990, pp. 1222–1230.

[24] A. Kaufmann and M. M. Gupta, *Introduction to Fuzzy Arithmetic Theory and Applications*, Van Nostrand Reinhold, New York.

[25] D. Dubois and H. Prade, "Fuzzy Sets—a Convenient Fiction for Modelling Vagueness and Possibility," *IEEE Transactions on Fuzzy Systems*, Vol. 2, No. 1, 1994, pp. 16–21.

[26] H.-C. Kuo and Y.-Y. Hsu, "A Heuristic Based Fuzzy Reasoning Approach for Distribution System Service Restoration," *IEEE Transactions on Power Delivery*, Vol. 9, No. 2, 1994, pp. 948–953.

[27] E. Economakos, "Application of Fuzzy Concepts to Power Demand Forecasting," *IEEE Transactions on Systems, Man and Cybernetics*, Vol. SMC-9, No. 10, 1979, pp. 651–657.

[28] H. J. Zimmermann, "Fuzzy Set Theory and Mathematical Programming," in *Fuzzy Sets Theory and Applications*, A. Jones, A. Kaufmann, and H. J. Zimmermann (Eds.), Reidel, Tokyo, pp. 99–114.

[29] R. Bellman and L. A. Zadeh, "Decision-Making in a Fuzzy Environment," *Management Sciences*, Vol. 17B, 1970, pp. 141–164.

[30] C. N. Lu, J. H. Teng, and W.-H. E. Liu, "Distribution State Estimation," *IEEE Transactions on Power Systems*, Vol. PWRS-10, 1995, pp. 229–240.

[31] N. Wilson, "Vagueness and Bayesian Probability," *IEEE Transactions on Fuzzy Systems*, Vol. 2, No. 1, 1994, pp. 16–21.

[32] M. Laviolelte and J. W. Seaman, "The Efficacy of Fuzzy Representations of Uncertainty," *IEEE Transactions on Fuzzy Systems*, Vol. 2, No. 1, 1994, pp. 4–37.

[33] G. T. Heydt, *Computer Analysis Methods for Power Systems*, Macmillan, New York.

V. Miranda

Instituto de Engenharia de Sistemas e Computadores Universidade Porto, Porto 4000 Portugal

Chapter 9

A Fuzzy Perspective of Power System Reliability

9.1. INTRODUCTION

The core of reliability analysis of power systems is the uncertainty that is characteristic of the failure-repair cycle of equipment. Typical models assume that the power system, as a whole, is fully repairable and that service disruption is caused by component failure. However, system function may be fully recovered by repair or replacement actions. Some of the underlying uncertainties are associated with "when" a failure occurs and "how long" it takes to recover from it. These uncertainties have been seen as having a particular nature, justified by the following:

- There is a perceived frame of repetition of events in the overall appreciation of the phenomena. This justifies the stochastic assumption and the use of probabilistic models.
- The number of events that may be considered included in the same type of controlled experiment is perceived as being potentially very high. Although one specific device may experience very few failures during its lifetime, a large number of (almost) identical devices are usually installed under very similar conditions. This justifies the assumption of the law of large numbers, which roughly states that the frequency of occurrence tends to be close to the probability of occurrence only for a large number of repetitions of events.

As a result, for mid- to long-term reliability assessment of power systems, these are assumed to be repairable and the analysis is based mainly on mean values of probability distributions. Of course, there are cases when this is not valid:

- For reliability assessment of a nuclear power station, for risk analysis, the system should not be considered fully repairable. Some conditions, based on sequences of failures, are clearly undesirable and must be considered as absorbing states due to their catastrophic consequences.
- In reliability assessment for short-term operational decisions in power systems, when the time ahead is, say, 1 h, mean values cannot be used to assess probability values. Instead, the full description of the probability distributions must be used.

We will concentrate in this chapter mainly on reliability assessment of power systems for mid- or long-term purposes, based on the mean values of probability distributions. We will also accept the general assumption that these distributions are exponential, at least during the useful life of the components.

Fuzzy reliability analysis is a necessary framework whenever at least one of the many input data is described by a fuzzy model. Fuzzy reliability assessment of a power network depends not only on the reliability indices of the components (which help in describing the events) but also on power and energy consumption characteristics (which help in evaluating the consequences of the events).

If we have *only a fuzzy description of*, then we are dealing with type I fuzzy reliability assessment. A fuzzy load is a consequence of incomplete knowledge about consumption behavior. It is typical of planning studies: Who can tell for sure what the load will be 10 years from now? And it may be typical also in distribution networks, whenever consumers are organized into typical clusters or patterns of behavior: A consumption of industrial type, how much is it exactly?

A fuzzy load model may describe both a degree of uncertainty in its exact value and a linguistic description of its nature or possible range of values. Simple expressions such as "more or less 10 MW" or more elaborate declarations such as "load will not be under 8 MW or above 12 MW, but the best estimate is between 10 and 11 MW, with little uncertainty" translate this type of imprecision in load definition.

If one is not sure of load values, then system reliability indices such as mean power disconnected are no longer described by crisp numbers. The uncertainty in the knowledge of the load values must be reflected in the resulting uncertain system indices. Fuzzy models allow one to clearly see this relationship between uncertain data and results.

In many cases, the notion of representing component reliability indices (such as failure rate or mean repair time) by crisp numbers must be challenged. For instance, much reliability data are obtained by analogy from data bases associated with equipment that is not exactly the one under analysis, either because it was not installed under the same conditions or just because some new types of equipment are being foreseen. Also, repair times depend not only on the components themselves but also on other systematic factors that include company efficiency.

Consequently, it is only natural that some uncertainty be associated with component indices, and this uncertainty is not of the probabilistic type. In fact, we are dealing with a twofold dimension of uncertainty, getting at a hybrid model that connects stochastic and fuzzy uncertainties: stochastic because we are still dealing with a failure-repair cycle and fuzzy because we cannot accurately describe all the conditions of the "experiments" that would lead to a pure probabilistic model. This leads to type II fuzzy reliability assessments, where *component reliability indices are fuzzy*.

Extending the concept, we could also consider type III calculations. This would be the case for models that deal with fuzzy reliability in a decision-making environment.

The chapter is organized in three sections, corresponding to the three types of fuzzy reliability studies identified. This material has been introduced, in a systematic way, for the first time in ref. 1.

9.2. TYPE I POWER SYSTEM FUZZY RELIABILITY MODELS

9.2.1. Introduction

Power system reliability assessment has for long extended the analysis beyond continuity of supply. The impact on consumer supply has become an essential way of measuring the goodness of a system design and of comparing the merits of alternatives. The most classical measures adopted are the average power unserved (disconnected) and the average annual energy not supplied, and we will base our discussion mainly on these indices.

Type I fuzzy reliability calculations are related to the uncertainty in defining the power consumption—and therefore in measuring the impacts of failures of supply. Fuzziness of a load value will be translated into uncertainty about the actual consequences of failures.

First, it is easy to realize that if the load is uncertain, so will the loss of load probability (LOLP) index, which describes "how many times" the supply is affected by loss of capacity (generating or transmission): If the load has smaller values, the LOLP index will be lower. As we shall see, this relation between a changing load and a varying calculated LOLP index is far from being linear, but its characteristics are well captured by the fuzzy set description. Second, in very general terms, if a load is fuzzy, so is the value of a power disconnected following a random interruption of supply; this fuzzy value will be represented in the following text as P.

Given the unavailability U of supply (which is the probability of having no supply), in hours per year, one may in general apply the following equation to calculate an average annual fuzzy energy not supplied E:

$$E = UP \equiv \lambda r \tag{9.1}$$

with λ being the rate of interruption and r the average repair time.

There are two ideas that justify adopting fuzzy representations of loads in power system models for reliability analysis purposes. The first relates to the evaluation of a system design: the fuzzy LOLP index P or E allows one to understand how uncertainty in load prediction affects the reliability performance of a system. The second relates to the comparison between alternatives: Two designs may be differently affected, in reliability performance, by the uncertainty in load forecast.

In fact, introducing a fuzzy dimension in the power system models allows explicit evaluation of alternatives under a planning criterion that, although always considered in practice, seldom has received an adequate attention in terms of modeling. This criterion is referred to as robustness: It explains how much uncertainty in the future (translated into uncertain data) can a system design withstand. We will discuss this subject in a later section.

9.2.2. Hierarchical 1 Fuzzy LOLP Index

Definitions. At the hierarchical level 1, at the generating system level (denoted HL 1), the classical calculation of the LOLP index requires the following data: a capacity outage probability table and a cumulative load duration curve. The general formula for calculating the LOLP index of a system is

$$\text{LOLP} = \sum_i p(y_i)q(y_i) \tag{9.2}$$

where y_i = capacity outage (entry i in the table)

$p(y_i)$ = probability of capacity outage i

$q(y_i)$ = probability of load exceeding system available capacity after outage i

Suppose now that there exists some uncertainty in defining the load curve. A fuzzy description of a cumulative load curve would be one that defines, at every level α, an interval of confidence for such curve. If, for instance, a crisp load curve is given by $L = f(q)$, then one would have $q = f^{-1}(L)$. A fuzzy load curve could then be defined at a level α and based on the crisp function f by

$$L_\alpha = [(1 - \Delta_\alpha^-)f(q); \ (1 + \Delta_\alpha^+)f(q)] \tag{9.3}$$

with Δ_α^- and Δ_α^+ being two non-strictly monotonic decreasing functions with α. Conversely, we may have a fuzzy description of the probability q through

$$q_\alpha = \left[\max\left\{0, f^{-1}\left(\frac{1}{1 - \Delta_\alpha^-}L\right)\right\}; \ \min\left\{1, f^{-1}\left(\frac{1}{1 + \Delta_\alpha^+}L\right)\right\}\right] \tag{9.4}$$

This is illustrated in Figure 9.1.

A Numerical Example. Consider a generating system with four similar 10-MW units, each with a forced outage rate (FOR) of 0.9. Take a crisp cumulative load duration curve, described by the equation

$$L = 28.846 - 19.25q$$

The table of capacity outage probabilities is

MW	Probability
0	0.6561
10	0.2916
20	0.0486
30	0.0036
40	0.0001

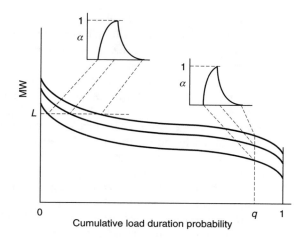

Figure 9.1 Fuzzy description of a cumulative load duration curve.

Let a fuzzy load curve be defined in an interval $L \pm 8\%$:

$$L_\alpha = [(1 - (0.08 - 0.08\alpha))(28.846 - 19.25q); (1 + (0.08 - 0.08\alpha))(28.846 - 19.25q)]$$

Of course, when $\alpha = 1$, this interval condenses into the crisp equation above. This fuzzy load curve is depicted in Figure 9.2, together with the possible capacity outages.

The fuzzy LOLP calculation is derived from

$$\text{Fuzzy LOLP} = \sum_i p(C - L_i)\tilde{q}(L_i) \tag{9.5}$$

where C = system installed capacity
$C - L_i$ = capacity outage
$p(C - L_i)$ = probability of capacity outage of $C - L_i$, a crisp value
$\tilde{q}$ = Fuzzy description of probability of load exceeding available capacity

In Eq. (9.5), we have the addition of fuzzy numbers supported on [0; 1] multiplied by positive constants with domain [0, 1]. As all values are positive, based on interval arithmetic applied to the respective intervals of confidence, calculating the

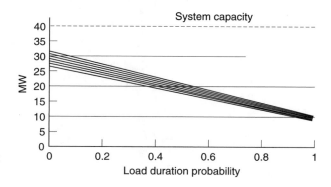

Figure 9.2 Fuzzy load curve and how it is intersected by the capacity outages.

fuzzy result at every level α is straightforward. The upper extreme of an interval of confidence at level α will be just the product of the upper extremes of the intervals of confidence for all the fuzzy components (weighted by the probabilities), and similarly for the lower extremes.

This means that a practical computation of the fuzzy LOLP index based on the regular partition of the interval $[0, 1]$ in levels of confidence α may be done by proceeding to the classical LOLP index calculations for the lower and upper extremes at every α. The result for the numerical example is shown in Figure 9.3 calculated based on the α levels between 0 and 1 at steps of 0.1.

We see from Figure 9.4 that a small uncertainty, represented by a linear variation of the load predicted, has a profound effect on the reliability index uncertainty. In this case, if the uncertainty range in loads could be kept above the α value of approximately 0.5, the LOLP index would have a relatively small uncertainty. However, for larger uncertainties in load definition, the LOLP index becomes quite different and its uncertainty grows remarkably.

If the load curve is not represented as in Figure 9.4 but by a step diagram, the practical calculations must follow a different approach. We will illustrate with an example for the same four-generator system. Assume that the load duration curve is defined through three fuzzy intervals, with the extremes described by triangular fuzzy numbers:

$$L \in [(0, 0, 0); (7, 11, 13)] \Rightarrow q(L) = 1$$

$$L \in [(7, 11, 13); (19, 21, 23)] \Rightarrow q(L) = 0.5$$

$$L \in [(19, 21, 23); (27, 29, 31)] \Rightarrow q(L) = 0.2$$

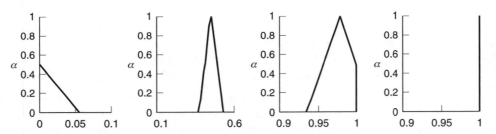

Figure 9.3 Contributions of each outage state to the fuzzy $q(L)$; to calculate the fuzzy LOLP index, each contribution $q(L_i)$ must be multiplied by $p(C - L_i)$, the probability of the outage state i.

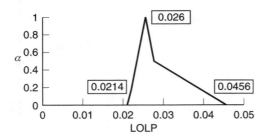

Figure 9.4 Fuzzy LOLP index for the example.

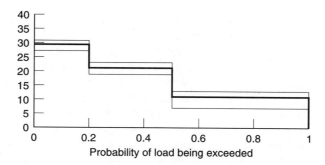

Figure 9.5 Fuzzy load step curve.

A traditional representation of such a load duration curve would be given (such as in Figure 9.5) in three fuzzy steps with triangular fuzzy numbers, represented as triplets as follows:

$$0 \leq q \leq 0.2 \Rightarrow L = (29, 31, 33)$$

$$0.2 \leq q \leq 0.5 \Rightarrow L = (19, 21, 23)$$

$$0.5 \leq q \leq 1 \Rightarrow L = (7, 11, 13)$$

However, the form presented in Figure 9.6 is more explicit on how the calculations must be performed. Figure 9.6 shows four cases of how an outage state interacts with two consecutive steps.

Assume that these steps have probabilities q_1 and q_2 such that $q_1 < q_2$. For an outage state, we can have, in general, the four cases A, B, C, and D. In cases A and D, the probability of the load exceeding the available capacity value L is a crisp number; but in cases B and C, it has a fuzzy representation, with two values having

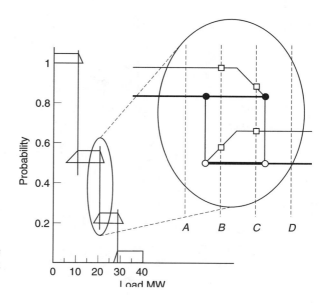

Figure 9.6 Fuzzy load duration curve and how an outage state may intersect this curve.

different possibility degrees. This fuzzy representation is not a fuzzy number but a related concept called a *fuzzy quantity*.

A fuzzy number is said to be convex, with the meaning that the successive α-cuts exhibit a nesting property and are represented by continuous intervals; but in a fuzzy quantity, an α-cut can be a union of nonadjacent intervals.

In the case of the fuzzy load duration curve such as in Figure 9.6, the fuzzy quantities representing the probability of the load exceeding a certain value for cases *B* and *C* are degenerate in the sense that they have only two possible values with different possibility degrees. These degrees are easily seen in Figure 9.6; the fuzzy quantities in cases *B* and *C* are sketched in Figure 9.7. In the numerical example we have been working on, for each outage state one would have the contributions for the LOLP index illustrated in Figures 9.7(c)–(f), obtained by multiplying the probability of each outage state by the fuzzy probability of load exceeding the available capacity at each stage.

"Adding" up all these contributions gives the fuzzy LOLP value depicted in Figure 9.8. In fact, it is not exactly an addition of fuzzy quantities, according to the

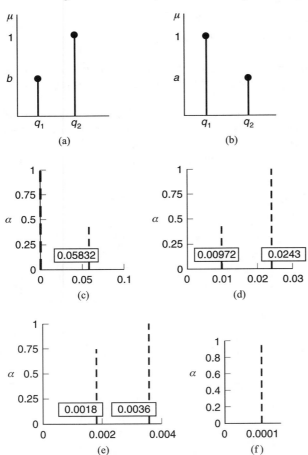

Figure 9.7 Fuzzy quantities representing the probability of load exceeding a given value in cases *B* (a) and *C* (b). Contribution of the outage state at 10 MW (c); at 20 MW (d); at 30 MW (e); and at 40 MW (f).

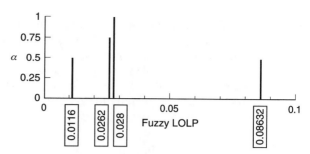

Figure 9.8 Fuzzy LOLP for the step diagram of Fig. 4.

rules of the sum of fuzzy quantities (Minkowsky's addition). Remember that we defined the fuzzy load diagram through expression (9.3), which is a parametric definition. The function itself is defined as fuzzy, but not its values independently. The instantiation of any value L, say, equal to the mode $+x$ percent for some probability q means that the whole curve at the mode $+x$ percent for all q values is instantiated. This is only convenient, for our purposes, because it avoids having to worry about physically meaningless (remember that load duration curves must be monotonically decreasing). The consequence in our calculations is that, to add up the contributions for the LOLP index, we cannot combine subdomains to the left of the mode with subdomains to the right of the mode $\alpha = 1$. This follows since combining these cases would mean that we were accepting, in a load step, a load lower than the mode and, in another step, a load greater than the mode, which contradicts the concept of fuzzy load duration curve defined.

The resulting fuzzy LOLP index for our example is given in Figure 9.8. It is defined in a discrete domain as a consequence of the discontinuities in the load duration curve.

9.2.3. Hierarchical 2 Fuzzy Reliability Assessment

We have shown that, at HL1, a "regular" fuzzy definition of the power system load may lead to very irregular shapes of reliability indices. Certainly we must expect also distortion of the shapes of membership functions, at hierarchical level 2 (HL2), which includes the generation and transmission systems. The constraints imposed by the network branches will have some sort of visible effect, too.

The reliability evaluation of a system design at HL2 may be done with a Monte Carlo (MC) approach. In fact, due to progress in the implementation of convergence acceleration techniques, MC simulation became more and more accepted as a practical tool. In this chapter, we will naturally proceed to describe how this technique can be used to derive type I fuzzy reliability indices, in a process that will be called fuzzy Monte Carlo (FMC).

In a classical MC, given a fixed or crisp load scenario, the process involves roughly the following:

(a) A scenario of outages is sampled.

(b) A special form of an optimal power flow (OPF) is run, to determine the minimum load disconnected.

(c) The results are combined to give the average answer sought.

In an FMC simulation, we follow along the same general lines. However, we need a special fuzzy OPF, or FOPF, to handle fuzzy power injections; and we need to aggregate the results respecting the rules for the operations with fuzzy numbers.

The MC reliability evaluation approach is based on linear models of the power system such as the DC model. This option is justified due to the heavy computation burden that an MC simulation requires. But the defenders of this option have always kept in mind that:

(a) These studies are mainly for planning purposes, where the load prediction is affected by uncertainty; the approximation of using the DC model could be justified in view of the lack of precision in data.

(b) An AC model would require much more data, with added uncertainty.

The idea to build an FMC has been first presented in ref. 2. But a practical FMC has only been possible after the development of an efficient FOPF implementing the DC model. In order to describe the FMC approach, we will have to summarize first some concepts behind solving a DC FOPF.

DC Fuzzy Power Flow. The DC power flow model can be used to obtain fuzzy descriptions of the bus angles and branch active power flows based on fuzzy load and generation data, and it is fully described, from a theoretical basis, in refs. 3 and 4. In summary, the resulting possibility distributions can be evaluated as follows:

(a) A deterministic power flow study is previously run and crisp bus angles and system branch active power flows are calculated based on the central values with membership $\alpha = 1$ of the fuzzy nodal power injection data.

(b) The bus angles and the branch active power flow possibility distributions are calculated by superimposing the possibility distributions of their deviations on the corresponding values obtained in a. The deviation possibility distributions of bus angles $\Delta\Theta$ are built using the DC model $[\mathbf{B}]$ matrix. The branch active power flow deviation possibility distributions $\Delta\mathbf{f_{ik}}$ are evaluated using the sensitivity coefficient matrix $[\mathbf{A}]$. The following are well-known equations, but as the vector $[\mathbf{DP}]$ is composed of fuzzy numbers, the operations must obey the properties of fuzzy arithmetic and the results are also fuzzy numbers:

$$[\Delta\Theta] = [\mathbf{B}]^{-1}[\Delta\mathbf{P}] \qquad [\Delta\mathbf{f_{ik}}] = [\mathbf{A}][\Delta\mathbf{P}] \qquad (9.6)$$

When solving the DC model in the deterministic case, one usually adopts the same node for a reference angle $\theta_{slack} = 0$ fixed as the slack bus.

In the fuzzy case, the preceding equations mean that the uncertainty of nodal injection in the slack bus depends on the uncertainties of the other injections. As no conditions are imposed on the uncertainties in $\Delta\mathbf{P}$, namely at the sources, or to the global data, the results obtained correspond to the widest distributions obtainable from the combination of all possible scenarios implicit in $\Delta\mathbf{P}$. Furthermore, the choice of the slack bus is not irrelevant, and different chosen buses will in general lead to different results, because those equations by themselves do not allow one to condition the uncertainty at the slack bus.

In terms of the definition of a fuzzy problem, the practical case arises when one makes statements about uncertainties at *all* nodes. The observation of the physical behavior of the system, in which generation is driven by consumption, requires that for every possible scenario of loads there must be at least one scenario of generations meeting the demand. There are thus several ways to define the fuzzy power flow problem:

(a) Declare uncertainties at all nodes but the slack bus and solve with the equations above.

(b) Declare uncertainties at all nodes but the slack bus and admit that total uncertainty in generation equals total uncertainty in consumption.

(c) Declare uncertainties at all nodes but the slack bus and declare for this node the maximum and minimum admissible generation values. This can be thought of as equivalent to describing the feasible power injections at this node by a rectangular membership function.

(d) Declare uncertainties at all nodes including the slack bus.

Cases b, c, and d cannot be solved directly. A procedure must then be adopted to correct the extreme values of branch flows given by the preceding equations. It is based on the technique of the dual simplex but can be executed by inspection of the problem features and is described in ref. 5.

An illustrative example is presented based on a network with only three lines, two source nodes, and one load node. All branch reactances have numerical value 1. Uncertainties will be described as rectangular membership functions, which allows also the reading of the results as an exercise in interval arithmetic. Load will be taken as described by a rectangular membership function in the interval [2, 5]. The next figures show the network.

First, a DC fuzzy load flow is solved by declaring a supply uncertainty of [1, 3] at node 2 and fixing the reference angle $\Theta = 0$ at node 1. The calculation of all branch flows and supply uncertainty at node 1, without taking into consideration its generating capacity limits of [min = 0, max = 5], would give the branch flows and slack bus injection shown in Figure 9.9.

As one can see, a negative supply at bus 1 would have to be considered, which is not admissible; therefore, one must take the supply uncertainty at this node as [0, 4] from the intersection of the result obtained and the capacity constraint. The calculations are continued and include the need for the corrective phase. The new results are shown in Figure 9.10.

In a new approach, the reference angle is fixed at node 1 (at zero value), an uncertainty in supply is declared at node 2 as [1, 3], and the global uncertainty in supply is now assumed equal to the uncertainty in demand. This fixes the uncertainty of supply at node 1 as [1, 2]. The calculations of branch flows need again the corrective phases and give as a result the ranges shown in Figure 9.11.

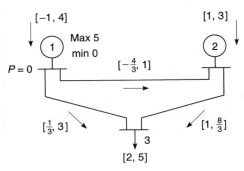

Figure 9.9 Fuzzy flows with no constraints imposed at the slack bus.

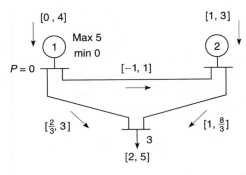

Figure 9.10 Fuzzy flows with generating capacity constraint at the slack bus.

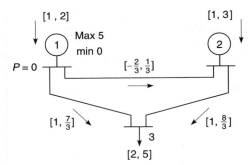

Figure 9.11 Fuzzy flows with uncertainty range fixed at the slack bus.

One notes that these successive steps do not affect the flow in some branches. This means that the uncertainty in their flows does not depend on the uncertainty of supply in some (more or less) distant nodes. One could also show that, in some cases, the extreme value of a branch flow could be obtained at a scenario where one of the generators or loads is not at any of its extreme possible values.

DC Fuzzy Optimal Power Flow. Fuzzy flows, as we have seen, are related to the idea of feasibility: All possible scenarios within the uncertainty ranges are taken into account. There is a rationale behind power generation that conditions the impact of the uncertainty in future loads on the uncertainty in dispatch decisions. In order to represent this operational feature, one must in some way try to "optimize" the uncertainty in generation costs. In reliability studies, the costs considered may be just associated with fictitious generators that simulate load disconnected. This means that we would like to take into account only those generation scenarios that make sense from an economic point of view (or security, or other). As this means combining optimization and power flows, it defines a FOPF. The full presentation of a DC FOPF model can be found in ref. 6.

The ideas behind a FOPF will be illustrated with the help of a simple example with three buses, two lines, and two generating stations (see Figs. 9.12–9.14). The load at bus 3 is assumed to be *more or less* 80 MW, described by the triangular possibility distribution shown in Figure 9.12. The generators at buses 1 and 2 are assumed to be able to operate between 0 and 70 MW. Therefore, one can calculate for each generator a possibility distribution for the feasible generation states, meaning that all possible scenarios where the possible loads are met have been considered and have been qualified by their degree of possibility; see Figure 9.12.

However, if we take into account that the generation cost of 1 MW is $1 at bus 1 and $2 at bus 2, we can also calculate the possibility distributions of the optimal generation scenarios: There is uncertainty in the optimal dispatch because the load is uncertain. In Figure 9.13 one can see that these possibility distributions are much narrower than those related to the feasible scenarios.

We can also obtain possibility distributions for load not supplied, due to any kind of system limitations, for instance, following a forced outage of a line such as shown in Figure 9.14. In this simple system, the outage of the line has the same effect as the outage of generator 2: The fuzzy consequences due to generator or branch outages are treated in the same way.

The expected power disconnected or average annual energy not supplied are indices generally adopted to describe system reliability. But if loads are uncertain, then one should also consider such indicators to be uncertain as well. System adequacy should also be addressed in uncertain terms when referred to forecast system load.

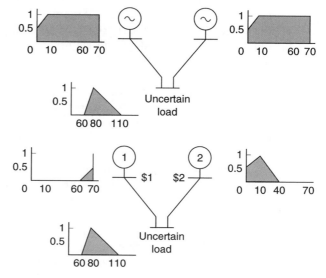

Figure 9.12 Possibility distributions for feasible generation scenarios.

Figure 9.13 Possibility distributions for optimal generation scenarios.

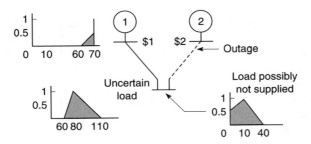

Figure 9.14 Possibility distribution for load not supplied in a scenario where a forced outage is assumed.

Contributions to Risk Analysis in Planning: Robustness, Exposure, and Hedging. As will be shown, FOPF can be used not only to calculate power injection and line flow possibility distributions (under the assumption that the system will be operated as economically as possible) but also to assess the robustness of the system performance and the degree of exposure to an uncertain future.

The FOPF models allow one to detect for which set of lumped and nested load scenarios the system will have feasible solutions while keeping track of the more economic ones. This set can be designated as one for which the system is *robust*. This is a concept from risk analysis whose general ideas are summarized in ref. 7.

A robustness index can be associated with this set when the uncertainties in data have received a fuzzy definition: It just takes the value $1 - \alpha$, where α is the lowest α-cut level for which the system still accommodates data uncertainties at that level. An *exposure* index is also readily available, taking the value α, and is associated with at least one system component (generation or transmission) causing a bottleneck in system flexibility to accommodate uncertainties in data.

As an example, take the system in Figure 9.12, which is similar to the one in Figure 9.9 but with a higher (while still uncertain) load. Load uncertainty is defined in such terms that any generation scenario cannot meet some load scenarios. In simple terms, this means that due to the uncertainty in load forecasting, we are not sure that the generating capacity will accommodate all the load. But above a certain level α, all load scenarios have a feasible generating combination. Therefore, the system is *robust* only if uncertainty in load is contained above α; below α, with a larger uncertainty, there are adverse scenarios, and in this case we must consider that the system is exposed to them. For the type of load uncertainty defined in this example, we have thus a robustness index of 0.5 and an exposure index of also 0.5.

This is valid not only for generating capacities but also for line power flow limits. In Figure 9.15, one would obtain a similar result if the maximum power limits were associated not with the generators but to the transmission lines that connect them with the load bus.

A *hedging* policy in planning would, based on the analysis of system exposure following distinct events, such as branch or generation outages, propose reinforcements so as to guarantee the system robustness in a larger set of nested scenarios resulting from data uncertainty.

In planning or operation environments it is useful to evaluate how increments in some branch or generation power constraints improve the security operational conditions of the system. This means that each increment in the capacity of any system

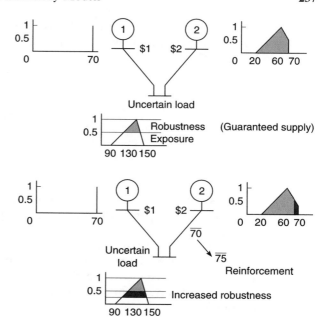

Figure 9.15 System is robust for load scenarios above $\alpha = 0.5$, i.e., for loads occurring in [110, 140] MW. If the uncertainty is larger, however, there is no guarantee that some load disconnection will not necessarily show.

Figure 9.16 Allowing power flow limit in line 2–3 to be increased from 70 to 75 MW improves the robustness index or decreases the exposure index.

component, namely transmission or distribution lines or substation expansion, comes associated with an improvement of the $1 - \alpha$ robustness index, which may be weighted against the investments that would be required in each case. This case is illustrated in Figure 9.16, where it is shown that an increase of flow capacity in line 2 from 70 to 75 MW reduces the exposure index from 0.5 to 0.25.

Another possible situation occurs when one wishes to reduce the exposure index by a certain amount. This can be achieved only by accepting improvements in some power flow or generation limits. The objective would then be to determine a set of such increments so that the specified exposure or robustness indices become guaranteed.

Both approaches can be readily used to rank the different generation expansion strategies according to their cost or to the reduction of the exposure index.

Combining Fuzzy and Probabilistic Assessments in Quality-of-Supply Assessment. In reliability assessment at HL 2, a sampled combination of outages will lead to a system configuration that perhaps will force some load to be disconnected. When fuzzy numbers represent the loads, we have seen that the power disconnected will also have a fuzzy representation. A reliability study will combine the effects of several combinations of outages to derive a representation of the average load disconnected, which in this case will also be fuzzy. This combined approach was first suggested in ref. 2, in which for every sampled scenario, in a typical probabilistic MC process, a fuzzy analysis would be performed. This means that outages would still be represented as probabilistic events and treated as such, but forecasted loads would be treated as uncertain (fuzzy) and therefore a fuzzy load flow would be required in each scenario.

Illustrating the ideas behind the technique, assume that for the system of Figure 9.12 one would have a fully reliable transmission system and the following probabilities:

Scenario	Probability
No failure	0.7
Failure of generator 1	0.1
Failure of generator 2	0.2
Failure of generators 1 and 2	0.0

Consider two cases: case *A* corresponds to the uncertain load of Figures 9.12 and 9.13; in case *B*, the load is still a triangular fuzzy number, with characteristic points (50, 60, 80).

In Figures 9.17 and 9.18 one can see the possibility distribution of the mean probabilistic value of the load disconnected associated with system operation in cases *A* and *B*, weighted from the possibility distributions calculated for all four scenarios. In a way, it can be interpreted as a discrete random walk of a possibility distribution.

Exposure and Robustness as Reliability Indices. We have seen that the exposure index quantifies the possibility of facing a load scenario where one cannot avoid disconnecting the load. The robustness index quantifies the degree of confidence that this costly measure will not be necessary given the uncertainty ranges defined by their fuzzy description.

Within a reliability study, a variety of system states affected by their probability of occurrence must be considered. Therefore, a lumped measure of the impact of outages in the expected system exposure or robustness can also be weighted by the same probabilities. In our example, the global system exposure index can easily be calculated; for example,

$$\text{Global } \alpha \text{ [case } A] = 0.7 \times 0 + 0.1 \times 1 + 0.2 \times 1 + 0 \times 1 = 0.3$$

$$\text{Global } \alpha \text{ [case } B] = 0.7 \times 0 + 0.1 \times 0.5 + 0.2 \times 0.5 + 0 \times 1 = 0.15$$

This index Global α is fuzzy and probabilistic at the same time, in the sense that the exposure of the system under adverse load scenarios (which are uncertain) is moderated by the probability of them really happening. Therefore, if an exposure index must in some way be related to the regret one would feel for having taken the decision of keeping the system with given characteristics, the Global α can be seen as an average expected regret.

Fuzzy Monte Carlo Simulation. In general, MC simulations are used to calculate reliability indices by randomly sampling system component outages. In an FMC, load uncertainties are not of the probabilistic type: They are defined by fuzzy numbers. In this case, load values are not sampled: instead, a FOPF is run for each outage simulation. Therefore, the main difference in computational burden between an FMC and a classical MC with crisp loads derives from the difference in

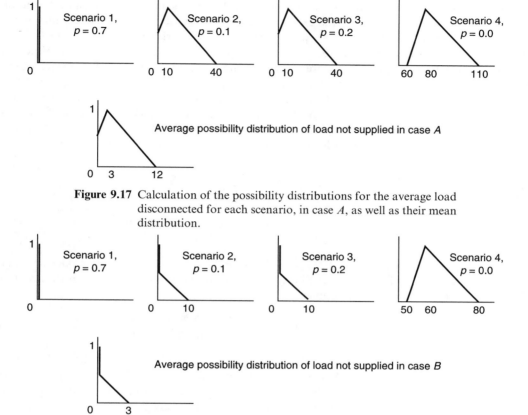

Figure 9.17 Calculation of the possibility distributions for the average load disconnected for each scenario, in case A, as well as their mean distribution.

Figure 9.18 Calculation of the possibility distributions for the average load disconnected for each scenario, in case B, as well as their mean distribution.

execution time between a DC OPF and a DC FOPF. As the latter takes in general about 1.4 times longer, the FMC process remains at a level of efficiency similar to the classical crisp MC, while returning much more valuable information [8].

FUZZY INDICES. Let us consider a function F whose expected value $E(F)$ is searched for. If $p(x)$ represents the probability of a system state x and X is the set of all possible system states; then $E(F)$ is given by (9.7). This expression can hardly be used in practical situations; instead, one gets and analyzes a sample of N system states and, for each state, monitors the function F. In this case, an estimate $\hat{E}(F)$ is given by (9.8):

$$E(F) = \sum_{x_i \in X} F(x_i)p(x_i) \tag{9.7}$$

$$\hat{E}(F) = \sum_{i=1}^{N} F(x_i) \tag{9.8}$$

Let us now assume that we run a FOPF for each sampled outage state x_i and obtain a value $F(x_i)$. Then, the expected possibility distribution $\hat{E}(F)$ is calculated using a fuzzified version of (9.8), where the addition is performed according to fuzzy arithmetic rules. So, using (9.8) as a crisp expression, we can estimate the expected values of robustness or exposure indices; using a fuzzified version of (9.8), we obtain an estimation of the expected power not supplied (PNS) possibility distribution.

With this Monte Carlo methodology integrating FOPF, we can derive also an estimate of the uncertainty affecting the system LOLP. The evaluation of the α-cut of LOLP, or the interval of confidence LOLP_α, can be done by estimating its extreme values, $\text{LOLP}_\alpha^{\min}$ and $\text{LOLP}_\alpha^{\max}$, with the expressions (9.9) and (9.10):

$$\text{LOLP}_\alpha^{\min} = \frac{1}{N} \sum_{i=1}^{N} I(\text{PNS}_\alpha(x_i)^{\min}) \tag{9.9}$$

$$\text{LOLP}_\alpha^{\max} = \frac{1}{N} \sum_{i=1}^{N} I(\text{PNS}_\alpha(x_i)^{\max}) \tag{9.10}$$

In these expressions, $\text{PNS}_\alpha(x_i)^{\min}$ and $\text{PNS}_\alpha(x_i)^{\max}$ represent the extreme values of the α-cut of the PNS possibility distributions associated with system state x_i and I is an indicator function given by

$$I(\text{PNS}) = \begin{cases} 1.0 & \text{if } \text{PNS} \neq 0.0 \\ 0.0 & \text{if } \text{PNS} = 0.0 \end{cases} \tag{9.11}$$

CONVERGENCE EVALUATION. Monitoring a so-called uncertainty coefficient allows one to control the convergence of a Monte Carlo simulation process. As indicated in ref. 9, for instance, this coefficient is given by

$$\beta^2 = \frac{V(\text{PNS})}{N \cdot E(\text{PNS})^2} \tag{9.12}$$

where $V(\text{PNS})$ stands for the variance of PNS. During the process, $E(\text{PNS})$ and $V(\text{PNS})$ are replaced by their current estimates, given by (9.8) and (9.13). One considers having obtained convergence if the calculated β value gets below some specified value $\beta^{\max}$; otherwise, more system states have to be sampled and analyzed:

$$\bar{V}(\text{PNS}) = \frac{1}{N-1} \sum_{i=1}^{N} [\text{PNS}(x_i) - E(\text{PNS})]^2 \tag{9.13}$$

$$N = \frac{V(\text{PNS})}{\beta E(\text{PNS})^2} \tag{9.14}$$

Expression (9.14) is especially interesting as it shows that we can reduce the sample cardinality N by adopting adequate reducing variance techniques for a given β.

In an MC simulation process integrating the FOPF algorithm, this convergence control criterion may be applied considering, for each sampled state, the PNS value obtained by the initial deterministic DC OPF study of the FOPF algorithm.

A more sophisticated control may also use as control parameters the extreme values of the α-cut at level 0 of the definition of the fuzzy PNS.

CONVERGENCE ACCELERATION TECHNIQUES. The convergence acceleration problem has already been addressed [9] in simulation processes where loads are represented by crisp values. We will now describe the application of antithetic sampling and control variable techniques when we have fuzzy numbers representing load uncertainties.

(a) *Antithetic sampling.* The antithetic sampling technique is based on the identification, for each sampled system state, of the so-called antithetic state. Consider a pseudo-random-number sequence as in (9.15) associated with the sampling process. In this sequence, u_i is the pseudo-random-number related to system component i. The corresponding antithetic sequence, given by (9.16), is used to identify the antithetic state:

$$u_1, u_2, u_3, \ldots, u_n \tag{9.15}$$

$$1.0 - u_1, \ 1.0 - u_2, \ 1.0 - u_3, \ldots, 1.0 - u_n \tag{9.16}$$

A number of pairs of system states may therefore be analyzed, so that a new function can be defined by

$$F(x) = \tfrac{1}{2}[F_1(x) + F_2(x)] \tag{9.17}$$

In this expression, F_1 is the value assumed by the function to be estimated for the sampled state while F_2 represents the value related to the antithetic state. If functions F_1 and F_2 are negatively correlated, then (9.18) will be satisfied and a reduction of the variance will be achieved:

$$V(F) < \tfrac{1}{4}[V(F_1) + V(F_2)] \tag{9.18}$$

This technique may be used in the FMC simulation provided that each sampled state and the corresponding antithetic one are analyzed using the FOPF algorithm. Expression (9.17) is replaced by a fuzzified version where addition is done according to the fuzzy arithmetic rules.

(b) *Control variable technique.* This technique is based on the approximation of the function F to be estimated by a so-called control or regression function Z. We define a residual function $Q(x)$ by (9.19) such that a new estimator $F^*(x)$, given by (9.20), can be used:

$$Q(x) = F(x) - Z(x) \tag{9.19}$$

$$F^*(x) = F(x) - Z(x) + E(Z) \tag{9.20}$$

If Z and F are strongly correlated, that is, if Z is a good approximation of F, the residues and the variance of F^* will be small. If we calculate $E(Z)$ by an analytical method, we may obtain an estimate of $E(F)$ using

$$\bar{E}(F) = \bar{E}(F^*) = E(Z) + \frac{1}{N} \sum_{i=1}^{N} [F^*(x_i) - Z(x_i)] \tag{9.21}$$

When we wish to estimate the expected PNS for a composite system, the PNS due only to generator outages (PNS_g) is often selected as a regression function. This choice is very convenient because we know how to calculate its expected value, $E(PNS_g)$, through building a capacity outage probability table.

With fuzzy loads, some care must be taken; expression (9.19) must be rewritten as

$$F(x) = Z(x) + Q(x) \tag{9.22}$$

Because we assume that the fuzziness of F derives from superimposing a fuzzy residue Q to a fuzzy Z that partially explains the system behavior, $Q(x)$ must be obtained by deconvolution and not subtraction. The fuzzy equivalent to (9.21) is

$$\bar{E}(PNS) = E(PNS_g) + \frac{1}{N} \sum_{i=1}^{N} [PNS(x_i) \ominus PNS_g(x_i)] \tag{9.21a}$$

where we wish to note that we are using fuzzy arithmetic operations: $+$ and $\cdot$ denote fuzzy addition, θ represents a deconvolution, and $PNS_g(x_i)$ is the fuzzy number representing the PNS related to state x_i due only to generation outages.

The membership function of $PNS_g(x_i)$ can be calculated from the outage capacity probability table using

$$PNS_g(x_i) = f(P_L - [P_g^{max} - P_g^{out}(x_i)]) \tag{9.23}$$

where P_L = fuzzy system total active load
P_g^{max} = system total installed generation capacity
$P_g^{out}(x_i)$ = addition of the out of service generator capacities in state x_i
$[P_g^{max} - P_g^{out}(x_i)]$ = available generating capacity in state x_i
f = function of fuzzy variable that has the effect of setting to zero the membership degree of negative values of the support set of $PNS_g(x_i)$

Given a fuzzy number X with membership function μ_X, a new fuzzy number Y with membership function μ_Y is built by

$$f(X) = Y: \begin{cases} a < 0 \Rightarrow \mu_Y(a) = 0 \\ a \geq 0 \Rightarrow \mu_Y(a) = \mu_X(a) \end{cases} \quad a \in \mathcal{R} \tag{9.24}$$

Finally, once the system capacity outage probability table is built, the $E(PNS_g)$ possibility distribution can be readily evaluated by (9.25) where $PNS_g(x_i)$ is given by (9.23), $p(x_i)$ is the probability of state x_i, and $\sum$ denotes fuzzy addition:

$$E(PNS_g) = \sum_{i=1}^{N} PNS_g(x_i) p(x_i) \tag{9.25}$$

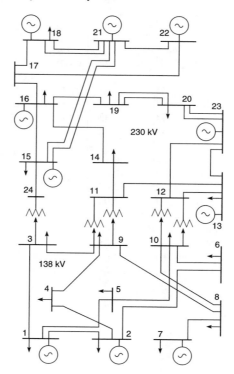

Figure 9.19 IEEE RTS network.

Example. Let us examine some practical effects of applying the FMC approach to the 24-bus, 38-branch, 32-generator network presented in Figure 9.19. This network is based on the IEEE Reliability Test System (RTS) (500 MVA was adopted as a power base). We have fuzzified the loads and increased their value to stress the system: The central value of the 1.0 α-level of each fuzzy load distribution corresponds to the RTS value multiplied by 1.8. The generator characteristics are presented in Table 9.1. We also defined two sets of load possibility distributions:

> *First set*: Trapezoidal fuzzy numbers describe load uncertainties. The extreme values of the 0.0 and 1.0 cuts correspond to (0.9, 0.95, 1.05, 1.1) of the central value (a maximum uncertainty range of 10%).
>
> *Second set*: Triangular fuzzy numbers describe load uncertainties. The extreme values of their 0.0 cut are 0.925 and 1.075 of the central value (a maximum uncertainty range of 7.5%).

The branch data of the RTS have also been changed, namely the generator limits (the original maximum limits have been doubled). Finally, the generator and branch outage rates of the RTS network were used in the reliability calculations. See Table 9.2.

TABLE 9.1   GENERATOR CHARACTERISTICS

Bus Number	Generator	P_g^{max} (MW)	Inc. cost ($/MWh)	FOR	Bus Number	Generator	P_g^{max} (MW)	Inc. cost ($/MWh)	FOR
1	1	40.0	3.0	0.1	15	3	24.0	2.0	0.02
1	2	40.0	3.0	0.1	15	4	24.0	2.0	0.02
1	3	152.0	4.0	0.02	15	5	24.0	2.0	0.02
1	4	152.0	4.0	0.02	15	6	310.0	6.0	0.04
2	1	40.0	3.0	0.1	16	1	310.0	5.5	0.04
2	2	40.0	3.0	0.1	18	1	800.0	9.0	0.12
2	3	152.0	4.0	0.02	21	1	800.0	8.0	0.12
2	4	152.0	4.0	0.02	22	1	100.0	2.0	0.01
7	1	200.0	5.0	0.04	22	2	100.0	2.0	0.01
7	2	200.0	5.0	0.04	22	3	100.0	2.0	0.01
7	2	200.0	5.0	0.04	22	4	100.0	2.0	0.01
13	1	394.0	6.0	0.05	22	5	100.0	2.0	0.01
13	2	394.0	6.0	0.05	22	6	100.0	2.0	0.01
13	3	394.0	6.0	0.05	23	1	310.0	5.0	0.04
15	1	24.0	2.0	0.02	23	2	310.0	5.0	0.04
15	2	24.0	2.0	0.02	23	3	700.0	7.0	0.08

TABLE 9.2   BRANCH CHARACTERISTICS

Extreme Buses	x (pu)	P^{max} (MW)	FOR	Extreme Buses	x (pu)	P^{max} (MW)	FOR
1, 2	0.0139	175	0.00044	12, 13	0.0476	500	0.00050
1, 3	0.2112	175	0.00059	12, 13	0.0966	500	0.00065
1, 5	0.0845	175	0.00038	13, 23	0.0865	500	0.00062
2, 4	0.1267	175	0.00045	14, 16	0.0389	500	0.00048
2, 6	0.1920	175	0.00055	15, 16	0.0073	500	0.00041
3, 9	0.1190	175	0.00043	15, 21	0.0490	500	0.00052
3, 24	0.0839	400	0.00175	15, 21	0.0490	500	0.00052
4, 9	0.1037	175	0.00041	15, 24	0.0519	500	0.00052
5, 10	0.0883	175	0.00039	16, 17	0.0259	500	0.00044
6, 10	0.0605	175	0.00132	16, 19	0.0231	500	0.00043
7, 8	0.0614	175	0.00034	17, 18	0.0144	500	0.00040
8, 9	0.1651	175	0.00050	17, 22	0.1053	500	0.00068
8, 10	0.1651	175	0.00050	18, 21	0.0259	500	0.00044
9, 11	0.0839	400	0.00175	18, 21	0.0259	500	0.00044
9, 12	0.0839	400	0.00175	19, 20	0.0396	500	0.00048
10, 11	0.0839	400	0.00175	19, 20	0.0396	500	0.00048
10, 12	0.0839	400	0.00175	20, 23	0.0216	500	0.00043
11, 13	0.0476	500	0.00050	20, 23	0.0216	500	0.00043
11, 14	0.0418	500	0.00049	21, 22	0.0678	500	0.00057

SAMPLING STATES: BASE CASE. The base case corresponds to the operation of all components. For this case and for either fuzzy load set, no load disconnection has been found. The state exposure is therefore 1.0, and its robustness is 0.

SAMPLING STATES: OVERLAPPING OUTAGES OF LINE 16 AND GENERATORS 13/3 AND 22/5. In this case, again for the trapezoidal load set, the system exposure has value 1.0 and robustness 0.0 (it is another way of saying that this scenario implies load disconnection for the best estimate of loads, described at a level 1). The resulting PNS possibility distribution is sketched in Figure 9.20.

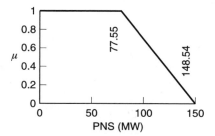

Figure 9.20 The PNS if line 16 and generators 13/3 and 22/5 are out of service (trapezoidal load set).

SAMPLING STATES: OVERLAPPING OUTAGE OF GENERATORS 1/1, 2/2, AND 21/1. For the trapezoidal fuzzy load set, the system showed no capacity to accommodate the load for uncertainties associated to α-cuts below 0.36. The state robustness and exposure indices are 0.64 and 0.36. Figure 9.21 displays a sketch of the state PNS possibility distribution, obtained with a FOPF.

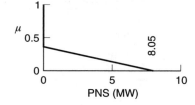

Figure 9.21 The PNS if generators 1/1, 2/2, and 21/1 are out of service (trapezoidal load set).

FUZZY MONTE CARLO ANALYSIS. Fuzzy Monte Carlo simulations were run for the two load sets using both antithetic sampling and control variable variance reduction techniques and specifying 10% for the uncertainty coefficient b. Figures 9.22 and 9.23 show the expected PNS possibility distribution and the LOLP possibility distribution for the triangular fuzzy loads, and Figures 9.24 and 9.25 show the same results for the trapezoidal fuzzy loads.

Table 9.3 presents the results of the calculation of the expected global exposure and robustness indices for the two sets of fuzzy loads. One may see that the larger uncertainty associated with the trapezoidal loads leads to a higher value of the exposure index, meaning that, for such a load forecast, the planner is facing a riskier decision. This comparison, made here between two uncertain load forecasts as an illustration, is necessary when assessing the merits of different system expansion plans in different future scenarios.

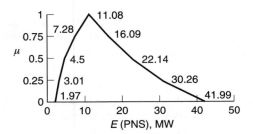

Figure 9.22 An E(PNS) possibility distribution (triangular load set).

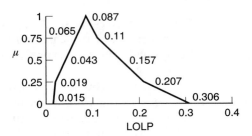

Figure 9.23 A LOLP possibility distribution (triangular load set).

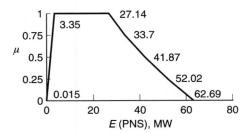

Figure 9.24 An E(PNS) possibility distribution (trapezoidal load set).

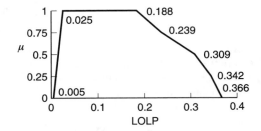

Figure 9.25 A LOLP possibility distribution (trapezoidal load set).

TABLE 9.3 EXPECTED GLOBAL ROBUSTNESS AND EXPOSURE INDICES

Load ↓ Robustness	Indices →	Exposure
Triangular load set	0.161	0.839
Trapezoidal load set	0.288	0.712

Both MC simulations converged after analyzing 3258 states without convergence acceleration and 716 states with the control variable technique. This reduction of the computing effort to 22% confirms the known advantages of the convergence acceleration techniques. We could also confirm that the antithetic scheme had a negligible effect in variance reduction, while the control variable method proved very efficient.

A fuzzy simulation and a crisp simulation converge after analyzing the same number of system states, because convergence is monitored for the sequence of PNS values obtained from the initial deterministic OPF study performed in each FOPF run. This sequence is the same in the two essays as in both cases the load central values are equal.

The FOPF study is very efficient as it only takes on average 40% longer than a classical crisp DC OPF. This is an important feature when analyzing a large number of system states as happens in an MC process: the FMC simulation also takes 1.4 of the computation time of a classical crisp load approach.

INTERPRETATION OF SYSTEM FUZZY RELIABILITY INDICES. The interpretation of the fuzzy indices given as a result of an FMC is enlightening. Take, for instance, the fuzzy E(PNS): Its membership function is directly associated with the membership functions of the loads. For any α-cut, one has an interval of uncertainty about the exact load values; the corresponding α-cut in the fuzzy E(PNS) gives the corresponding range of uncertainty of the expected power not supplied index. Figure 9.26 displays this correspondence for the $\alpha = 0.5$ level of the triangular fuzzy load set.

The FMC and the fuzzy indices reveal some system design weaknesses. For instance, take the triangular fuzzy load set: Its maximum range of uncertainty (at α-level 0.0) is $\pm 7.5\%$. However, the corresponding range of uncertainty for the fuzzy E(PNS), related to its central value, is $(-77\%, +278\%)$; for the LOLP, its range is $(-82\%, +252\%)$. The expected exposure and robustness indices combine possible load scenarios (described by fuzzy numbers) with system states (originated by random component outages). They are measures of system adequacy to meet the demand. A higher value of expected exposure index means that for more combinations of contingencies and load scenarios it is likely that some load disconnection must take place.

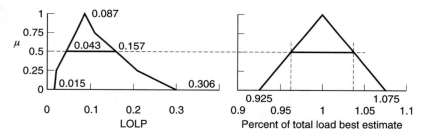

Figure 9.26 Correspondence between load uncertainty range and LOLP uncertainty range.

9.3. TYPE II POWER SYSTEM FUZZY RELIABILITY MODELS

9.3.1. Basic Fuzzy Reliability Concepts

The discipline of reliability engineering depends on the analysis of system behavior, taking in account failures of the equipment. Classical reliability studies are based on two assumptions: the *probabilistic assumption*, which states that the system behavior can be fully described and understood through probabilistic, and the *binary state assumption*, which requires that the state of an equipment, of failure or functioning, be completely defined. These two assumptions form the basis of the so-called PROBIST model: under this conventional view, the system performs its assigned functions during some exposure period under a given known environment.

But the consideration of some fuzzy concepts leads to the following reliability models:

- **PROBIST** model: assuming the probabilistic assumption and the binary state assumption
- **PROFUST** model: keeping the probabilistic assumption but introducing a fuzzy state assumption
- **POSBIST** model: introducing a possibilistic assumption as to the description of the events and the laws governing their repetition, together with the binary state assumption
- **POSFUST** model: combining the possibilistic assumption and the fuzzy state assumption

Some authors [10] have addressed the general subject of fuzzy reliability within this model frame of reference. This section will be devoted mainly to the **PROFUST** model; however, it is necessary to clarify the meaning of the fuzzy state assumption. The first interpretation of a fuzzy state assumption is that one cannot precisely define the state of a component, namely the meaning of system failure, as if the borders between the functioning state and the failed state were not precise, leading to the definition of a fuzzy success and a fuzzy failure state. But, in fact, another interpretation is equally valid: One can define exactly what the operating and the failed states are, but one is unable to define precisely how the transition occurs, namely how often.

In the **PROFUST** model, the states are well defined but the probability of finding a particular component at a given state is given by a fuzzy number (instead of a crisp number). However, because the probability assumption is retained, particular probabilities of actual outcomes must still add up to 1 in the whole state space. This introduces a factor of dependence between fuzzy probability values that is not found in other fuzzy models.

Furthermore, the probabilistic assumption requires the following:

1. An event is precisely defined.
2. A frame of repetition of events is precisely described.
3. For practical purposes, a large size of collected data is available.

The **PROFUST** model seems to be very adequate in many applications in the power system field. It does not require abandoning all the tools and concepts that have been built in the last two decades. In many cases, it may just be presented as an extension of them; it may get a rather straightforward interpretation as a special type of sensitivity analysis to uncertainty in data, which is in general cherished by engineers. It requires more fuzzy arithmetic than fuzzy logic, which helps by being more readily understood and accepted by engineers familiar with traditional tools; due to these characteristics, it can be truly argued that what is being proposed is not the replacement of old faithful probabilistic models but just their enhancement.

An easy way of introducing the PROFUST model is by questioning the practical interpretation of simple concepts such as failure rate or mean repair time.

9.3.2. A Fuzzy Exponential Distribution

In classical probability models usually adopted in power systems, the reliability function is given by

$$R(t) = \exp\left(-\int_0^t \lambda(u)\,du\right) \tag{9.26}$$

where $\lambda(t)$ is the instantaneous hazard rate or failure rate. If $\lambda(u)$ is a constant λ and therefore independent of time, the reliability function becomes the very well known

$$R(\tau) = e^{-\lambda\tau} \tag{9.27}$$

Here $R(t)$ gives the probability of a component surviving a time t in a constant failure rate environment.

The unreliability, or the probability of finding the component in the failed state during time t, is of course

$$Q(\tau) = 1 - e^{-\lambda\tau} \tag{9.28}$$

In power system studies, the raw data are usually constituted by the frequency of failures and the mean repair time. These data may be found in several databases. However, it is typical of power system components to have small failure rate values and small repair times (compared to functioning times). Therefore, it is classical to approximate the frequency of failures with the failure rate, and therefore, we will assume that the raw data extracted from databases are failure rates.

But a database is a collection of information about pieces of equipment possibly installed in very diverse conditions, or of somewhat different technological generations, or operated by utilities with different views about maintenance or quality. Besides, when planning a power system, one may only apply failure rates, by analogy, to new equipment that will be built in the future (remember the technological progress). In fact, failure rates will change over time because of the aging of equipment or because of rehabilitation actions.

It is not surprising that we may accept, for a failure rate value relative to some type of equipment (say, a breaker or a transformer), instead of a crisp number such as 0.01 failures/year, an interval of confidence such as [0.008, 0.012] failures/year or even a fuzzy number. Recall that the designation "interval of confidence" does not relate to any classical statistical concepts but to the discourse of the "fuzzy set community." In this sense, an interval of confidence α corresponds to the cut set at level α defined in relation to the membership function of a fuzzy set.

As one knows, a fuzzy number may be seen as an aggregation of nested intervals of confidence; for example, $\lambda_\alpha = [0.01 - (0.002(1 - \alpha)),\ 0.01 + (0.002(1 - \alpha))]$ represents a triangular fuzzy number, with λ_α being the interval of confidence at level α, as defined. This fuzzy failure rate is represented in Figure 9.27.

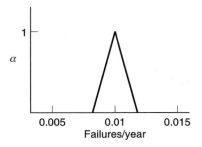

Figure 9.27 A fuzzy failure rate.

We can now consider the extension of the reliability function $R(t)$ to the fuzzy case [11]. At a fixed α, λ_α defines an interval of confidence $\lambda_\alpha = [\lambda_\alpha^-, \lambda_\alpha^+]$; for every t, the boundaries of such an interval define

$$R_\alpha(\tau) = [R_\alpha^+ = e^{-\lambda_\alpha^+ \tau}; \ R_\alpha^- = e^{-\lambda_\alpha^- \tau}] \tag{9.29}$$

being obvious that

$$\lambda_1^2 \lambda_2 \Rightarrow e^{-\lambda_2 \tau^2} e^{-\lambda_1 \tau} \tag{9.30}$$

We have therefore, at each level α, an interval of confidence for the reliability or survival function delimited by a lower survival law R_α^+ and an upper survival law R_α^-. The consideration of all α-levels leads to the definition of a fuzzy reliability function $R_\alpha(t)$, such as suggested in Figure 9.28.

The mean time to failure of a classical reliability function is given by

$$\text{MTTF} = \int_0^\infty t R(t) \, dt = \frac{1}{\lambda} \tag{9.31}$$

Taking into account Eqs. (9.29)–(9.31), it is easy to define a fuzzy MTTF for the fuzzy reliability function $R_\alpha(t)$, based on the interval of confidence representation

$$\text{MTTF}_\alpha = \left[\frac{1}{\lambda_\alpha^+}; \frac{1}{\lambda_\alpha^-} \right] \tag{9.32}$$

This definition is straightforward due to the monotonic characteristics of the functions involved. Besides, all λ values are nonnegative, and therefore there are no problems in calculating the intervals of confidence in Eq. (9.32).

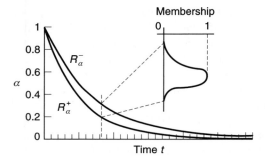

Figure 9.28 Fuzzy reliability function and an illustration of its membership values at a certain t.

In a similar fashion, if the exponential assumption is accepted for repair actions, we may define a mean time to repair $r_{.}$ in relation with repair rates μ_α^- and μ_α^+ such as

$$r_\alpha = \left[\frac{1}{\mu_\alpha^+}; \frac{1}{\mu_\alpha^-}\right] \tag{9.33}$$

In power system reliability engineering, mean repair times are usually also raw data. Equation (9.33) shows that a fuzzy definition of the mean repair time may be thought of as built over an underlying exponential fuzzy distribution of repair times, which depend on repair rates μ_α also being fuzzy. Furthermore, Eq. (9.33) means that the fuzzy numbers r_α and μ_α are the inverse of each other:

$$r_\alpha = \frac{1}{\mu_\alpha} \tag{9.34}$$

This operation is valid because the fuzzy numbers $r_{.}$ and $\mu_{.}$ are defined on the positive axis of the real line.

9.3.3. Series and Parallel Systems

These definitions allow us to calculate the fuzzy reliability of series and parallel systems. The fuzzy reliability of a serial system of n components is given by

$$R^{\text{ser}} = R_1 \cdot R_2 \cdots R_n = \prod_{i=1}^{n} R_i \tag{9.35}$$

We have here the product of fuzzy numbers, all defined on the positive axis. In terms of intervals of confidence, if each $R_a = [R_a;^-; R_a;^+]$, we have

$$R_a;^{\text{ser}} = [R_{a;1}^- \cdot R_{a;2}^- \cdots R_{a;n}^-; R_{a;1}^+ \cdot R_{a;2}^+ \cdots R_{a;n}^+] \tag{9.36}$$

An example illustrates this calculation. Consider a series of five similar components, each one with a probability of survival or reliability described as a triangular fuzzy number, depicted in Figure 9.29 and described in terms of a triplet as [0.91; 0.93; 0.95] or in terms of intervals of confidence as

$$R_\alpha = [0.93 - (0.02(1 - \alpha)); 0.93 + (0.02(1 - \alpha))]$$

The product of such five fuzzy numbers is illustrated in Figure 9.29. It is *not* a triangular fuzzy number, but surely such an approximation would seem justified in this case.

The fuzzy reliability of a parallel system of n components is given by

$$R^{\text{par}} = 1 - \prod_{i=1}^{n} (1 - R_i) \tag{9.37}$$

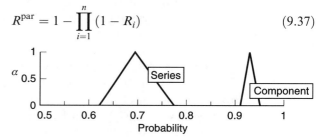

Figure 9.29 Fuzzy reliability of a five-component serial system.

The operations of subtraction and multiplication must be understood, of course, as in the context of fuzzy arithmetic, which derive from applying interval arithmetic to the intervals of confidence at every level α.

Consider a parallel connection of three similar components, each with a probability of survival or reliability described as a triangular fuzzy number, depicted in Figure 9.30 and described in terms of a triplet as [0.81; 0.85; 0.89] or in terms of intervals of confidence as

$$R_\alpha = [0.85 - (0.04(1 - \alpha)); 0.85 + (0.04(1 - \alpha))]$$

Applying Eq. (9.12), one reaches the result depicted in Figure 9.30. The result is again *not* a triangular fuzzy number, as one may check in Figure 9.31 (zooming in the system possibility distribution of Figure 9.30), but still here this would be a reasonable approximation.

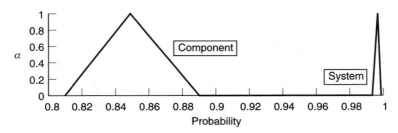

Figure 9.30 Fuzzy reliability of a single component and a three-component parallel system.

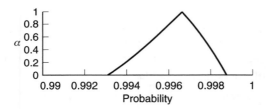

Figure 9.31 Fuzzy reliability of a three-component parallel system (zooming in Fig. 9.30).

9.3.4. Fuzzy Markov Models

From a power system point of view, Markov models are extremely important, because they form the basis of many simplified models used in practice. In the sense of the PROFUST fuzzy reliability model, it is possible to define Markovian models for the behavior of a component or a system, with the following characteristics:

(a) The state space is completely defined and crisp; it is clearly the definition of a failure or a success state, for example.

(b) The transitions between states are assumed as obeying the general probabilistic laws; namely, a Markov model with no memory is characterized by exponential distributions with constant rate.

(c) The definition of the transitions themselves is fuzzy; one is uncertain about the actual value of the transition rates and therefore describes them as fuzzy numbers.

Let us analyze a two-state Markov model of a component with a failure f and a success s state (Fig. 9.32) and fuzzy failure rate λ and a fuzzy repair rate μ. The objective is to try to assess the value of the (stationary) probabilities P_s and P_f of finding the component in either state, but as the raw data are fuzzy, these probability values will also be fuzzy.

For every instantiation of λ' and μ', the stochastic transition equations must apply. We have therefore

$$\begin{bmatrix} 1-\lambda & \mu \\ \lambda & 1-\mu \\ 1 & 1 \end{bmatrix} \begin{bmatrix} P_s \\ P_f \end{bmatrix} = \begin{bmatrix} P_s \\ P_f \\ 1 \end{bmatrix} \tag{9.38}$$

for the stationary probabilities in a discrete-time process and

$$\begin{bmatrix} -\lambda & \mu \\ \lambda & -\mu \\ 1 & 1 \end{bmatrix} \begin{bmatrix} P_s \\ P_f \end{bmatrix} = \begin{bmatrix} 0 \\ 0 \\ 1 \end{bmatrix} \tag{9.39}$$

for the limit probabilities in a continuous-time process. The result is well known from classical crisp models; it may be obtained after deleting the first equation of (9.39), for instance,

$$P_s = \frac{\mu}{\lambda + \mu} \qquad P_f = \frac{\lambda}{\lambda + \mu} \tag{9.40}$$

These expressions can be taken as a basis for a fuzzy model. However, it would be wrong to just replace in them the crisp rates λ and μ by fuzzy definitions: The result would display a much larger uncertainty than necessary, because one would be using *more than once* the same fuzzy variable in the calculations. The correct form of the expressions for the fuzzy values of P_s and P_f is

$$P_s = \frac{1}{1 + (1/\mu)\lambda} \qquad P_f = \frac{1}{1 + (1/\lambda)\mu} \tag{9.41}$$

or, if the mean repair time r is used as fuzzy raw data,

$$P_s = \frac{1}{1 + r\lambda} \qquad P_f = \frac{1}{1 + (1/\lambda)(1/r)} \tag{9.42}$$

A numerical example may be as follows: Assume that the failure rate and the mean repair time are uncertain and represented by triangular fuzzy numbers such as

$$\lambda \rightarrow \text{``more or less 0.1'' failures/year} \rightarrow [0.08; 0.1; 0.12]$$

$$r \rightarrow \text{``around 48 h''} \qquad\qquad \rightarrow [40; 48; 56]$$

Using the fuzzy equations (9.42), one gets the results depicted in Figures 9.32 and 9.33. Again we get fuzzy numbers looking very much triangular. The reader is

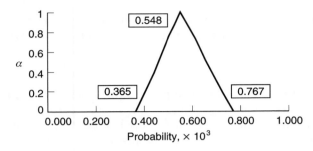

Figure 9.32 Fuzzy probability P_f of the failure state.

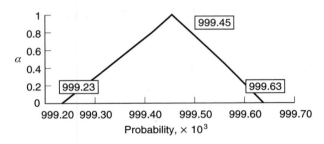

Figure 9.33 Fuzzy probability P_s of the success state.

invited to try and use Eq. (9.40) and compare the (wrong) results thus obtained with the correct ones. One may also notice that the fuzzy results obtained satisfy the fuzzy equation $P_s = 1 - P_f$, which seems only natural; however, it is not possible to write $P_s + P_f = 1$ and give it a fuzzy interpretation, because the right-hand side is an uncertain (fuzzy) number and the left-hand side is a certain (crisp) value.

In another example with a three-state model, a component other than failure and success states 1 and 3 can also be found in a derated state 2. From Figure 9.34, the stochastic transition equations are

$$
\begin{bmatrix}
-\lambda & 0 & \mu \\
\lambda & -\beta & 0 \\
0 & \beta & -\mu \\
1 & 1 & 1
\end{bmatrix}
\begin{bmatrix}
P_1 \\
P_2 \\
P_3
\end{bmatrix}
=
\begin{bmatrix}
0 \\
0 \\
0 \\
1
\end{bmatrix}
\tag{9.43}
$$

The solutions, arranged so that a fuzzy result may be found, are

$$
P_1 = \frac{1}{1 + \lambda(1/\beta + 1/\mu)} \qquad P_2 = \frac{1}{1 + \beta(1/\lambda + 1/\mu)} \qquad P_3 = \frac{1}{1 + \mu(1/\lambda + 1/\beta)}
\tag{9.44}
$$

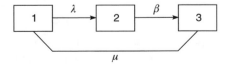

Figure 9.34 A three-state Markov model.

Assume that the failure rate λ, the transition rate β, and the mean repair time r are uncertain and represented by triangular fuzzy numbers such as

$$\lambda \rightarrow \text{"more or less } 0.2\text{" failures/year} \rightarrow [0.18; 0.2; 0.22]$$

$$\beta \rightarrow \text{"around" 1 transition/year} \qquad \rightarrow [0.6; 1; 1.4]$$

$$r \rightarrow \text{"approximately" 72 h} \qquad\quad \rightarrow [64; 72; 80]$$

The results from applying Eq. (9.44) are depicted in Figures 9.35–9.37.

There is no obvious equation relating the three fuzzy probabilities and the unity value that may be satisfied by these results. The reason for this is that the nonlinearity of the solution forces the extreme values of the state probabilities to be achieved with nonextreme values of some of the transition rates at every α interval of confidence. However, the central values of the fuzzy probabilities still add up to 1, as expected.

Assume now that states 1 and 2 are classified as success states and that one wishes to assess the value of the probability $P_{1 \cup 2}$ of finding the system in one of those states. This must not be calculated by the fuzzy addition of individual probabilities! In fuzzy terms,

$$P_{1 \cup 2} \neq P_1 + P_2$$

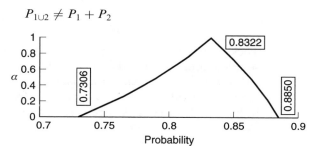

Figure 9.35 Fuzzy probability of success state 1.

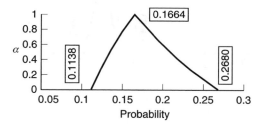

Figure 9.36 Fuzzy probability of derated state 2.

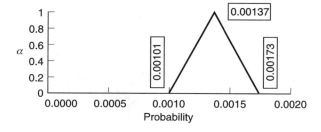

Figure 9.37 Fuzzy probability of failure state 3.

This operation would lead to a fuzzy number admitting the possibility of probability values greater than 1, which is nonsense. Instead, the correct calculation must be based on

$$P_{1\cup2} = 1 - P_3$$

Using the fuzzy arithmetic rules (or interval arithmetic rules applied to each α-level), the result for the numerical example discussed above is the one in Figure 9.38. It is easy to check that, departing from expressions (9.44), one may arrive at the result

$$P_{1\cup2} = 1 - P_3 = \left(1 + \frac{1/\mu}{1/\mu + 1/\beta}\right)^{-1} \tag{9.45}$$

Solving expression (9.45) according to the fuzzy arithmetic rules, it is easy to arrive at the same result, as depicted in Figure 9.34. The last example presented relates to a two-component, four-state model, such as in Figure 9.39. The fuzzy state probabilities in this case are given by the fuzzy equations

$$P_1 = \frac{1}{1 + \lambda_1/\mu_1} \frac{1}{1 + \lambda_2/\mu_2} \qquad P_2 = \frac{1}{1 + \mu_1/\lambda_1} \frac{1}{1 + \lambda_2/\mu_2}$$

$$P_3 = \frac{1}{1 + \lambda_1/\mu_1} \frac{1}{1 + \mu_2/\lambda_2} \qquad P_4 = \frac{1}{1 + \mu_1/\lambda_1} \frac{1}{1 + \mu_2/\lambda_2} \tag{9.46}$$

These examples show one way of solving some fuzzy Markov processes: The crisp solution equations for the limit stationary state probabilities may be fuzzified, *but only after being conveniently arranged*, such that each variable never gets, directly or indirectly, divided by itself or subtracted from itself. An easy way to guarantee this is to have each variable only represented once in an expression.

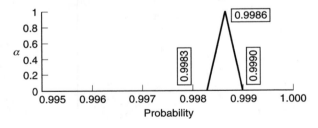

Figure 9.38 Fuzzy probability of states 1 and 2.

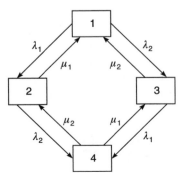

Figure 9.39 A two-component, four-state Markov model.

Unfortunately, this is easier said than done. Not only is it a hard job to solve analytically the stochastic transition equations—feasible only for small systems—but also in most cases it will not be possible to arrange the expressions in the desired way. When this is the case, numerical computations must be performed under another approach.

The basic principle that allows the generalization of crisp mathematical concepts into the fuzzy framework is the *extension principle*, whose definition we recall briefly. Consider an operation $\blacklozenge$ valid with real numbers such that $c = a \blacklozenge b$; its extension to fuzzy numbers is achieved by

$$\mu(c) = \sup\{\min\{\mu(a), \mu(b)\} | \forall a, b : c = a \blacklozenge b\}$$

This means that, in principle, if a pair (a, b) maps into a number c, c receives a membership degree equal to the minimum of a and b membership degrees; furthermore, if two pairs (a_1, b_1) and (a_2, b_2) map into the same c, then the maximum of the possible membership grades that would be given to c is chosen as the membership grade of c.

Naturally, for the interval of confidence of fuzzy numbers at a membership level α, there will only be elements c mapped from pairs (a, b), where a and b belong to that interval of confidence at level α. This results from the conjugation of the extension principle and the definition of fuzzy numbers. In fact:

(a) Suppose $\mu(a) < \alpha$; then $\mu(c) = \min\{\mu(a), \mu(b)\} < \alpha$, meaning that the pair (a, b) does *not* give evidence supporting c as belonging to the interval of confidence at level α.

(b) The same applies for $\mu(b) < \alpha$.

(c) Suppose $\mu(a) = \alpha' > \cdot$; this means that α belongs to an interval of confidence at degree α', which, by the definition of fuzzy numbers, is nested or included in the interval of confidence at degree α; therefore, a also belongs to the interval of confidence at level α.

(d) The same reasoning applies to b.

Therefore, the extremes of an interval of confidence at a certain level α must be searched among all possible combinations of values (a, b) belonging to intervals of confidence of the same degree.

This has a direct application to the reliability problem above, based on fuzzy Markov models. One possible way of solving the problem of determining the fuzzy probabilities of each state in a Markov process is by defining a nonlinear programming (NLP) formulation. In fact, if one has a problem with n states and m transition rates λ_j among the states, one may try and calculate, at each level α, the interval of confidence relative to the fuzzy probability P_k of a given state k at that level, defined as $P_k(\alpha) = [P_k;^-(a); P_k;^+(a)]$:

$$P_k;^-(a) = \min\{P_k(\lambda) | \forall \lambda_j : \lambda_j;^-(a) \leq \lambda_j \leq \lambda_j;^+(a)\} \tag{9.47}$$

$$P_k;^+(a) = \max\{P_k(\lambda) | \forall \lambda_j : \lambda_j;^-(a) \leq \lambda_j \leq \lambda_j;^+(a)\} \tag{9.48}$$

For a given Markov process, for the maximum and minimum values of the probability P_k of state k at level α, one will have therefore

$$P_k;^-(a) = \min P_k \qquad P_k;^+(a) = \max P_k \tag{9.49}$$

subject to

$$(\mathbf{M} - \mathbf{I})\mathbf{P} = \mathbf{0} \qquad [1 \ 1 \ \cdots \ 1]\mathbf{P} = 1$$

$$\lambda^-(\alpha) \leq \lambda \leq \lambda^+(\alpha) \tag{9.50}$$

where $\mathbf{P}$ is the vector of instantiated state probabilities, $\mathbf{M}$ is the transpose of the classical stochastic transition matrix, $\mathbf{I}$ is the identity matrix, λ is the vector of instantiated transition rates, and $\lambda^-(\alpha)$, $\lambda^+(\cdot)$ are the extrema of the intervals of confidence of the transition rates at level α.

In constraints (9.50) we find the equation $[1 \ 1 \ \cdots \ 1]\mathbf{P} = 1$. It means that the instantiated probability values must always add up to 1, and this is the direct consequence of the probabilistic assumption of the PROFUST approach. In particular, this formulation allows the calculation of the fuzzy probability of a set of states considered together in an aggregate state G. In this case, in the objective function (9.17), one only needs to replace P_k by $P_k = \sum_{j \in G} P_j$. This is not the same as first calculating each fuzzy P_j and then adding them to give (wrong) P_k.

Figure 9.40 illustrates the Markov space state diagram, from an example found elsewhere [12] that discusses the problem of the reliability of a generating transformer substation with two identical three-phase transformer banks and a spare.

Each bank is considered to have failed if any one of the three single-phase transformers in the bank fails. The bank can be returned to service upon repair or replacement of the failed transformer. One also assumes that no further failures will occur in the bank once it has been deenergized and removed from service.

Some of the crisp rates in the original example were as follows:

Failure: 0.1/year.

Repair: 12/year (equivalent to a repair time of 1 month).

Replacement: 183/year (equivalent to an installation time of about 2 days).

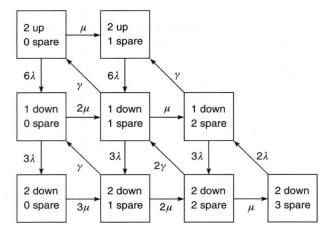

Figure 9.40 Two transformer banks with one spare; states numbered 1–9 from left to right and top to bottom.

Fuzzy triangular transition rates could instead be defined as follows:

Failure: [0.05; 0.1; 0.2]/year.

Repair: [4; 12; 20]/year (between about 18 days and 3 months; best estimate: 1 month).

Replacement: [120; 183; 183]/year (between about 2 and 3 days; best estimate: 2 days).

Figure 9.41 shows the fuzzy probabilities of states 1 and 2 and the joint consideration of $1 + 2$. Figure 9.42 shows the results for the aggregate states $1 + 2$ (both transformers up), $3 + 4 + 5$ (one up, one down), and $6 + 7 + 8 + 9$ (two down).

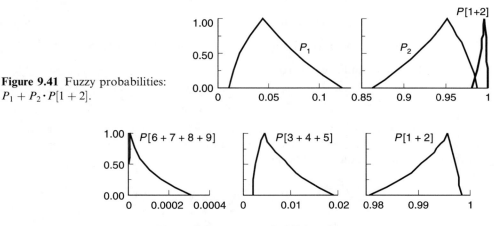

Figure 9.41 Fuzzy probabilities: $P_1 + P_2 \cdot P[1 + 2]$.

Figure 9.42 Fuzzy probabilities of aggregate states.

9.3.5. Frequency and Duration Models

Coming back to Eq. (9.41), one may see that, if the product of μ by $1/\lambda$ gives, at each level $\cdot$, an interval of confidence containing values much larger than 1, the probability of the failure state may approximately be given, even in fuzzy terms, by

$$P_f = \lambda r \qquad (9.51)$$

Based on this approximation, we will now proceed to present some approximate formulas for fuzzy calculations in parallel and serial systems. Most of these formulas are essentially the same used in classical power system reliability, arranged so that the extension principle may be respected.

Parallel Systems. Assume two components 1 and 2 in parallel, with fuzzy failure rates and repair times. The failure state is defined as the one in which both components have failed. The following fuzzy formulas apply:

$$\text{System failure rate} \rightarrow \lambda = \lambda_1 \lambda_2 (r_1 + r_2)$$

$$\text{System unavailability} \rightarrow U = \lambda_1 r_1 \lambda_2 r_2 \qquad (9.52)$$

$$\text{System mean repair time} \rightarrow r = \frac{1}{1/r_1 + 1/r_2}$$

Serial Systems. Consider two components 1 and 2 in series, with fuzzy failure rates and repair times. The failure state is defined as the union of two states in which any of the components have failed (we do not admit overlapping failures or neglect them). The following fuzzy formulas apply:

$$\text{System failure rate} \rightarrow \lambda = \lambda_1 + \lambda_2$$

$$\text{System unavailability} \rightarrow U = \lambda_1 r_1 + \lambda_2 r_2 \qquad (9.53)$$

$$\text{System mean repair time} \rightarrow \text{This is not straightforward.}$$

The crisp formula would result from

$$r = \frac{U}{\lambda} = \frac{\lambda_1}{\lambda_1 + \lambda_2} r_1 + \frac{\lambda_2}{\lambda_1 + \lambda_2} r_2$$

At every level α, we must find the upper and lower values for $r(\alpha) = [r^-(\alpha); r^+(\alpha)]$ within the bounds of the failure rates and the repair times; the lower bound, for instance, will be given by

$$r^-(\alpha) = \min \left(\frac{\lambda_1}{\lambda_1 + \lambda_2} r_1 + \frac{\lambda_2}{\lambda_1 + \lambda_2} r_2 \right)$$

subject to

$$\lambda^- \leq \lambda \leq \lambda^+ \qquad r^- \leq r \leq r^+$$

Now as all values are positive, the minimum of $r(\alpha)$ will be attained at the lower bounds of r_1 and r_2. If by chance these are equal, the values of λ_1 and λ_2 are irrelevant. If not, let us assume that $r_1^- < r_2^-$ or, which is the same, that $r_2^- = r_1^- + k$, with $k > 0$. Then,

$$r^-(\cdot) = \min \left(\frac{\lambda_1}{\lambda_1 + \lambda_2} r_1^- + \frac{\lambda_2}{\lambda_1 + \lambda_2} r_1^- + \frac{\lambda_2}{\lambda_1 + \lambda_2} k \right)$$

subject to

$$\lambda^- \leq \lambda \leq \lambda^+$$

or

$$r^-(\cdot) = \min \left(r_1^- + \frac{1}{\lambda_1/\lambda_2 + 1} k \right)$$

subject to

$$\lambda^- \leq \lambda \leq \lambda^+$$

Clearly, now, as r_1^- and k are constants, the minimum of r_α is obtained at the upper bound of λ_1 and the lower bound of λ_2. Therefore, we must write

$$r^-(\alpha) = r_1^-(\alpha) + \frac{1}{\lambda_1^+(\alpha)/\lambda_2^-(\alpha) + 1}[r_2^-(\alpha) - r_1^-(\alpha)] \qquad (9.54)$$

Similarly, whenever $r_1^+(\alpha) < r_2^+(\alpha)$,

$$r^+(\alpha) = r_1^+(\alpha) + \frac{1}{\lambda_1^-(\alpha)/\lambda_2^+(\alpha) + 1}[r_2^+(\alpha) - r_1^+(\alpha)] \qquad (9.55)$$

Equations (9.54) and (9.55) give directly the answers for the calculation of the fuzzy mean repair time under the conditions stated.

Deconvolution Approach. We have seen that the calculation of the mean repair time demands some extra effort. Formulas (9.54) and (9.55) are exact but may not be the most easy way to obtain the fuzzy value of that index. In fact, if one bears in mind that the fuzzy equation

$$U \approx \lambda r \qquad (9.56)$$

must hold (within the general approximations accepted), then an easy way of obtaining r from (9.56) is by the deconvolution of this expression. Recall that by deconvolution we mean finding the fuzzy number r that multiplied by the fuzzy λ, would give the fuzzy U; this is not the same as solving $r = U/\lambda$ by applying the rules of fuzzy arithmetic.

9.3.6. Conditional Probability: The Fuzzy Case

Within the PROFUST approach, conditional probability principles must apply. However, the operation rules are not straightforward. In general, a conditional probability approach may derive from expression such as ($\bar{B}$ is the complement of B)

$$P(A) = P(A|B)P(B) + P(A|\bar{B})P(\bar{B})$$
$$= P(A|B)P(B) + P(A|\bar{B})[1 - P(B)] \qquad (9.57)$$

To calculate the fuzzy description of $P(A)$, based on fuzzy descriptions of $P(B)$, $P(A|B)$, and $P(A|\bar{B})$, we must refer to the extension principle, as before, and the direct fuzzification of expression (9.57) is incorrect, because $P(B)$ appears twice, adding and subtracting. Therefore, the following approach can be followed, for each level α, where intervals of confidence $[P_\alpha^-; P_\alpha^+]$ are defined for $P(B)$, $P(A|B)$, and $P(A|\bar{B})$: A *lower bound* $P_\alpha^-(A)$ is defined by an NLP problem

$$\min(\cdot) = P(A|B)P(B) + P(A|B;^-)[1 - P(B)] \qquad (9.58)$$

subject to

$$P_\alpha^-(A|B) \le P(A|B) \le P_\alpha^+(A|B)$$

$$P_\alpha^-(B) \le P(B) \le P_\alpha^+(B)$$

$$P_\alpha^-(A|\bar{B}) \le P(A|\bar{B}) \le P_\alpha^+(A|\bar{B})$$

This formulation assumes that $P(A|B)$ and $P(A|\bar{B})$ are independent. From the objective function, we see that $\min \Phi \Rightarrow [P(A|B) = P_\alpha^-(A|B)] \wedge [P(A|\bar{B}) = P_\alpha^-(A|\bar{B})]$. An equivalent NLP problem is then

$$\min \Phi' = [P_\alpha^-(A|B) - P_\alpha^-(A|\bar{B})]P(B) + P_\alpha^-(A|\bar{B}) \qquad (9.59)$$

subject to

$$P_\alpha^-(B) \leq P(B) \leq P_\alpha^+(B)$$

Its solution depends on the relative values of $P_\alpha^-(A|B)$ and $P_\alpha^-(A|\bar{B})$. It may be summarized as

$$P_\alpha^-(A) = \begin{cases} [P_\alpha^-(A|B) - P_\alpha^-(A|\bar{B})]P_\alpha^-(B) + P_\alpha^-(A|\bar{B}) \Leftarrow [P_\alpha^-(A|B) \geq P_\alpha^-(A|\bar{B})] \\ [P_\alpha^-(A|B) - P_\alpha^-(A|\bar{B})]P_\alpha^+(B) + P_\alpha^-(A|\bar{B}) \Leftarrow [P_\alpha^-(A|B) < P_\alpha^-(A|\bar{B})] \end{cases} \quad (9.60)$$

An *upper bound* $P_\alpha^+(A)$ is defined in an analogous way:

$$P_\alpha^+(A) = \begin{cases} [P_\alpha^+(A|B) - P_\alpha^+(A|\bar{B})]P_\alpha^+(B) + P_\alpha^+(A|\bar{B}) \Leftarrow [P_\alpha^+(A|B) \geq P_\alpha^+(A|\bar{B})] \\ [P_\alpha^+(A|B) - P_\alpha^+(A|\bar{B})]P_\alpha^-(B) + P_\alpha^+(A|\bar{B}) \Leftarrow [P_\alpha^+(A|B) < P_\alpha^+(A|\bar{B})] \end{cases} \quad (9.61)$$

9.3.7. Minimum Cut Set Method with Fuzzy Indices

The minimum cut set method is a well-known approach to the assessment of the reliability of general systems, namely power systems. It is not an exact method, but within the relative values for component reliability indices usually found in power systems, it gives a satisfactory accuracy. The method is appropriate to find nodal, point, or local indices, and it is specially adequate for continuity assessment. It finds applications in the assessment of the design of power stations or substations and also in subtransmission or distribution systems. A relevant paper referring to this technique is, for example, ref. 13. We will discuss the consequences of having fuzzy component indices in the application of the minimum cut set method. This subject has been discussed elsewhere [14].

Here is a sketch of the minimum cut set basics:

(a) First, one must specify the node in a network where one wishes to find the continuity-related indices.

(b) Then, one must determine which component failure modes cause the supply to that node to be interrupted.

(c) Each component failure mode that causes supply interruption is called a "cut": If only one component causes the interruption, then we face a first-order cut; else when two components fail to cut the supply, we have a second-order cut; and so on.

(d) Reliability indices are calculated for each cut, applying, if necessary, formulas for components in parallel.

(e) The reliability calculations are then performed as if the cuts were independent entities in series; namely, the formulas for series of components are applied.

What difficulties might we expect from having fuzzy component reliability indices? The fact is that one expects to have, during the calculations, repeated entries of the same fuzzy indices: For instance, two second-order cuts may share one component; therefore, according to the method, its indices will contribute to the global indices in two ways. Can we directly apply the "fuzzified" formulas of the crisp method? The answer is yes (in general, with precautions). And the reason is that, in the formulas for series and for parallel components, for the failure rate and the unavailability, we only find additions and multiplications. Therefore, a variable will never divide itself or subtract from itself: Check the expressions for λ and U in (9.51) and (9.52).

We will also have a fuzzy index for the average annual energy not supplied (ENS). Recall that in a crisp model we may write a relation between the unavailability U and the average power disconnected PNS as

$$\text{ENS} = U \cdot \text{PNS} \tag{9.62}$$

In a fuzzy reliability model of type II, PNS is a crisp value. In the minimum cut set method, each cut is associated with a crisp value of load disconnected: It does not matter if the frequency or the repair times vary, the consequence of the outage is losing a certain amount of power. This is valid even if we consider a model with partial cuts, meaning that a certain outage mode may lead to only the partial loss of the supply at a certain load point. Therefore, ENS comes just from the multiplication of a crisp number PNS by a fuzzy number U.

An Example. As an example, see Figure 9.43, where on the left an electrical scheme is drawn; on the right lies an equivalent block diagram adequate for reliability modeling. Assume that the failure rates and repair times of blocks 1–4 are given through specifying a best estimate and a range of uncertainty; this allows one to define triangular fuzzy numbers to represent those uncertain indices.

Load L can be assumed as given by a load curve such as 10%, 20 MW; 40%, 15 MW; and 100%, 10 MW. This diagram is depicted in Figure 9.44. Assuming full redundancy of the parallel branches, the average load disconnected is 12.5 MW. It is a crisp value.

The fuzzy reliability indices of components 1–4 are summarized in Table 9.4.

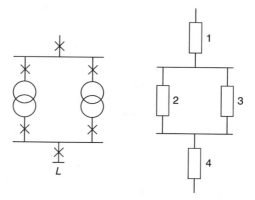

Figure 9.43 Electric diagram and equivalent block diagram for a reliability study.

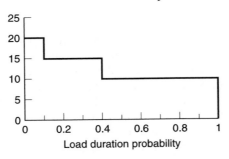

Figure 9.44 Load duration curve for the numerical example.

TABLE 9.4 FUZZY RELIABILITY INDICES

Component	Failure Rate (per year)		Mean Repair Time (h)	
	Uncertainty Range	Best Estimate	Uncertainty Range	Best Estimate
1	0.008–0.015	0.01	10–30	20
2	0.001–0.003	0.002	190–220	200
3	0.001–0.003	0.002	220–280	250
4	0.008–0.0015	0.01	10–30	20

In this example and referring to the block diagram in Figure 9.43, we find two first-order cuts (blocks 1 and 4) and one second-order cut (overlapping outage of blocks 2 and 3). The calculations proceed as follows:

1. Calculate the fuzzy λ and U indices of the second-order cut using Eq. (9.52).
2. Calculate the fuzzy r index of the second-order cut using the deconvolution approach of Eq. (9.56).
3. Calculate the fuzzy λ and U indices for point L, applying Eq. (9.53) to the series of the two first-order cuts and second-order cut.
4. Calculate the fuzzy r index for load point L using the deconvolution approach of Eq. (9.56).

In Figure 9.45 one may find a representation of the possibility distributions for the failure rate, the mean repair time, and the unavailability for load point L. Observe how these distributions are still very approximately triangular as a consequence of all fuzzy data being also triangular.

A more detailed (with more information content) calculation can be made by load level. In Figure 9.46, the possibility distributions of the mean annual energy not supplied are represented for each load level multiplied by each probability of occurrence. We may also see the possibility distribution for the total average ENS at L.

Therefore, with the precautions described, finding a fuzzy version of a minimum cut set method is quite straightforward. When it is possible to accept a triangular approximation for the successive results of the calculations, all one needs to do is to calculate three crisp models (for the minimum and maximum values of the uncertainty ranges and for the best estimate). The most relevant contribution of

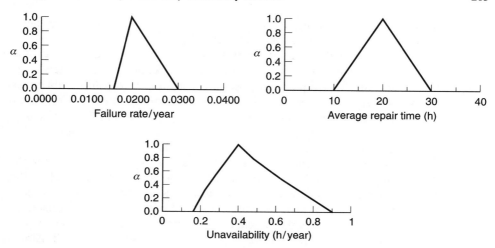

Figure 9.45 Fuzzy indices at load L, in the example.

having a fuzzy model is, in this case, the way the results are displayed and the useful interpretations one may derive from them. They underline the correspondence between the degree of uncertainty in the actual data and the resulting uncertainty range in the system indices.

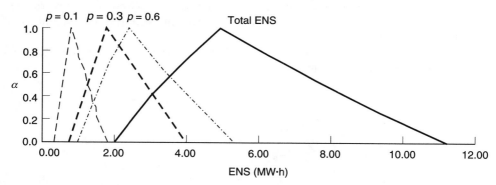

Figure 9.46 Possibility distributions at each load level with probability p and the total value for the fuzzy index average value of annual energy not supplied at L.

9.4. TYPE III POWER SYSTEM FUZZY RELIABILITY MODELS

9.4.1. Introduction

Type III models for fuzzy reliability calculations in power systems deal with decision-making problems. Types II and I have so far required only arithmetic operations with fuzzy numbers; but in type III calculations we will also be concerned with ranking fuzzy options. These models are typical of planning activities and combine together technical and economical factors.

We have already seen how to perform a technical evaluation of a solution when some data are uncertain. But surely many of the economic parameters that one must take in account for planning purposes are also subject to considerable uncertainties. Among these, we may refer to the cost of equipment, to interest rates, and, of course, to the costs of the power disconnected or the energy not supplied. Each of these types of data has a role in quantifying the feasibility of a system design.

In planning, when deciding among alternatives, the investment criterion is usually subject to minimization. But if the costs are uncertain (as they really are, depending on so many uncontrolled factors, including international politics) would not one wish to know how good a decision is, that is, to know whether a ranking of alternatives would not change if other values for the costs were considered?

Figure 9.47 displays the typical decision-making dilemma dealing with an uncertain definition of the merits of an alternative. Assume that a decision maker is comparing the cost of two alternative plans denoted by A and B. If crisp calculations have been made, it would seem that plan A should be preferred because it is less expensive. However, a fuzzy modeling of the uncertainties in data could lead to the situation depicted in Figure 9.47. At once the decision maker realizes that, below α_1, A may exceed B, and this becomes obvious below α_2; that is, if the decision maker fears uncertainties in data larger than the ones defined at the α_1-level, then he or she must weigh the risk of making a choice that may be regretted later. On the other hand, if uncertainties could be constrained to remain above α_1, then the decision in favor of plan A would be the best whatever instantiation of the uncertain values would occur.

Incidentally, we are again considering a fuzzy interpretation of the concepts of robustness and exposure, already discussed.

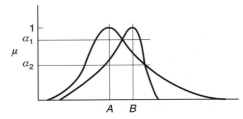

Figure 9.47 Plan A is estimated to cost less than plan B but perhaps is riskier.

9.4.2. Ranking Fuzzy Numbers

In many cases, one not only wishes to have information on the compared risks among decisions but also needs to rank fuzzy numbers according to some preference criterion. There are several proposals for obtaining a linear ordering of fuzzy numbers. There is no unique order that may be defined without ambiguity. One process that has been proposed is referred to as total distance criterion (TDC) in ref. 15 and removal in ref. 16. Another process, very much used in fuzzy control applications, is based on the concept of center of mass of the fuzzy number. For simplicity, we will refer only to fuzzy numbers with the support on the positive axis.

Removal, or TDC. The membership function of a fuzzy number defined over $x \in R$ may be divided into two parts, namely a function $L(x)$ on the left side of the mode m of the fuzzy number and a function $R(x)$ on the right side. Let the support of this number be $[a, b]$ so that $L(a) = 0$ and $R(b) = 0$ and $L(m) = R(m) = 1$. The TDC, or removal, of a nonnegative fuzzy number with respect to 0 is given by the average of the sum of the areas between the 0 axis and both L and R functions such as

$$\text{Rem}(A) = \frac{1}{2}\left(m - \int_a^m L(x)\,dx + m + \int_m^b R(x)\,dx\right) \tag{9.63a}$$

Rearranging yields

$$\text{Rem}(A) = m + \frac{1}{2}\left(\int_m^b R(x)\,dx - \int_a^m L(x)\,dx\right) \tag{9.63b}$$

Equivalent to this form is the TDC definition, which is based on the integration along the membership axis of the arithmetic mean value of the α-level sets of the fuzzy set considered:

$$\text{TDC} = \int_0^1 \frac{R^{-1}(\alpha) + L^{-1}(\alpha)}{2}\,d\alpha$$

where R^{-1} and L^{-1} correspond to the inverse functions of R and L. It is easy to show that under these definitions the TDC and Rem indices are equivalent.

For a triangular fuzzy number A, represented by the triplet (a, m, b), one has

$$\text{Rem}(A) = \tfrac{1}{2}(a + 2m + b) \tag{9.64}$$

Inspired by the Hurwicz criterion for ranking intervals, a generalized Hurwicz criterion (GHC) for ranking fuzzy sets has recently been proposed [17]. It has a clear relation to the Rem criterion or TDC and depends on a parameter δ that is defined externally:

$$\text{GHC} = \int_0^1 [\delta R^{-1}(\alpha) + (1 - \delta)L^{-1}(\alpha)]\,d\alpha$$

For $\delta = \tfrac{1}{2}$, GDC becomes equal to the Rem criterion or TDC. The interest of this formulation is that different types of risk aversion by decision makers may be translated into the ranking of fuzzy numbers by adequately setting the δ parameter, giving more or less weight to the left or right behavior of the fuzzy number.

Center of Mass G. The center of mass G of a fuzzy number A with membership function $\mu_A(x)$ is given by

$$G(A) = \frac{\displaystyle\int_a^b x\mu(x)\,dx}{\displaystyle\int_a^b \mu(x)\,dx} \tag{9.65}$$

For the triangular fuzzy number $A(a, m, b)$, one has

$$G(A) = \tfrac{1}{3}(a + m + b) \tag{9.66}$$

One practical difference between the removal criterion and the center of mass for establishing a ranking of fuzzy numbers is that the center of mass gives slightly more weight to what happens at low membership values than does the removal. As a numerical example, try to compare the removal and the center-of-mass values for the triangular numbers $A(1, 10, 10)$ and $B(5, 5, 14)$, depicted in Figure 9.48, we reach the following contradictory conclusions:

$$\mathrm{Rem}(A) = \frac{31}{4} > \mathrm{Rem}(B) = \frac{29}{4} \Rightarrow B < A$$

$$G(A) = \frac{21}{3} < G(B) = \frac{24}{3} \Rightarrow A < B$$

which means that the center-of-mass technique is more risk adverse than the removal technique.

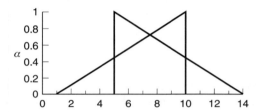

Figure 9.48 Two triangular fuzzy numbers $A(1, 10, 10)$ and $B(5, 5, 14)$.

9.4.3. Ranking Decisions: The Regret Criterion

In decision-making processes, in many problem contexts, ranking solutions according to their attribute values may not be the more close procedure to human thinking. One important question that direct ranking may not answer is: "How much will one regret by choosing one alternative instead of another?"

In problems with crisp objective function values, the ordering of solutions according to those values provides the answer to that question. However, when the alternatives are characterized by fuzzy objective function values, it is not so simple: Checking Figure 9.48, we may see that at some α-levels, B is always less than A, but at lower α-levels, B may exceed A.

With no loss of generality, let us assume a context of minimization. We now define a fuzzy number $\mathrm{Reg}(A|B)$, meaning the regret felt by a decision maker when he or she chooses A instead of B; using the interval of confidence notation,

$$\mathrm{Reg}_\alpha(A|B) = [\{\max\{0; (A_\alpha^- - B_\alpha^+)\} \max\{0; (A_\alpha^+ - B_\alpha^-)\}] \tag{9.67}$$

Of course, if one chooses A and A results better than B, no regret will be felt; that is why expression (9.67) does not allow for negative regrets.

In Figure 9.49, we find two fuzzy numbers, precisely $\mathrm{Reg}(A|B)$ and $\mathrm{Reg}(B|A)$. The decision dilemma is now deciding between these two regrets. In principle, the

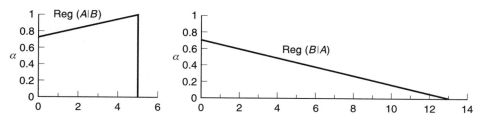

Figure 9.49 Fuzzy regrets, referring to Figure 9.45. Left: regret for choosing A instead of B; right: regret for choosing B instead of A.

smaller the potential regret, the better. We can now go back to the removal or the center-of-mass techniques and rank the two fuzzy regrets. In the case of Figure 9.49, both criteria place $\text{Reg}(B|A) < \text{Reg}(A|B)$; therefore, decision B would perhaps be preferable to decision A, in this pairwise comparison.

The relevant point to be realized is that decisions about fuzzy-valued solutions must be weighed in pairwise comparisons, while solution fuzzy values may be ranked based on individual defuzzified values.

Both the G and the Rem criteria can establish a ranking of fuzzy values, because they exhibit a transitive property: If A ranks better than B and B better than C, then A ranks better than C. Preferences based on this ranking are therefore transitive. It remains to be proved for the new regret ranking criterion that it leads also to a transitivity in preferences.

9.4.4. Fuzzy Dominance

We now discuss the concept of dominance in multicriteria problems. This discussion is relevant due to the fact that in many power system planning problems, typically two criteria (at least) are weighed against each other: costs (investment + operation) versus reliability.

The concept of an efficient, dominant, or Pareto-optimal set of solutions in a multicriteria environment is well understood and is at the root of most decision-aiding models. In short, a solution is said to dominate another in a pairwise comparison if it is at least as good in all criteria and strictly better in at least one of them (this means that a rational decision maker would always prefer the dominant to the dominated solution). Those solutions for which one cannot find any others that dominate them form, precisely, the nondominated set.

In Figure 9.50 we have represented three solutions in a two-attribute space—to give it a power system flavor. We identified the objectives as minimizing cost and minimizing the power not supplied. We can easily see that C is dominated by A and B, and these are nondominated, in a context of minimization.

The fuzzy case is somewhat more complicated. In Figure 9.51 we have three solutions, A, B, and C, with fuzzy values on each attribute (say, a fuzzy cost and a fuzzy power not supplied as a consequence of uncertainties in data). To make it more evident, assume that this fuzziness is just represented by intervals at $\alpha = 0$ (giving a

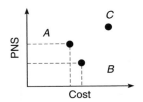

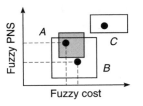

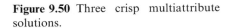

Figure 9.50 Three crisp multiattribute solutions.

Figure 9.51 Three fuzzy multiattribute solutions.

rectangle) and a point at $\alpha = 1$, and suppose that these points would represent the center of mass of each fuzzy solution.

If a decision maker evaluates the solutions only based on the centers of mass, he or she will reach the same conclusions as in the previous example. However, the fuzzy representation shows that the uncertainties in data do not allow such a simplistic conclusion: For example, at level $\alpha = 0$, there is the possibility that in some instances of A and B, B may dominate A (while still both dominate C in every case). There are some instances of B, at some α-levels close to zero, at which in fact B would be the absolute best solution.

Definitions of dominance between solutions with several degrees of strengths are therefore possible. The definition of a nondominated border is no longer a trivial task. Given a border of fuzzy solutions in a multiple attribute space, we can nevertheless apply, for each criterion, expression (9.67) in successive pairwise comparisons of neighboring solutions and determine, under the regret criterion, if they are in fact mutually nondominated. Trade-offs based on defuzzified regrets may also be calculated.

This illustrates one of the important contributions of fuzzy modeling to the understanding of the problems dealt with in decision making and in power systems in particular. In contrast with crisp models, where the concepts of solutions and decisions are not distinguished, *in fuzzy models there is a clear separation between the (fuzzy) solution values and the decisions* that must be made. This distinction derives from the evaluation of risk, resulting from the simultaneous consideration of multiple possible scenarios possible by the fuzzy representation of uncertainty.

9.4.5. Application of Type III Calculations: Switching Device Location

We will illustrate the type III models with some examples from distribution networks. Let us consider the problem of deciding, in a distribution network and based on reliability criteria, where best to locate switching devices that may help in isolating faulty areas and recovering the supply to unaffected areas.

In a single-criterion model, one could calculate a cost for the interruptions of supply and then add it to the investment cost, with the objective of minimizing a global cost investment and reliability. In many countries it has been possible to establish values for different types of consumers for the kilowatt-hours not delivered and the kilowatts disconnected. For the purpose of being added to investment costs, these values must be adequately capitalized through the adoption of suitable rates and year horizon. If g and h are such values, then the cost D of an outage mode could be approximately evaluated, in general terms, by

$$D = \lambda(gL + hE) \quad \text{or} \quad D = \lambda(gL + hrL) \quad \text{or} \quad D = (g\lambda + hU)L \qquad (9.68)$$

with L the average load disconnected and E the average annual energy not supplied.

Of course, these g and h parameters are strong candidates to be taken as fuzzy numbers. The diversity in values resulting from different surveys is just a confirmation of this need of explicitly recognizing the fuzziness inherent in the ideas of the cost of kilowatts disconnected or kilowatt-hours not supplied. But we may also assume that the reliability indices λ and U may be fuzzy as well as the load L. The cost D of an outage mode therefore becomes fuzzy. And if one includes in the model the uncertainty about equipment costs, then we have an optimization model with all numerical data described by fuzzy numbers!

9.4.6. Radial Networks

A crisp model to search for the "best" solution in the switching device location problem in radial networks may be found elsewhere [18]. The underlying concept is dynamic programming. The network is consistently bottom-up reduced until a final "optimal" decision may be taken, allowing a top-down process that sets up the options of which type of switching equipment to install at every branch.

The trade-off is the following: More devices will reduce the average power disconnected and the average annual energy not supplied, resulting from outages in the lines; but more devices also add some more unreliability to the network, and besides they have a cost. This algorithm may be used with fuzzy data [19], provided that only sums and multiplications are kept in the formulas. We assume that the radial network does not have access to alternative supply sources, other than the substation. We reproduce below the updated formulas that allow the determination of an optimal solution. In these formulas, the following definitions hold:

L_j	Average load dependent on branch j	May be fuzzy
$\lambda_j(\lambda_p)$	Failure rate of branch j (of device type p)	May be fuzzy
$U_j(U_p)$	Unavailability of branch j (of device type p)	May be fuzzy
S_p	Switching time for device type p	May be fuzzy
V_p	Cost of device type p	May be fuzzy

The types of devices considered are as follows:

o No device installed

s A manually operated isolator, with switching time S_s

t A telecommanded switch, with switching time S_t

a An IAR (interrupter autorecloser) or a fuse, with switching open time taken as $S_a = 0$

In the case of an active fault occurring beyond an IAR, the action of the IAR can be described like this:

- Whenever a fault occurs, it is cleared at the substation and the whole network is disconnected from the supply
- Activated by a relay sensitive to the no-voltage condition, the IAR opens.
- At the substation, within a prefixed timing, supply is connected again under the command of a reclosing relay.
- With voltage restored, the IAR closes.
- At the substation, the circuit breaker opens again, because the fault is being fed again.
- Again under the no-voltage condition, the IAR opens and *remains* open.
- At the substation, the circuit breaker recloses again, this time successfully.

Therefore, the action of the IAR is, in fact, to isolate faulty subnetworks and allow service restoration for the rest of the system, from a reliability analysis point of view, we will deal with an IAR as with a fuse.

Contribution of a Single Branch. Consider the possibility of including a device in a branch j knowing that on the path between this branch and the root (the substation) the nearest manual switch is located in branch n. The nearest telecontrolled switch is in branch m and the nearest recloser interrupter (IAR) is at branch k (see Fig. 9.52). Furthermore, one has $1 < k < m < n < j$ in the sense that k is closer to the root, labeled 1, than m, and so on.

Then, the following decisions may be considered, leading to the contribution of branch j to form a global composite investment-reliability cost D_{jx} (where $x \in \{o, s, t, a\}$):

i. No device at j:

$$D_{jo} = g\lambda_j L_k + h\lambda_j S_t L_k + h\lambda_j S_s L_m + hU_j L_n$$

ii. A manual switch at j:

$$D_{js} = g\lambda_j L_k + h\lambda_j S_t L_k + h\lambda_j S_s L_m + hU_j L_j + g\lambda_s L_k + h\lambda_s S_t L_k$$
$$+ h\lambda_s S_s L_m + hU_s L_n + V_s$$

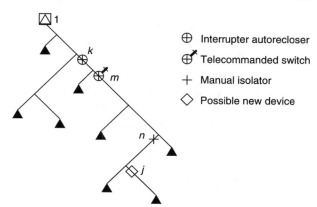

Figure 9.52 Radial network: deciding about installing or not a device at j, with the nearest devices at n, m, and k, in the path between the branch j and the root 1.

iii. A telecontrolled switch at j:

$$D_{jt} = g\lambda_j L_k + h\lambda_j S_t L_k + hU_j L_j + g\lambda_t L_k + h\lambda_t S_t L_k + h\lambda_t S_s L_m + hU_t L_n + V_t$$

iv. An automatic device (such as an IAR) at j:

$$D_{ja} = g\lambda_j L_j + hU_j L_j + g\lambda_a L_k + h\lambda_a S_t L_k + h\lambda_a S_s L_m + hU_a L_n + V_a$$

If no manual switch exists at location n, for any n such that $m < n < j$, the formulas are transformed into:

i'. $D'_{jo} = D_{jo} - h\lambda_j S_s L_m$

ii'. $D'_{js} = D_{js} - h\lambda_s S_s L_m$

iii'. $D'_{jt} = D_{jt} - h\lambda_t S_s L_m$

iv'. $D'_{ja} = D_{ja} - h\lambda_a S_s L_m$

If no telecontrolled switch exists at m, for any m such that $k < m < n$, the formulas are transformed such that $L_m = L_k$ and still:

i''. $D''_{jo} = D_{jo} - h\lambda_j S_t L_k$

ii''. $D''_{js} = D_{js} - h\lambda_j S_t L_k - h\lambda_s S_t L_k$

iii''. $D''_{jt} = D_{jt} - h\lambda_t S_t L_k$

iv''. $D''_{ja} = D_{ja} - h\lambda_a S_t L_k$

If neither a telecontrolled switch exists at m nor a manual switch exists at location n, for any n and m such that $k < m < n < j$, the formulas are transformed such that $L_m = L_k$ and still:

i'''. $D'''_{jo} = D'_{jo} - h\lambda_j S_t L_k$

ii'''. $D'''_{js} = D'_{js} - h\lambda_j S_t L_k - h\lambda_s S_t L_k$

iii'''. $D'''_{jt} = D'_{jt} - h\lambda_t S_t L_k$

iv'''$_{ja}$. $D'''_{ja} = D'_{ja} - h\lambda_a S_t L_k$

These "formula transformations" do not mean that we will perform actual subtractions; instead, they mean that the terms affected with a negative sign are withdrawn from the formulas (this is important in a fuzzy context).

General Algorithm. The algorithm for finding the optimal solution can be defined by a function OPTIM being recursively called, beginning at the first branch (called root) emerging from the substation:

```
function OPTIM (branch)
```

```
if branch has descendants then
```

$$\text{OPTIM} = \min\{D_{\text{branch},x}\} \mid x \cdot \{o, s, t, a\}\} + \sum \text{OPTIM (descendants of branch)};$$

```
else branch has no descendants and
```

$$\text{OPTIM} = \min\{D_{\text{branch},x} \mid x \cdot \{o, s, t, a\}\};$$

```
main
```

```
Solution=OPTIM(root).
```

In a fuzzy context, the min operation in the OPTIM function requires a ranking of the alternatives. The min expression must receive this interpretation: It denotes the preferred alternative under some ranking criterion, including the regret criterion.

The above algorithm has been built on the basis of the average power not supplied, L; this has been evaluated at every branch i of the network as L_i. The algorithm may also be adopted if, instead of L_i, one would have in all expressions the value NC_i, number of customers dependent on branch i. In this case, g and h should be interpreted as weights. The effect of the unreliability of a branch i would be measured by $(g \cdot NC_i + h \cdot rNC_i)$, which would make it related to the SAIFI and SAIDI (and CAIDI) indices adopted by many utilities as indicators of distribution network reliability level. Here SAIFI is system average interruption frequency index, or average annual number of outages; SAIDI is system average interruption duration index, or average customer minutes outage per year; and CAIDI is customer average interruption duration index, or average outage duration.

9.4.7. Location of the Open Point of an Interconnection

The model for an interconnection line between two substations ("left" and "right") is represented in Figure 9.53. The line is divided into several sections, or branches, having radial subnetworks in the dependence of each. A branch k (possibly with a dependent radial subnetwork) is represented by values such as average load L_k, failure rate λ_k, and unavailability U_k, or number of customers NC_k. For the sake of generality, we assume that different kinds of consumers may be attached to different branches, so that for branch k specific values g_k and h_k apply.

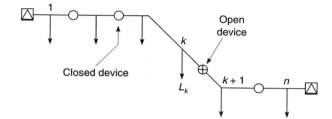

Figure 9.53 Model for an interconnection line.

The nodes between these branches are taken as candidate locations for switching devices. One open point must be chosen either at one of the substations or somewhere in the middle. Although not mandatory, we will consider (without loss of generality) that the operation of this normally open device can be telecommanded, while the operation of the other switching devices will remain manual.

Following a fault, switching actions can be divided into two kinds: the ones that isolate the faulty zone and the ones that allow the restart of supply to some branches (even by an alternative path). One must consider switching times S_m for manual action and S_t for telecontrolled switching. The combination of these times will be referred to as the average switching time S.

Assume that all the nodes have a switch taken as completely reliable and one needs to decide which one to keep in the open state during normal operation. If the switch between branches k and $k+1$ is open (the others remaining closed), in a first approach one can say that the equivalent reliability cost D' is given by

$$D' = \sum_{i=1}^{k} \lambda_i \sum_{i=1}^{k} (g_i + h_i S)L_i + \sum_{i=1}^{k} h_i U_i L_i + \sum_{i=k+1}^{n} \lambda_i \sum_{i=k+1}^{n} (g_i + h_i S)L_i + \sum_{i=k+1}^{n} h_i U_i L_i$$

(9.69)

The first term expresses the impact of power disconnected from a fault occurring in any branch from 1 to k and the energy not delivered during switching actions. The second term expresses the energy not delivered to consumers connected to the faulted section of the network, after being isolated and during repair time. The third and fourth terms express the same for faults occurring at the other side of the open point.

An optimization problem in terms of outage costs would then be to find k such that this expression leads to a minimum. It can be shown that this problem can be transformed by introducing binary variables x, y giving as a basic formulation

$$\text{Minimize} \quad \Phi = \sum_{i=1}^{n} x_i \lambda_i \sum_{i=1}^{n} (g_i + h_i S) x_i L_i$$

$$+ \sum_{i=k+1}^{n} y_i \lambda_i \sum_{i=k+1}^{n} (g_i + h_i S) y_i L_i + \sum_{i=1}^{n} h_i U_i L_i$$

$$\text{Subject to} \quad x_i + y_i = 1 \quad i = 1, \ldots, n \qquad (9.70)$$

$$x_i \geq x_j \quad \forall i < j$$

$$x_i, y_i \in \{0, 1\}, \quad i = 1, \ldots, n$$

Fixing $x_i = 1$ means that branch i will be connected to the "left" supplying source; fixing $y_i = 1$ means that branch i will be connected to the "right" supplying source.

The first restriction means that a branch can only be connected to one of the substations. The second restriction means that a branch cannot be connected to the left substation unless all other branches between it and the substation are too. The same condition for the right substation is implicit.

This is an integer quadratic problem with linear constraints; for crisp data, its optimum either is at the boundaries ($x_1 = 0$ or $y_n = 0$) or may be found by iterative bisection (two integer optima may exist; therefore the first and second best solutions should be searched for). A more detailed description can be found in ref. 19, namely in taking care of what happens at branches adjacent to the substations.

As the number of possible open-switch locations is not likely to be very high in most practical cases, one may also solve the problem for every location and compare the values thus obtained. The parameters in the objective function of (9.70) can be taken as fuzzy, and therefore, in order to reach a decision, a comparison between fuzzy numbers must be done following any of the rules previously described.

9.4.8. Optimal Number of Devices in an Interconnection Line

Finding an optimal balance between reliability and the investment to be made in switching devices is a more complicated problem (it can be seen as combinatorial). Assume that we have, as an objective function, the sum of the cost (which may be fuzzy) of a specific configuration of switching devices installed in the line and the "cost" of the unreliability associated with that configuration. We propose a "greedy" heuristic that tries to reach a good compromise. In short:

```
1. Assume that all branches are separated by switches.

2. REPEAT:

   2.1. Locate the two best solutions for the open point of
        the interconnection.

   2.2. Evaluate the marginal advantage of suppressing each
        switch, one at a time. If no benefit is found, STOP and
        take the present solution as the best one.

   2.3. Suppress the device with greatest negative marginal
        value, and build a single branch from the two pre-
        viously separated ones.

   UNTIL no devices can be eliminated.
```

Eliminating a device between branches j and $j+1$ leads to saving its cost ($V_{j/j+1}$) at the expense of reducing reliability. In the general case, it can be shown that this means adding $\Delta D' = h_j U_{j+1} L_j + h_{j+1} U_j L_{j+1}$ and reducing $V_{j/j+1}$ to the objective function (through a deconvolution operation if one is dealing with fuzzy numbers).

In a complete formulation of this problem, one must pay attention to some small details such as suppressing the location of the normally open device as a candidate location (cost evaluations must be based on the second-best solution) or suppressing a device adjacent to a substation. Details may be found in the literature [19].

The evaluation of the marginal advantage of suppressing an open point is based on the comparison of the values of the objective function for the two cases (with and without a device). This comparison may be made following the rules described previously. An application of these techniques has been reported elsewhere [19].

9.5. CONCLUSIONS

This chapter has given a summary of features of fuzzy reliability models. One would like to offer the following conclusions:

- Reliability models deal with a certain type of uncertainty and try to extract coherent conclusions from the assumed random behavior of many characteristics of the problems, namely in power systems.
- Reliability models use, as raw material, a diversified set of data that are themselves contaminated with another type of uncertainty, which may get an adequate representation with fuzzy sets.
- It is, however, possible to build hybrid models that respect the probabilistic assumptions and at the same time include the "doubts" about the actual value of indices or other data.
- The fuzzification of traditional reliability models must be done with care, and under the rules established by the extension principle, otherwise one may get an undesirable propagation of errors and erroneous results with "too large" possibility distributions.
- This fuzzification leads to the extension of the concept of component indices (e.g., failure rate, mean repair time) and system indices (e.g., power disconnected, energy not supplied) to fuzzy indices, with their associated possibility distributions.
- If the uncertainty in the raw data is described by means of triangular (or trapezoidal) fuzzy numbers, then the results may be expected as being also triangular, with a fairly good approximation.
- The additional computing effort for many fuzzy reliability models is not excessive and may be negligible in many cases.

- The fuzzy model outputs include the results for the crisp data case (which means that the results of a classical study are directly obtained, even if one has only available a fuzzy tool).
- But the fuzzy models have the property of supplying to the system analyst a considerable amount of extra information, only extracted from classical models with a considerable amount of work.
- One of the important conclusions that fuzzy reliability models allow one to reach is about the direct link between a specific range of uncertainty in data and the corresponding range in the results.
- Fuzzy reliability models may therefore also be seen as an economical way of performing some types of sensitivity analyses.
- This allowed the definition of new reliability indices—exposure, robustness—that extract their meaning from an interpretation of the membership functions or possibility distribution results.
- Furthermore, fuzzy reliability models may also be combined or included in decision models, namely in power system planning.
- A ranking of decisions is no longer possible in classical terms, because of the fuzziness of attribute or objective function values.
- But new concepts like "regret" may be adopted in pairwise comparisons to assess the risks of opting for one or the other of two alternatives.

The subjects dealt with in this chapter are far from being thoroughly investigated; this suggests many avenues to be explored in the future.

References

[1] V. Miranda, "Fuzzy Reliability Analysis of Power Systems," in *Proceedings of the Twelfth Power System Computation Conference PSCC'96*, Vol. 1, Dresden, Germany, Aug. 1996, pp. 558–566.

[2] V. Miranda and L. M. V. G. Pinto, "A Model to Consider Uncertainties in Power System Operation" (in Portuguese), in *Proceedings of SNPTEE*, Rio de Janeiro, Brazil, Oct. 1991.

[3] V. Miranda, M. Matos, and J. T. Saraiva, "Fuzzy Load Flow—New Algorithms Incorporating Uncertain Generation and Load Representation," in *Proceedings of the Tenth Power System Computation Conference PSCC'90*, Graz, Austria, Aug. 1990.

[4] V. Miranda, "Fuzzy Flows in Gas and Electricity Networks," in *Proceedings of the European Simulation Multiconference—ESM'91*, Erik Mosekilde (ed.), Technical University of Denmark, Copenhagen, Denmark, 1991.

[5] V. Miranda, M. Matos, and J. T. Saraiva, "Fuzzy Flows in Linear Networks with Corrections for Some Dependencies in Nodal Uncertainties," *Investigacão Operational/Operations Research Review*, Vol. 12, No. 2, 1992.

[6] J. T. Saraiva, V. Miranda, and L. M. V. G. Pinto, "Impact on Some Planning Decisions from a Fuzzy Modeling of Power Systems," *IEEE Transactions on Power Systems*, Vol. 9, No. 2, 1994.

[7] H. M. Merril and A. J. Wood, "Risk and Uncertainty in Power System Planning," paper presented at the Tenth Power System Computation Conference PSCC'90, Graz, Austria, Aug. 1990.

[8] J. T. Saraiva, V. Miranda, and L. M. V. G. Pinto, "Generation-Transmission Power System Reliability Evaluation by Monte Carlo Simulation Assuming a Fuzzy Load Description," *IEEE Transactions on Power Systems*, Vol. 11, No. 2, 1996.

[9] Electric Power Research Institute (EPRI), "Users Guide to the Over/Under Capacity Planning Model," Report EA-1117, EPRI, Palo Alto, CA, July 1979.

[10] K.-Y. Cai, "Fuzzy Variables as a Basic for a Theory of Fuzzy Reliability in the Possibility Context," *Fuzzy Sets and Systems*, Vol. 42, 1991, pp. 145–172.

[11] A. Kaufmann and M. M. Gupta, *Fuzzy Mathematical Models in Engineering and Management Sciences*, North-Holland, Amsterdam, 1988.

[12] R. Billinton and R. N. Allan, *Reliability Evaluation of Power Systems*, Pitman, London, 1984.

[13] R. N. Allan, R. Billinton, and M. F. Oliveira, "An Efficient Algorithm for Deducing the Minimal Cuts and the Reliability Indices of a General Network Configuration," *IEEE Transactions on Reliability*, Vol. R-25, 1976, pp. 226–233.

[14] V. Miranda, "Reliability Calculations in Power Systems with Fuzzy Indices" (in Portuguese), in *Proceedings of the First Spanish-Portuguese Workshop in Electrical Engineering*, Vigo, Spain, July 1990.

[15] R. R. Yager, "A Procedure for Ordering Fuzzy Subsets over the Unit Interval," *Information Science*, Vol. 50, 1981, pp. 143–151.

[16] A. Kaufmann and M. M. Gupta, *Fuzzy Mathematical Models in Engineering and Management Sciences*, North-Holland, Amsterdam 1988.

[17] J. J. Saade, "A Unifying Approach to Defuzzification and Comparison of the Outputs of Fuzzy Controllers," *IEEE Transactions on Fuzzy Systems*, Vol. 4, No. 3, 1996, 227–237.

[18] V. Miranda, M. F. Oliveira, and A. A. Vale, "Optimal Location of Switching Devices in Distribution Networks," in *Proceedings of CIRED 83*, Vol. 6, AIM, Liège, Belgium, 1983.

[19] V. Miranda and M. Matos, "Distribution System Planning with Fuzzy Models and Techniques," *Proceedings of CIRED 89*, Vol. 6, IEE, Brighton, United Kingdom, 1989.

H. Matsumoto
Hitachi Research Laboratories
832-2 Horiguchi, Hitachinaka-shi
Ibaraki-ken
312 Japan

Chapter 10

Operation Support Expert System for Startup Schedule Optimization in Fossil Power Plants

10.1 INTRODUCTION

Recent growth in both capacity and complexity of fossil power plants demands not only advances in machine design but also in plant operational strategies. As the power generation share of nuclear plants increases, fossil plants undergo heavy-duty operation cycles involving frequent startup and shutdown and fast load following. Furthermore, the fossil plants are required to meet stringent environmental regulations regarding air pollution [1]. Consequently, reducing the frequency of startup and achieving safe and economical operation under environmental regulations and reliable machine life management have become important issues [2].

Optimizing the operation of generators is required not only for economical reasons but also for the associated environmental advantages, especially with the growth in capacity and complexity of gas and steam turbine combined-cycle power plants. The latter are receiving a good deal of attention because they are characterized by higher efficiency, faster startup, lower environmental pollution, shorter construction time, and smaller site space requirements. As a result, finding ways of reducing startup time without shortening machine life while satisfying environmental constraints is an important area of plant control.

This chapter describes the application of fuzzy theory in fossil power plant operation and two expert systems included in an operational support system for a conventional one-through-type supercritical pressure fossil power plant [3] and a gas and steam turbine combined-cycle power plant [4]. One of the expert systems used for the conventional plant performs the functions of startup scheduling optimization

280

and on-line schedule modification using fuzzy reasoning to assist plant operators. The second expert system for the combined-cycle power plant optimizes the startup schedule to provide quick and economical plant startup while observing the requirements of NO_x emission regulations and reliable machine life management.

10.2. EXPECTATIONS OF FUZZY THEORY IN PLANT STARTUP

10.2.1. Plant Startup Problem

The startup schedule optimization problem can be defined as that of minimizing the time needed for startup while meeting constraints imposed on machine life consumption, safety of operation, and environmental regulations. The faster the startup, the greater is the stress developed in the steam turbine and the boiler structure and the more serious is its effect on reducing equipment service life. Moreover, faster gas turbine startup is associated with higher NO_x emissions. The operation of fossil power plants requires not only minimizing the startup time but also maximizing the flexibility of its operation. The startup schedule should be executed as accurately as possible even when unexpected events take place during startup sessions. In such a case, the startup schedule should be modified flexibly and appropriately to accommodate plant dynamics. The complexity and nonlinearity of plant dynamics make methods such as nonlinear programming too time consuming and impractical for getting the optimal solution.

10.2.2. State of the Art

To reduce startup time, state-of-the-art techniques are used in off-line scheduling and on-line control. In the off-line scheduling method developed for simple steam cycle plants, a startup schedule is generated utilizing a metal matching chart using steam and metal temperatures just before startup [5]. Conventional control systems of combined-cycle plants basically involve the same method.

In the on-line control method developed for startup of steam turbines, speed-up and load-up rates are periodically optimized using the estimated rotor stresses from measured temperatures of turbine casings [6]. To improve the accuracy and the stability of stress control, a turbine automatic startup system (HRASS) was developed [7]. This system uses a plant dynamics model that consists of an autoregressive boiler model and a turbine heat transfer model.

It has been observed that using off-line scheduling methods, startup is achieved but takes longer. However, using on-line control methods reduces startup time but completion is less reliable. It is important to address these two problems simultaneously. But this is difficult to realize within the scope of conventional computer control technology, because of the complicated physical relationships among process variables in power plants. We focus our attention on a synergetic approach that combines dynamics simulation and knowledge engineering to provide for an improved startup schedule.

10.2.3. The Expert Way of Thinking

The startup schedule optimization problem is considered along with turbine stress dynamics. Figure 10.1 shows two example relationships between the startup schedule and stress dynamics. After the boiler is started, the turbine is rolled off, and its speed increases toward the rated speed. Subsequently, the generator is parallelled to the power system and loaded up toward a target level to complete the startup session. The solid curve represents the case after optimization. Optimality is achieved when the startup schedule is achieved in the least possible time without violating any stress limit operational constraint.

If a stress pattern is prespecified, then an experienced operator or plant engineer can make appropriate schedule modifications. Such actions may include reducing the time during which stress margins are exceeded. In general, knowledge of elements of the plant process dynamics such as turbine and boiler stress and NO_x emission allows experienced operators to modify the startup schedule easily. They can recognize temporal and spatial causal patterns and employ a reasoning process that is independent from the process dynamic. The reasoning focuses on causes and their effects. They also evaluate the process dynamics qualitatively and recognize the relation between the startup schedule and its effect on process dynamics. Then, they can modify the schedule appropriately and efficiently using qualitative reasoning based on long-term experience. The question now is how to translate this into an automatic computer-based procedure.

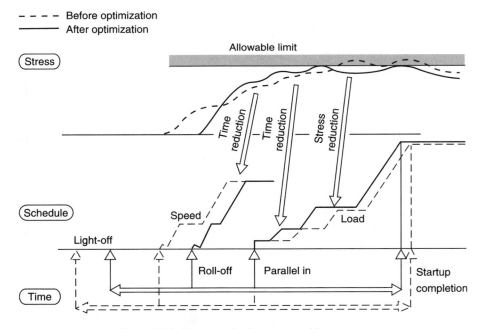

Figure 10.1 Startup schedule and turbine stress.

10.2.4. Expectations of Fuzzy Theory as a New Paradigm

Conventional optimization methods such as nonlinear programming in conjunction with dynamics simulation are too time consuming and impractical for obtaining optimum solutions by ordinary computers because plant operations are dynamic, large scale, and nonlinear. Moreover, conventional mathematical methods are not able to handle subjective information effectively. Therefore, an integration of dynamics simulation and modeling of human expertise such as heuristic, fuzzy, and qualitative thinking processes (as shown in Figure 10.2) is expected to provide faster solutions.

In this approach, fuzzy theory provides many possibilities as a new paradigm. Fuzzy theory models the thinking process of experts starting with qualitatively evaluating process variables and proceeding to the final decision on quantitative schedule modification. Qualitative evaluation of process variables such as *positive big margin* and *negative small margin* can be put in a computer model using fuzzy representations utilizing the concept of membership functions. All process variables needed to be evaluated are obtained through quantitative dynamics simulations. Feature extraction of a process response pattern such as *positive small margin after negative big margin* and *negative small margin after negative medium margins* can be represented by fuzzy logic.

The method of startup schedule modification relies on fragmentary knowledge such as *if stress response pattern is A, then turbine startup schedule parameter X should be modified* and *if the NOX response pattern is B, then gas turbine startup parameter Y should be modified*, which can be represented by fuzzy rules. Qualitative schedule modifications resulting from different aspects of startup dynamics can be simulated

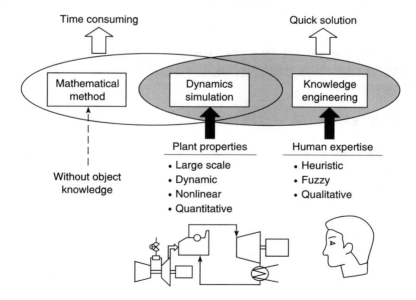

Figure 10.2 System framework.

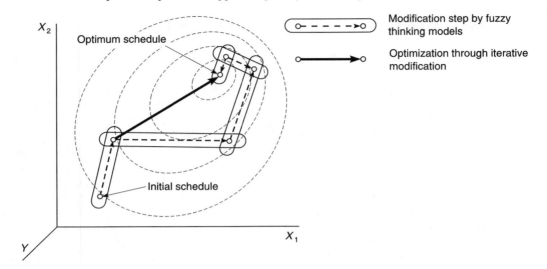

Figure 10.3 Optimization by expert qualitative knowledge.

using fuzzy reasoning. The final decision on quantitative schedule modification rate for each schedule parameter can be realized using defuzzification algorithms. Figure 10.3 shows the optimization process. The fuzzy models of expertise described earlier are used to modify the startup schedule. The optimal schedule will be found in more than one modification step using the models in combination with dynamics simulation. The most remarkable feature of this approach is that a quantitative optimum solution can be expected from a combination of qualitative knowledge and dynamics simulations.

10.3. OPERATION SUPPORT AND STARTUP SCHEDULE OPTIMIZATION FOR CONVENTIONAL FOSSIL POWER PLANTS

10.3.1. System Design Concept

Objectives. This expert system is designed to realize the following four major objectives:

- Reduce startup time.
- Reduce schedule execution error.
- Improve operational flexibility.
- Reduce the operators' work burden.

The startup schedule should be as short as possible and executed accurately. It should allow flexible modification when unexpected schedule errors and contingencies take place during startup sessions.

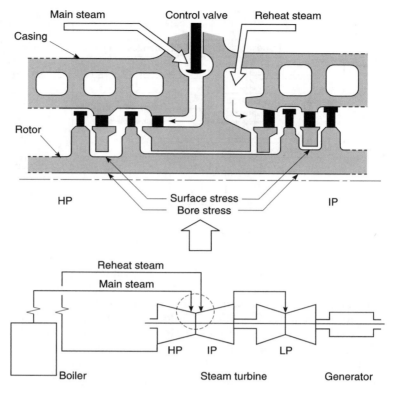

Figure 10.4 Locations of critical stress. HP, IP, LP: high, intermediate, and low pressure.

Location of Critical Stress. Turbine thermal stress is an especially decisive factor in reducing startup time. Stress occurs on the rotor surface and rotor bore at the labyrinth packings behind the first stage of the high-pressure turbine (HPT) and intermediate-pressure turbine (IPT), as shown in Figure 10.4. In these areas, a heavy heat flux appears from the surface toward the bore during the startup period as a result of leaked steam having a high temperature and high velocity. Thus, the faster the startup, the greater the stress developed and the more serious is its effect on shortening turbine service life. It is very important to know and manage stresses associated with startup schedules.

10.3.2. Functional Structure of the System

Outline of the System. The functional structure of this system is shown in Figure 10.5. A schedule that can minimize the time required for startup and deviation of the startup completion time is created from the startup scheduling part. The operators can be supported in making startup schedules and guided in operations to cope with unexpected schedule errors and stress monitoring and prediction control

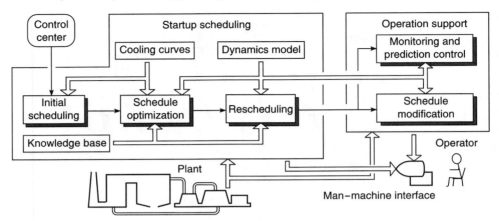

Figure 10.5 Functional structure of the system.

by the operation support part. The deviation represents a time difference between the time established by the control center and the actual startup completion time. A mathematical model of plant dynamics is introduced to reflect the uncertainties that depend on the initial conditions. Following the initiation of plant startup according to the schedule, the operators can be guided to modify the schedule appropriately when required after the turbine speed and load are altered by unexpected causes. In the major parts of the system, fuzzy reasoning is introduced to represent knowledge and the way experts think and to utilize them to improve computational abilities.

Startup Scheduling. The startup scheduling part consists of functions for initial scheduling prior to schedule optimization, during schedule optimization, and rescheduling just before turbine roll-off. Quantitative startup characteristics of the plant that are assumed to be started in accordance with a given schedule can be calculated based on a dynamic model. The initial schedule is created according to the initial conditions of the plant using cooling curves. The initial schedule can speed the convergence of schedule optimization. In the optimization process through fuzzy reasoning, turbine thermal stresses obtained from the dynamic model are qualitatively evaluated, and the schedule is modified. After the modification, startup dynamics are calculated and evaluated again. Through this iterative process, fast convergence to the optimum schedule makes it possible to predict startup characteristics accurately using the dynamics model involving complex relations. In the same manner, fuzzy reasoning is applied to the rescheduling function.

Operation Support. The turbine stresses are periodically predicted and controlled by the function for monitoring and prediction control during the actual startup process. The turbine speed or load is held when an unexpected excess stress appears. The unexpected speed or load hold is released after the stress returns to within allowable limits. Next, the startup schedule is appropriately modified through the function for schedule modification. The modification aims to minimize the error of startup completion time between the original schedule and the schedule after the unexpected hold occurred. The modification rate is determined using a stress prediction algorithm using the dynamic model.

Man–Machine Interface. The system is provided with a user-friendly man–machine interface using interactive graphic displays. The displays support the operators with information showing trends of controlled process variables, the processes of schedule optimization, schedule execution, schedule modification, and schedule optimization rules stored in the knowledge base.

10.3.3. Algorithm of Startup Scheduling

Initial Scheduling. As the initial schedule is used for the first step in the iterative optimization process, it should be as close to the optimum point as possible. Here, the plant stoppage period designates the time difference between the parallel in time established by the control center and the parallel off time when the generator is separated from the power system. Subsequently, the two specific startup modes are selected from four modes that are predefined in accordance with the length of the stoppage period. The predefined four modes are the cold mode, warm mode, hot mode, and very hot mode. Schedule parameters that define each startup pattern from the boiler light-off to the load-up completion are given by the following:

t_1	Boiler startup time (from light-off to roll-off)	t_4	Third speed hold time
		t_5	First load hold time
t_2	First speed hold time	t_6	Second load hold time
t_3	Second speed hold time	t_7	Third load hold time

In addition, speed-up and load-up rates are specified. The two steps of the schedule parameters that belong to the selected two modes are linearly interpolated or extrapolated according to the stoppage period. After determining the initial schedule parameter, the boiler light-off time can be calculated. Next, initial process values corresponding to boiler light-off time are estimated using the cooling curves. The initial values used here are superheater outlet steam temperature, reheater outlet steam temperature, main pipe metal temperature, reheat pipe metal temperature, water wall inlet liquid temperature, economizer inlet liquid temperature, high-pressure turbine metal temperature, intermediate-pressure turbine metal temperature, and main steam pressure.

Estimation of the initial process values is as follows. The process values at the present time T_{op} when the startup schedule is going to be made are estimated using the cooling curves equation, given by

$$T_{op} = (T_{SD} - T_A) \exp\left(-\frac{t_{op} - t_{SD}}{t_C}\right) + T_A \qquad (10.1)$$

where T_{SD} = temperature at parallel off time
t_{op} = present time
t_C = cooling time constants
T_A = ambient temperature
t_{SD} = parallel off time

The process values at the boiler light-off T_{IG} are estimated as

$$T_{IG} = (T_{SD} - T_A) \exp\left(-\frac{t_{IG} - t_{SD}}{t_C}\right) + T_A \qquad (10.2)$$

The process values at the boiler light-off T_{IG} can also be estimated without consideration of T_{SD} and t_{SD} using Eqs. (10.1) and (10.2) as

$$T_{IG} = (T_{OP} - T_A) \exp\left(-\frac{t_{IG} - t_{OP}}{t_C}\right) + T_A \qquad (10.3)$$

The main steam pressure is obtained in the same manner as described by Eq. (10.3).

Schedule Optimization. Figure 10.6 shows the procedure for schedule optimization. Schedule parameters t_i are time periods that define the startup pattern from boiler light-off to startup completion. The optimum schedule defines the shortest schedule that makes the summation of t_i a minimum, keeping the turbine stresses within the allowable limits during the whole startup process. Conventional optimization algorithms that need hundreds of iterative calculations for optimization of seven parameters are not practical. A heuristic optimization applying fuzzy reasoning is attempted to obtain fast convergence. By knowing the stress pattern that

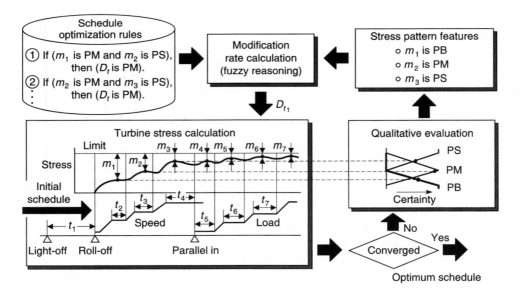

Figure 10.6 Schedule optimization procedure.

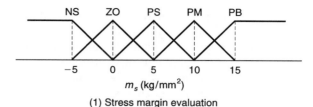

(1) Stress margin evaluation

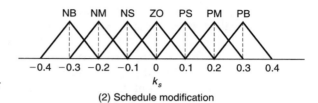

Figure 10.7 Membership functions for schedule optimization rules.

(2) Schedule modification

develops in the metal of the turbine rotors, experienced operators can modify the startup schedule easily. Such heuristic knowledge is used for the schedule optimization process.

The startup process is divided into seven time sections, as shown in Figure 10.6. The minimum stress margins m_i for each section are calculated. The margins are used to obtain the stress pattern. The schedule optimization rules give fragmentary knowledge for determining schedule modification rates in relation to stress patterns. This system deals with a case in which the parallel in time is established by the control center. But this system can also easily respond to other time commands, such as turbine roll-off time, startup completion time, and so on.

Corresponding to the initial schedule, the turbine stress is obtained from the dynamic model. The stress margins are evaluated qualitatively using membership functions to extract qualitative features of the stress pattern. There are 150 rules used for determining the qualitative modification rates of the schedule parameters D_{ti} in qualitative relations with $MS(j)$ or $MB(j)$. Here, $MS(j)$ and $MB(j)$ are defined as smaller stress margins of the rotor surfaces and bores between the high-pressure turbine and the intermediate-pressure turbine, respectively. Next, each stress margin is classified into a qualitative expression using membership functions. These qualitative features of stress patterns are compared to the schedule optimization rules by fuzzy reasoning. An example of the rules is as follows:

IF (MS(5) is PB and MS(6) is PM)

THEN (D_{t4} is NM and D_{t5} is NM and D_{t6} is NS)

The symbols PB, PM, PS, ZO, NS, NM, NB are attached to the membership functions shown in Figure 10.7. They have the following qualitative meanings:

PB Positive big   NB Negative big

PM Positive medium   NM Negative medium

PS Positive small   NS Negative small

ZO Zero

Following pattern matching of the qualitative features of stress pattern and schedule optimization rules, qualitative schedule modification rates are defined using membership functions. The plant startup characteristics, according to the modified schedule, can be predicted again through the dynamics model. Through the iterative manner described earlier, fast convergence to the optimum schedule can be expected. Convergence is asserted when the reduction rate of startup time by schedule modification becomes sufficiently small.

Table 10.1 shows some example schedule optimization rules that are actually used by the fuzzy reasoning. The rules in the table are used for determining the modification rates of three schedule parameters in qualitative relations with two stress margins. Taking rule 51, for instance, if the stress margins MS(5) and MS(6) are PB and NS, respectively then the modification rates D_{t4}, D_{t5}, and D_{t6} are defined as NM, NS, and PS, respectively (Figure 10.8).

The boiler light-off time may be changed in accordance with each iterative optimization process. The new initial process values T_{IG} at the new boiler light-off time t'_{IG} are estimated in the same manner as that of Eq. (10.3). The values are represented as

$$T_{IG'} = [T_{OP} - T_A] \exp\left(-\frac{t_{IG'} - t_{OP}}{t_C}\right) + T_A \qquad (10.4)$$

In addition to hold times of speed and load, changing rates of speed and load can also be introduced as extended modification parameters for the schedule optimization already described.

Thermal stresses of the turbine rotors are generated by temperature differences between the rotor surface and the bore. If the rotor temperature distributions are known, the stresses can be calculated. But it is difficult to measure the rotor temperatures directly, and the corresponding casing temperatures do not always coincide with the rotor temperatures. Furthermore, even measurement of the steam temperature behind the turbine first stage is difficult because of its mechanical structure. Considering these difficulties in measuring the temperatures, a new calculation method for the dynamics model of the rotor stresses is developed. Figure 10.9 shows the main steps in the calculation. Both current stresses and future stresses are obtained through the dynamic model. A detailed calculation procedure of the model can be found in ref. 7.

Rescheduling. The optimum schedule is intended to keep the stresses within their specified limits. Even then, it is possible that the actual process dynamics deviate from their predicted values. Because of these deviations, an unscheduled speed hold or load hold might be caused by excess stresses generated after the turbine is rolled off. In order to avoid this phenomenon and to maintain the optimum schedule, the startup schedule is reviewed by the function for rescheduling just before turbine roll-off. The schedule optimization procedure described earlier is adapted to this function in the same manner using current status as initial conditions for the dynamic model.

TABLE 10.1 EXAMPLE SCHEDULE OPTIMIZATION RULES

Stress margin $M_s(5)$	NS					ZO					PS					PM					PB				
$M_s(6)$	NS	ZO	PS	PM	PB	NS	ZO	PS	PM	PB	NS	ZO	PS	PM	PB	NS	ZO	PS	PM	PB	NS	ZO	PS	PM	PB
Schedule modification rate D_{14}	PS	PS	PS	PS	PS	ZO	ZO	ZO	ZO	ZO	ZO	ZO	ZO	ZO	ZO	NS	ZO	NS	NS	NS	NM	NM	NM	NM	PB
D_{15}	PB	PM	PM	PM	PS	PS	PS	PS	PS	ZO	PS	ZO	ZO	ZO	NS	ZO	NS	NS	NS	NM	NS	NM	NM	NM	PM
D_{16}	PB	PM	PM	PS	PS	PM	PM	PS	PS	ZO	PM	PS	PS	ZO	ZO	PS	PS	NS	ZO	NS	PS	ZO	ZO	NS	NS
Rule no.	11	12	13	14	15	21	22	23	24	25	31	32	33	34	35	41	42	43	44	45	51	52	53	54	55

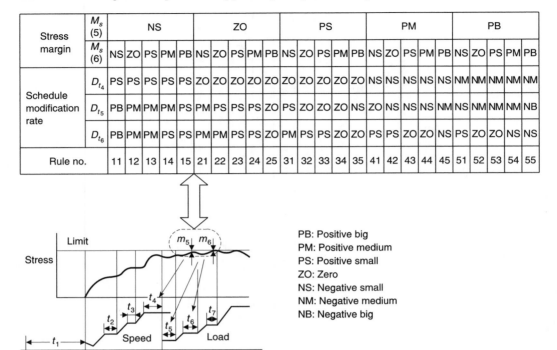

Stress margin	M_s (5)	NS					ZO					PS					PM					PB				
	M_s (6)	NS	ZO	PS	PM	PB	NS	ZO	PS	PM	PB	NS	ZO	PS	PM	PB	NS	ZO	PS	PM	PB	NS	ZO	PS	PM	PB
Schedule modification rate	D_{t_4}	PS	PS	PS	PS	PS	ZO	ZO	ZO	ZO	ZO	ZO	ZO	ZO	ZO	ZO	NS	NS	NS	NS	NS	NM	NM	NM	NM	NM
	D_{t_5}	PB	PM	PM	PM	PS	PM	PS	PS	PS	ZO	PS	ZO	ZO	ZO	NS	ZO	NS	NS	NS	NM	NS	NM	NM	NM	NB
	D_{t_6}	PB	PM	PM	PS	PS	PM	PM	PS	PS	ZO	PM	PS	PS	ZO	ZO	PS	PS	ZO	ZO	NS	PS	ZO	ZO	NS	NS
Rule no.		11	12	13	14	15	21	22	23	24	25	31	32	33	34	35	41	42	43	44	45	51	52	53	54	55

PB: Positive big
PM: Positive medium
PS: Positive small
ZO: Zero
NS: Negative small
NM: Negative medium
NB: Negative big

Figure 10.8 Example schedule optimization rules.

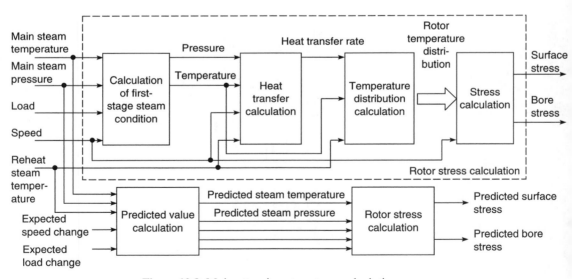

Figure 10.9 Main steps in rotor stress calculation.

10.3.4. Operation Support Algorithm

Stress Monitoring and Control. In the stress monitoring and control function, either speed or load is controlled along with the startup schedule, as well as simultaneous stress monitoring of predicted stresses. When one of the predicted stresses reaches a limit, speed or load-hold control is applied. In this way, the predictive stress calculation is effective in avoiding excess stresses. During the unexpected speed or load hold, stresses are reduced enough to allow raising speed or load again. Subsequently, the startup schedule is appropriately modified by the function for schedule modification.

Schedule Modification. An unexpected hold of speed or load is caused by anomalous conditions of the turbine such as differential expansion, vibratory level, or condenser vacuum level. After recovering from the anomalous condition, the original schedule with the unexpected hold is released and appropriately modified according to the following steps:

1. Stress dynamics are predicted with the assumption that the turbine speed or load is raised along the modified schedule which neglects the unexpected hold time t_{UH}, as shown by the dotted line t_{UH} in Figure 10.10. The minimum stress margin S_1 during the whole startup process is calculated.
2. Stress dynamics are predicted assuming that the startup session is continued along the schedule which is equally shifted as the unexpected hold time t_{UH}, as shown by the broken line. The minimum stress margin S_2 is calculated.
3. If the stress margin S_1 is positive, the schedule used in step 1 is adopted for actual use. On the other hand, when S_1 is negative, the startup schedule is

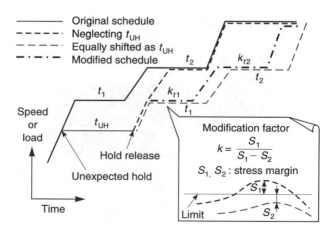

Figure 10.10 Schedule modification.

created using a modification factor k obtained by interpolation as follows:

$$k = \frac{S_1}{S_1 - S_2} \qquad k > 0 \qquad (10.5)$$

The results of the modification are shown by the chain line in Figure 10.10. The modified schedule can be expected to make the stress margin positive and a minimum. Consequently, the error of the startup completion time is expected to be a minimum.

10.3.5. Hardware System

The expert system is installed in hardware, as one of the applications of an expert system shell, HITREX, the Hitachi Tree-Based Real-time Expert System [8]. HITREX is specially tailored for fossil power plant engineers, with full-system support for easy installation and automatic linkage to process variables. The system ensures efficient plant operation in a wide range of system configurations from large to small. It consists of a control computer (HIDIC-90/25) for on-line use and an engineering workstation (ES330) for off-line maintenance of the knowledge base (Figure 10.11). Implementation of the fuzzy rules is fast and easy through an on-screen matrix in which logical relations can be simply defined. Input variables used in fuzzy reasoning can be easily assigned and modified through interactive screen dialogue.

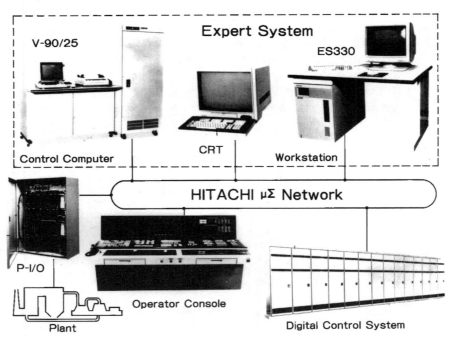

Figure 10.11 Power plant control system.

10.3.6. Simulation Study

Hardware System for Simulation. A simulator developed to verify and evaluate the system consists of a control computer (HIDIC-90/25) and an engineering workstation (ES330) that is the same as the expert system mentioned earlier. Causes of schedule errors can be entered into the dynamics model installed in the control computer through the keyboard of the engineering workstation.

Schedule Optimization and Rescheduling. A typical example of the simulated schedule optimization process is shown in Figure 10.12. In this figure, numbers correspond to the iteration steps, where the seventh is the converged optimum schedule. Time required for startup is prolonged by 24 min compared with the initial schedule in order to reduce the stresses to within limits. The modification parameters adopted in this case are hold times of speed and load. In the figure, only the larger stresses between HPT and IPT are shown, but actually, in the simulation, four stresses are taken into consideration.

Figure 10.13 shows a simulated rescheduling process just before the turbine roll-off. The time delay of startup completion is reduced to 4 min even though the time for turbine roll-off and parallel in time are delayed by 5 and 7 min, respectively. Simulation results prove that fuzzy reasoning works effectively for schedule optimization and accurate startup.

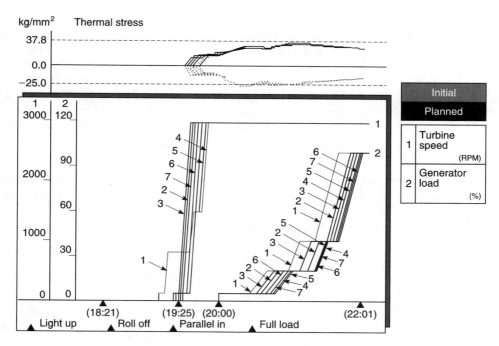

Figure 10.12 Schedule optimization process.

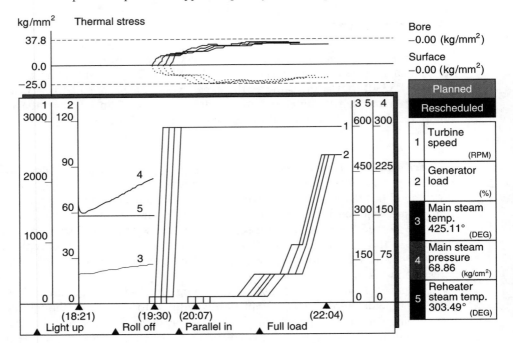

Figure 10.13 Rescheduling process.

Schedule Modification. Figure 10.14 shows a simulated schedule modification when the parallel in time is delayed by an anomalous step change in the main steam temperature by 10 min after the turbine reaches rated speed. The time delay of startup completion is reduced to only 3 min even though the parallel in time is delayed by 10 min, because of some stress margins that exist during the loading-up process. Simulation results under the assumption that deviations from normal temperatures of the main steam, the reheat steam, and the turbine metal exist prove that fuzzy reasoning works effectively for rescheduling and accurate startup.

Time Reduction for Startup. A comparison between conventional methods and the proposed method for the required startup time evaluated under the same stress level developed at the maximum point is necessary. Here, the startup time is evaluated using the following steps:

1. The maximum stress level is obtained through a simulation of the conventional method, which introduces a metal matching chart to create startup schedules as described in ref. 5.
2. The optimum schedule is obtained through a simulation of the proposed method. In this step, speed and load-hold times are modified as schedule parameters under a stress limit is the same level as the maximum stress obtained in step 1.

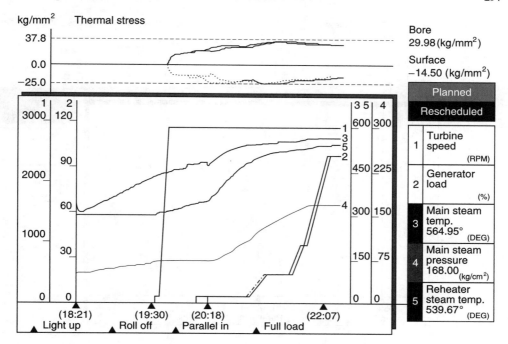

Figure 10.14 Schedule modification.

3. The optimum schedule is obtained through the extended proposed method. In this step, in addition to speed and load-hold times, changing rates for speed and load are also modified under the same stress limit used in step 2.

As for the first step, a startup schedule and stress characteristics are obtained through a simulation of the conventional method used in a startup case for the warm mode (Figure 10.15). In this case, the maximum stresses of the rotor surface and the rotor bore are −31.29 and 36.54 kg/mm, respectively. The required startup time is 287 min.

As for the next step, the optimum schedule is obtained by the proposed method, which modifies speed and load-hold times as schedule parameters when the stress limit for the rotor surface is −31.29 kg/mm (Figure 10.16). But the rotor bore stress is limited to 37.8 kg/mm, which is the result of an elasticity limit of the rotor metal. The boiler light-off time in this case is reduced by 7 min, and the startup completion time is also cut by 21 min. As a result, the proposed method reduces the startup time by 14 min compared with the conventional method, yielding a reduction rate of 5%.

Even in this case, some stress margins still remain during the loading-up part. It seems that there is some room for time reduction in this part if the changing rates of speed and load are introduced as modification parameters for schedule optimization. Figure 10.17 shows the improved startup characteristics along with the extended proposed method. Making good use of the stress limits, the startup time is reduced by 54 min. The time reduction rate compared to the conventional method is 23.7%.

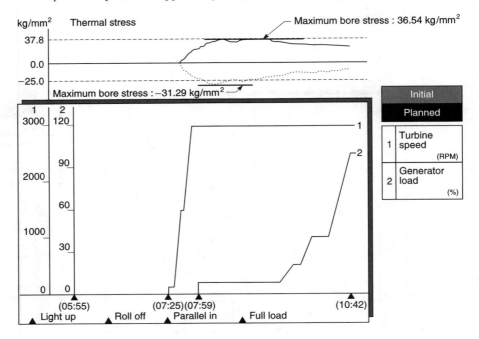

Figure 10.15 Schedule obtained through conventional method.

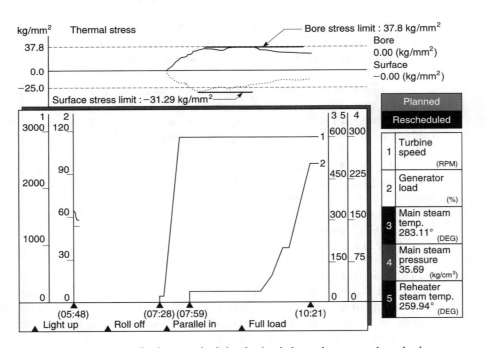

Figure 10.16 Optimum schedule obtained through proposed method.

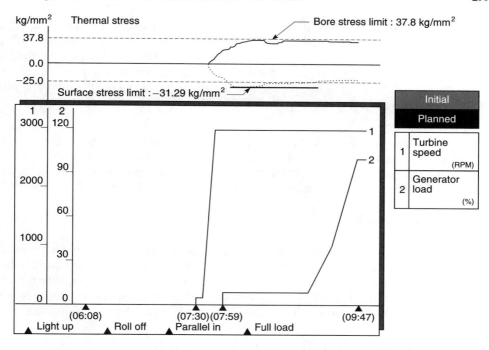

Figure 10.17 Optimum schedule obtained through extended proposed method.

Simulations of the schedule optimization in hot and cold modes are made in the same manner as in the warm mode described earlier. The simulation results are shown in Table 10.2. The startup times are reduced considerably by the extended

TABLE 10.2 COMPARISON OF REQUIRED STARTUP TIME

Startup Mode (Stoppage Period)	Hot (8 h)	Warm (32 h)	Cold (150 h)
Startup time (min)			
Conventional method, t_a	164	287	565
Proposed method, t_b	164	273	488
Extended proposed method, t_c	150	219	399

Time reduction rate TRR (%)

Proposed method:

$$\text{TRR} = \frac{t_a - t_b}{t_a} \times 100$$

	0.0	5.0	14.0

Extended proposed method:

$$\text{TRR} = \frac{t_a - t_c}{t_a} \times 100$$

	10.4	23.7	29.4

proposed method for all startup modes. But no reduction is obtained for the hot mode by the proposed method. The reason is that a suboptimum schedule is obtained for this mode by the conventional method. Both times for light-off and startup completion are not modified even though the time for roll-off and parallel in are slightly modified. The simulated plant has a one-through-type supercritical pressure and a rated capacity of 1000 MW.

Computation Time. The computation time depends on the simulated time range of the startup process and the number of iterations of the schedule optimization. Using a control computer (HIDIC-V90125) in the simulation study, the CPU time is about 70 s for the hot mode, 170 s for the warm mode, and 320 s for the cold mode.

10.3.7. Conclusions

An operation support expert system based on on-line dynamics simulation and fuzzy reasoning for startup schedule optimization in fossil power plants is discussed. This system introduces a function for iterative schedule optimization with fuzzy reasoning and a plant dynamics model and a function for accommodating unexpected schedule errors. Simulations with this system show the following:

- Plants were started quickly and accurately.
- Operators prepared the optimum startup schedule, even under unusual process conditions.
- Operators' work was lessened in monitoring and executing startup schedules.
- Operators were provided with functions for on-line assessment and off-line learning of the plant dynamics.
- Very fast convergence of the schedule optimization existed compared with conventional operations research techniques.

This system has advanced capabilities for early prediction of schedule errors caused by excess stresses and for providing operators with guidance for plant preventive operation. This system is believed to be capable of meeting new needs in connection with medium-load operations of fossil power plants.

10.4. STARTUP SCHEDULE OPTIMIZATION FOR COMBINED-CYCLE POWER PLANT

10.4.1. System Design Concept

Operational Constraints. There are many operational constraints on combined-cycle power plants. Stresses that develop in the steam turbine and heat recovery steam generator (HRSG) and NO_x from the HRSG are especially decisive factors for reducing startup time (Figure 10.18).

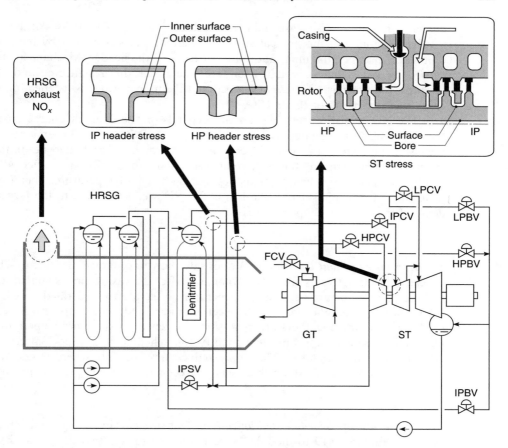

FCV: Fuel control valve
HPCV: High-pressure control valve
IPCV: Intermediate-pressure control valve
LPCV: Low-pressure control valve

HPBV: High-pressure bypass valve
IPBV: Intermediate-pressure bypass valve
LPBV: Low-pressure bypass valve
IPSV: Intermadiate-pressure stop valve

Figure 10.18 Plant configuration and operational constraints.

The turbine stresses occur on the rotor surfaces and rotor bores at the labyrinth packings behind the first stage of the HPT and IPT. In these areas, a heavy heat flux appears from the surface toward the bore during startup as a result of leaked steam having a high temperature and high velocity. The stresses of the steam turbine involve thermal and mechanical stresses. Thermal stress is caused by temperature difference between the surface and bore of the rotor. Mechanical stress is caused by centrifugal force in the rotor. Therefore, the faster the turbine startup, the greater the stress developed and the more serious is its effect on shortening the service life of the turbine.

The HRSG stresses take place on the inner and outer surfaces at the headers of the high-pressure superheater outlet and the reheater outlet due to the fast and wide changes of steam temperature and pressure during startup. The stresses of the HRSG involve thermal and mechanical stresses. Thermal stress is caused by temperature differences between the inner and outer surfaces of the header. Mechanical stress is caused by the steam pressure. Therefore, the faster the gas turbine and HRSG startups, the greater the stress developed and the more serious is its effect on shortening the service life of the HRSG.

The NO_x emission rate of the plant depends on the dynamic characteristics of the gas turbine and the denitrifier installed in the HRSG. The NO_x emissions from the gas turbine are a function of speed and load, which generally depend on the fuel input rate. Furthermore, the NO_x reduction rate of the denitrifier is affected by its catalyst temperature, which strongly depends on flue gas temperature. Therefore, the faster the gas turbine startup, the less is the NO_x reduction due to the larger time delay of the catalyst temperature rise.

Expertise in Plant Operation. We consider how to optimize the startup schedule in relation to the stresses and NO_x dynamics. After the gas turbine is ignited, the gas turbine, the steam turbine, and the generator accelerate toward rated speed. Next, the generator is parallelled with the electric system and loaded up toward a target load level, completing the startup session. In general, experienced operators and plant engineers can appropriately modify the schedule if a stress pattern and a NO_x pattern are available. They may reduce time periods where stress or NO_x margins exist and may extend them where excess stress or NO_x levels appear.

Strategies of Schedule Optimization. This system introduces the strategies shown in Figure 10.19 to modify the schedule parameters that designate the plant startup pattern. The startup schedules for the gas turbine and steam turbine are modified by predicting stresses of the steam turbine and HRSG and the NO_x through the dynamic models.

The schedule of the gas turbine is modified by the gas turbine primal scheduling (GTPS) module, which considers NO_x dynamics, and gas turbine global tuning (GTGT), which considers the stress dynamics. The schedule for the steam turbine is modified by the steam turbine primal scheduling (STPS) module, which considers the stress dynamics, and steam turbine local tuning (STLT), which considers NO_x dynamics. The GTPS primary function is to modify the gas turbine startup schedule, because of the direct effects on the NO_x emission. The STPS primary function is to modify the steam turbine startup schedule, because of the direct effects on the stresses. The GTGT tunes the gas turbine startup schedule globally, because of the indirect effects on the stresses through the heat exchangers in the HRSG. The STLT tunes the steam turbine startup schedule locally, because of the indirect effects on the NO_x emission through the heat exchangers. Earlier experience with plant operations is introduced in these strategies using fuzzy reasoning to represent knowledge and the way experts think and to utilize the strategies to reach the optimum schedule quickly.

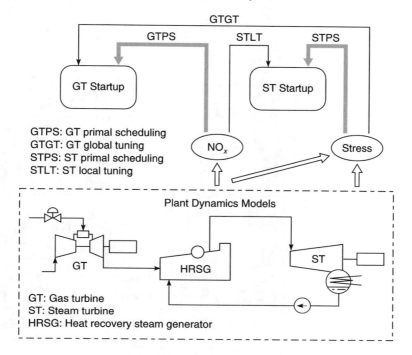

Figure 10.19 Strategies of schedule optimization.

10.4.2. Schedule Optimization Algorithm

Functional Structure of the System. The functional structure of the system and configuration of the plant are shown in Figure 10.20. The system consists of functions for startup schedule optimization and machine control. The latter executes schedules generated by the schedule optimization function. The main process of schedule optimization consists of functions for schedule assumption, dynamic models, dynamic evaluation, convergence judgment, schedule modification, schedule presentation, and optimum schedule setting.

The system is studied on a plant that has three pressure-staged reheat-type units with 235.7 MW rated capacity. The denitrifier is installed between two high-pressure evaporators. The speed and load of the gas turbine are regulated by the fuel control valve (FCV). The steam turbine consists of an HPT, an IPT, and a low-pressure turbine (LPT). The steam inlet flows to the turbines are regulated by a high-pressure control valve (HPCV), an intermediate-pressure control valve (IPCV), and a low-pressure control valve (LPCV). The turbine bypass steam flows are regulated by a high-pressure bypass valve (HPBV), an intermediate-pressure bypass valve (IPBV), and a low-pressure bypass valve (LPBV). The temperature of the steam from the intermediate-pressure drum is adjusted to the HPT outlet steam temperature by an intermediate-pressure stop valve (IPSV).

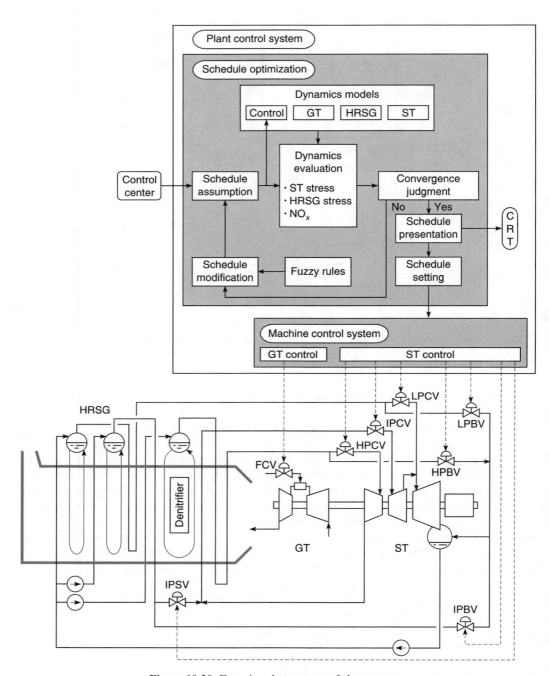

Figure 10.20 Functional structure of the system.

The control center of the power system specifies a startup completion time and a target load level as a startup demand. The schedule assumption module generates an initial schedule along with the command and sends it to the dynamic models. In the dynamic models, the plant startup process is simulated as if the gas turbine, the HRSG, and the steam turbine were started in accordance with the initial schedule. The stresses of the steam turbine and the HRSG and NO_x are also simulated with the model and forwarded to the dynamic evaluation module. This module calculates margins for the operational constraints and startup time. Next, the convergence judgment module decides whether the schedule has reached the optimum point or not. The optimum point means the shortest schedule without exceeding any stress or NO_x limits. Therefore, the initial schedule may not be the optimum, so the evaluated margins must be forwarded to the next module for schedule modification. The schedule is modified in this module using fuzzy rules prepared through the strategies of the schedule optimization described earlier. Next, the modified startup schedule is sent to the dynamic model again. Through the iterative manner described earlier, convergence to the optimum schedule can be expected. Convergence of the optimization is judged when the reduction rate of startup time obtained from the function for schedule modification becomes sufficiently small.

After finding the optimum schedule, the schedule parameters are sent to the module for machine control by the optimum schedule setting module. The iterative optimization search process is presented on a CRT display through the schedule presentation module as a man–machine interface.

Schedule Parameters. Schedule parameters $X_1, \ldots, X_{28}$, which designate the startup patterns of the plant, are shown in Figure 10.21. Among these, 12 parameters $(X_1, \ldots, X_{12})$ are selected as optimizing objects due to their effect on the operational constraints and, consequently, on the time reduction for startup. The other parameters for control and operating time are defined in a basic engineering phase.

Conventional optimization algorithms need hundreds of iterations for 12 parameter and are time consuming and not practical. A heuristic optimization applying fuzzy reasoning is attempted to obtain rapid convergence. By knowing the stress and NO_x patterns, experienced operators can modify the schedule parameters easily. Such heuristic knowledge is used for the schedule optimization process.

Plant Dynamic Models. A dynamic model of the HRSG is produced by considering transient heat and mass balances in all heat exchangers, drums, piping, and gas ducts. The gas turbine and steam turbine are simulated as static models, except for the steam turbine stresses, because of quicker responses compared with those of the HRSG to study startup characteristics.

Thermal stresses of the steam turbine are generated by temperature differences between the rotor surface and the bore. If the rotor temperature distribution is known, stress can be calculated. However, it is difficult to measure the temperature distributions directly, and the corresponding casing temperatures do not always coincide with the rotor temperatures. Furthermore, even measuring the steam temperature behind the turbine first stage is difficult because of its mechanical structure.

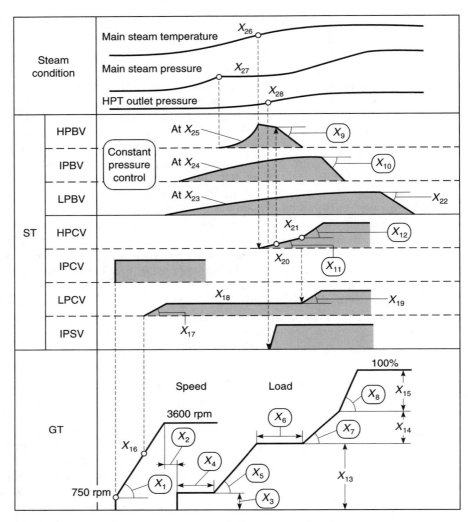

Object parameters of schedule optimization:

X_1: Changing rate of speed

X_2: Hold time at rated speed

X_3: Initial load of GT

X_4: Hold time at initial load of GT

X_5: First-stage load-up rate of GT

X_6: Hold time at second-stage load of GT

X_7: Second-stage load-up rate of GT

X_8: Third-stage load-up rate of GT

X_9: Closing speed of HPBV

X_{10}: Closing speed of IPBV

X_{11}: First-stage closing speed of HPCV

X_{12}: Second-stage closing speed of HPCV

Figure 10.21 Plant startup schedule parameters.

Considering these difficulties in measuring the temperature, dynamic models of the steam turbine rotor stresses are developed. The main calculation steps of the models involve estimation of steam conditions and heat transfer coefficients at the rotor surfaces using steam conditions (i.e., temperature and pressure) and at the inlets of the control valves and computing temperature distributions and stresses at the rotor surfaces and bores. The rotor stresses combine the thermal stresses with the mechanical stresses, which are generated by the centrifugal force corresponding to the rotating speed. Details of the model can be found elsewhere [7].

Thermal stresses of the HRSG can be calculated in a manner similar to that for the turbine stresses. The main calculation steps involve estimating heat transfer coefficients at the inner surface of the headers using the steam flow rates and steam conditions (i.e., temperature and pressure) and calculation of the temperature distributions and stresses of inner and outer surfaces of the headers. The stresses of the HRSG include the thermal stresses and the mechanical stresses, which are generated by steam pressure.

The denitrifier model consists of an equipment model and its controller mode, as shown in Figure 10.22. The NO_x from the gas turbine is reduced by a complicated denitrification process involving dynamics of temperature distributions of the flue gas and catalyst and distribution of absorbed ammonia (NH_3) in the catalyst. Therefore, the equipment model is divided into 10 sections in the direction of gas flow. Each section has calculation modules for heat balance and the denitrification reaction. Instantaneous values of the NO_x emission and NH_3 leakage and the moving average of NO_x emission are obtained from this equipment model. The controller model is used to simulate control actions that determine the NH_3 injection rate. All the input variables of this denitrification model come from the dynamic model of the HRSG.

Evaluation of Startup Dynamics. The whole startup period is divided into time sections to evaluate the startup dynamics: five sections for the stresses and six for NO_x, as shown in Figure 10.23. The whole startup period t_8 is defined as follows:

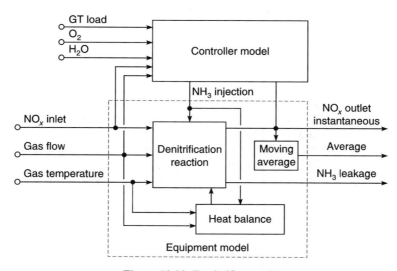

Figure 10.22 Denitrifier model.

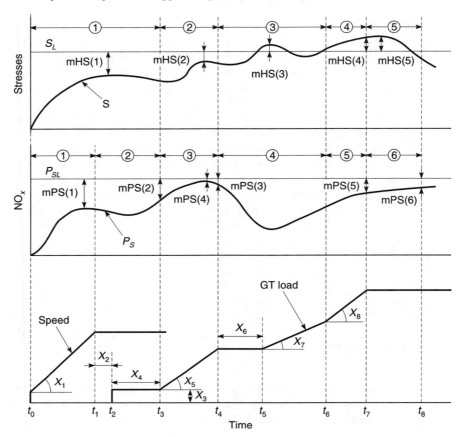

Figure 10.23 Time sections for dynamics evaluation.

$$t_8 = \max(t_{81}, t_{82}) \tag{10.6}$$

where $t_{81} = t_7 + t_e$
   $t_{82} = t_9 + t_e$
   t_7 = time needed for gas turbine startup, or time from gas turbine ignition to reach rated load
   t_9 = time needed for plant startup, or time from gas turbine ignition to reach 97% of steam turbine load
   t_e = time extension for evaluation (1800 s assumed in this system)

The margin m for each variable is generally defined as

$$m = S_L - S \tag{10.7}$$

where S = calculated variable through dynamics models
   S_L = operational limit

Minimum stress margins calculated for each time section $(i = 1, \ldots, 5)$ about four locations of critical stress are designated as follows:

mHS(i) Minimum stress margin at HPT rotor surface

mIS(i) Minimum stress margin at IPT rotor surface

mHB(i) Minimum stress margin at HPT rotor bore

mIB(i) Minimum stress margin at IPT rotor bore

The stress margins of the HRSG can be evaluated in a manner similar to that for the turbine stresses. The locations of critical HRSG stresses are the outer surfaces of headers of the high-pressure superheater and reheater. Minimum stress margins calculated for each time section $(i = 1, \ldots, 5)$ are designated as follows:

mHHD(i) Minimum stress margin at the outer surface of the high-pressure superheater header

mIHD(i) Minimum stress margin at outer surface of the reheater header

The NO$_x$ margins for each time section $(i = 1, \ldots, 6)$ can be evaluated in a manner similar to the turbine stresses. In this case, margins for the instantaneous value and moving average are calculated. Minimum NO$_x$ margins calculated for each time section are designated as follows:

MPS(i) Minimum NO$_x$ margin for instantaneous value

mPA(i) Minimum NO$_x$ margin for moving average

The critical values of the margins should be noted in the startup of the plant without exceeding any operational limits. Therefore, the respective smallest values are selected from the turbine stress margins, the HRSG stress margins, and the NO$_x$ margins as follows:

$$m_T = \min(-\mathrm{mHS}, \mathrm{mHB}, -\mathrm{mIS}, \mathrm{mIB}) \tag{10.8}$$

$$m_B = \min(\mathrm{mHHD}, \mathrm{mIHD}) \tag{10.9}$$

$$m_P = \min(\mathrm{mPS}, \mathrm{mPA}) \tag{10.10}$$

These values are used in fuzzy reasoning.

Schedule Modification by Fuzzy Reasoning. The schedule modification algorithm using multifuzzy reasoning is structured as shown in Figure 10.24 based on the preceding system concept. Fuzzy reasoning for the STPS and GTGT regulates the turbine and HRSG stresses, and STLT and GTPS regulate the NO$_x$ emission. The fuzzy reasoning module provides candidates $[\Delta X_T(i), \Delta X_B(i)(i = 1, 4, \ldots, 12),$ $\Delta X_P(i)(i = 1, \ldots, 11)]$ for modification rates $\Delta X(i)(i = 1, \ldots, 12)$ of the schedule parameters $X(i)(i = 1, \ldots, 12)$. Here, the subscripts T, B, and P mean their candidates obtained from the margins m_T, m_B, and m_P, respectively. Final modification rates $\Delta X(i)(i = 1, \ldots, 12)$ are selected from these candidates through priority decision gates that consist of high-value gates and low-value gates. These gates select safer

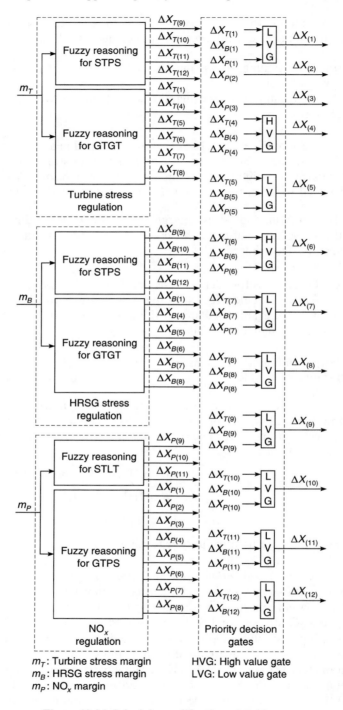

Figure 10.24 Schedule modification algorithm.

values to keep the process variables within the limits. The larger the parameters $X(4)$ and $X(6)$ become, the safer the startup is, because these two parameters correspond to load-hold periods of the gas turbine. The other parameters speed the startup if they are reduced.

The margins m_T, m_B, and m_P are evaluated qualitatively using membership functions, as shown in Figures 10.25(a1) and (b1), to extract qualitative features of the dynamic characteristics of the plant. The membership functions shown in Figures 10.25(a2) and (b2) are used to determine schedule modification coefficients k_S and k_G. The k_S and k_G in Figure 10.25(a2) are determined by fuzzy reasoning for STPS and GTGT using MT and MB, respectively. The k_S and k_G in Figure 10.25(b2) are determined by the fuzzy reasoning for GTPS and STLT, respectively, using m_P. The symbols PB, PS, ZO, NS, and NB are attached to the membership functions. They have the same qualitative meanings as given in Section 10.3.3.

The extracted qualitative features of the dynamic characteristics of the plant are compared with fuzzy rules. The example rules shown in Tables 10.3, parts (a)–(d) are used by fuzzy reasoning for STPS, GTGT, GTPS, and STLT, respectively. Rules in the tables give fragmentary knowledge for determining schedule modification rates in relation to stress and NO_x patterns. The effect of schedule parameters on startup dynamic is important for making the fuzzy rule tables efficient and of minimum size. The rules in Table 10.3, part (a), are used for determining the schedule modification coefficients $k_S(1)$, $k_S(2)$, $k_S(3)$, and $k_S(4)$ in qualitative relations with two steam turbine stress margins out of $m_T(2)$, $m_T(3)$, $m_T(4)$, and $m_T(5)$. The rules in Table 10.3, part (b), are used for determining the schedule modification coefficients $k_G(1)$, $k_G(4)$,

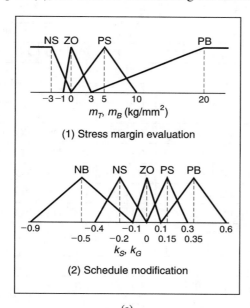

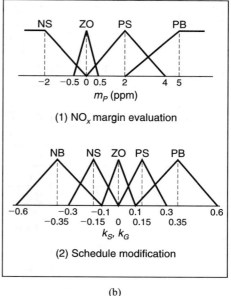

(a)   (b)

Figure 10.25 Membership functions for (a) STPS and GTGT and (b) GTPS and STLT.

TABLE 10.3   EXAMPLE RULE TABLES FOR FUZZY REASONING

(a) Example Rule Table for STPS

① $m_T(4)$	NS				ZO				PS				PB			
② $m_T(5)$	NS	ZO	PS	PB	NS	ZO	PS	PB	NS	ZO	PS	PB	NS	ZO	PS	PB
① $k_S(1)$	NB	NS	ZO	ZO	NS	ZO	ZO	PS	NS	ZO	PS	PS	ZO	ZO	PS	PB
② $k_S(2)$	NB	NS	ZO	ZO	NS	ZO	ZO	PS	NS	ZO	PS	PS	ZO	ZO	PS	PB
③ $k_S(3)$	NB	NS	NS	NS	NS	ZO	ZO	ZO	ZO	ZO	PS	PS	ZO	ZO	PS	PB
④ $k_S(4)$	NB	NS	NS	ZO	NS	ZO	ZO	PS	NS	ZO	PS	PS	ZO	ZO	PS	PB

(b) Example Rule Table for GTGT

① $m_T(2)$	NS				ZO				PS				PB			
② $m_T(3)$	NS	ZO	PS	PB	NS	ZO	PS	PB	NS	ZO	PS	PB	NS	ZO	PS	PB
① $k_G(1)$																
④ $k_G(4)$	PB	PS	PS	PS	PS	ZO	ZO	NS	PS	ZO	NS	NS	PS	ZO	NS	NS
⑤ $k_G(5)$	NB	NB	NS	NS	NS	ZO	ZO	ZO	NS	ZO	PS	PS	NS	ZO	PS	PS
⑥ $k_G(6)$	PB	PB	PS	PS	PS	ZO	ZO	ZO	PS	ZO	NS	NS	PS	ZO	NS	NS
⑦ $k_G(7)$	NB	NS	ZO	ZO	NS	ZO	ZO	PS	NS	ZO	PS	PS	NS	ZO	PS	PB
⑧ $k_G(8)$																

(c) Example Rule Table for GTPS

① $m_P(4)$	NS				ZO				PS				PB			
② $m_P(5)$	NS	ZO	PS	PB	NS	ZO	PS	PB	NS	ZO	PS	PB	NS	ZO	PS	PB
① $k_G(1)$																
② $k_G(2)$																
③ $k_G(3)$																
④ $k_G(4)$	PB	PS	PS	PS	PS	ZO	ZO	ZO	PS	ZO	NS	NS	PS	ZO	NS	NB
⑤ $k_G(5)$	NB	NS	NS	NS	NS	ZO	ZO	ZO	NS	ZO	PS	PS	NS	ZO	PS	PB
⑥ $k_G(6)$	PB	PS	PS	PS	PS	ZO	ZO	ZO	PS	ZO	NS	NS	PS	ZO	NS	NB
⑦ $k_G(7)$	NB	NS	NS	NS	NS	ZO	ZO	ZO	NS	ZO	PS	PS	NS	ZO	PS	PB
⑧ $k_G(8)$	NB	NS	ZO	PS	NS	ZO	PS	PS	NS	ZO	PS	PB	NS	ZO	PS	PB

(d) Example Rule Table for STLT

① $m_P(2)$	NS				ZO				PS				PB			
② $m_P(3)$	NS	ZO	PS	PB	NS	ZO	PS	PB	NS	ZO	PS	PB	NS	ZO	PS	PB
① $k_S(1)$	NS	ZO	ZO	PS	NS	ZO	ZO	PS	NS	ZO	PS	PB	NS	ZO	PS	PB
② $k_S(2)$	NB	NS	NS	NS	ZO	ZO	ZO	ZO	ZO	ZO	PS	PS	ZO	ZO	PS	PB
③ $k_S(3)$	NB	NS	NS	NS	ZO	ZO	ZO	ZO	ZO	ZO	PS	PS	ZO	ZO	PS	PB

$k_G(5)$, $k_G(6)$, $k_G(7)$, and $k_G(8)$ in qualitative relations with two steam turbine stress margins out of $m_T(1)$, $m_T(2)$, $m_T(3)$, $m_T(4)$, and $m_T(5)$. These two rule tables are similarly used for the STPS and GTGT in the functions of HRSG stress regulation.

The rules in Table 10.3, part (c), are used for determining the schedule modification coefficients $k_G(1)$, $k_G(2)$, $k_G(3)$, $k_G(4)$, $k_G(5)$, $k_G(6)$, $k_G(7)$, and $k_G(8)$ in qualitative relations with two NO_x margins out of $m_P(1)$, $m_P(2)$, $m_P(3)$, $m_P(4)$, $m_P(5)$, and $m_P(6)$. The rules in Table 10.3, part (d), are used for determining the schedule modification coefficients $k_S(1)$, $k_S(2)$, and $k_S(3)$ in quantitative relations with two NO_x margins out of $m_P(1)$, $m_P(2)$, and $m_P(3)$. The null parts in these tables mean that the relations between the schedule parameters and the operational margins are negligibly small.

Plural membership functions are selected for each schedule modification coefficient after pattern matching of the qualitative features of the startup dynamics with these rule tables. Following this, a single value for each schedule modification coefficient is defined by calculation of the balanced point of the selected membership functions. An example of the calculation for schedule modification coefficient k_S is shown in Figure 10.26. In this case, the value of the balanced point k_{SG} is actually used to regulate the steam turbine stresses. The balanced point k_{SG} is defined by

$$k_{SG} = \frac{\sum_{i=1}^{4} W(i)k_S(i)}{\sum_{i=1}^{4} W(i)} = 0.0273 \qquad (10.11)$$

Here, $W(i)(i = 1, \ldots, 4)$ are the weights of the membership functions. Next, the candidate for schedule modification rate ΔX_T is defined by

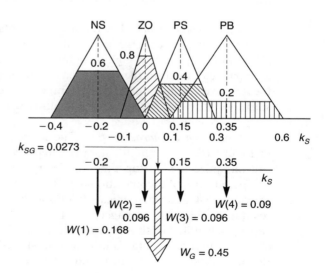

Figure 10.26 Example calculation of schedule modification coefficient K_s.

$$\Delta x_T = \alpha k_{SG}[X_{max} - X_{min}] \tag{10.12}$$

Here, α, X_{max}, and X_{min} are a tuning coefficient and the maximum and minimum allowable limits of the schedule parameter, respectively. The tuning coefficient α for each schedule parameter is tuned through a simulation study described next. The X_{max} and X_{min} are uniquely designated for each schedule parameter in view of safe and practical plant operation.

10.4.3. Simulation Study

Values of Operational Constraints. It is necessary to compare startup characteristics between conventional methods and the proposed method under the same stress and NO_x levels developed at the maximum point. The startup characteristics are evaluated in the following steps:

1. The maximum stress and NO_x levels are obtained through a simulation of the conventional method, which introduces a metal matching chart to produce startup schedules as described elsewhere [5].
2. The optimization process is developed through a simulation of the proposed method. In this step, 12 schedule parameters are modified under operational constraints that are at the same level as the maximum values obtained in step 1.

A simulation result for the conventional method is shown in Figure 10.27. The required startup time is 108 min. In this case, the maximum values of the rotor surface stress, the rotor bore stress, the HRSG header stress, and NO_x emission are -31.2 kg/mm, 36.8 kg/mm, 13.5 kg/mm, and 8.3 ppm, respectively. These values are used as the operational constraints in the subsequent simulation study.

Tuning Coefficients. The tuning coefficients $\alpha(I)(I = 1, \ldots, 12)$ are tuned based on the stability of convergence of schedule optimization. Table 10.4 shows the tuning steps. Two steps are enough to obtain stable convergence. Three of the coefficients are changed based on the convergence stabilities of the schedule parameters during the iterative optimization process. Simulation results are shown in Figure 10.28. The startup energy loss is reduced significantly with startup time.

Convergence Characteristics. The convergence characteristics of step 2 is shown in Figure 10.29. The ninth iterative step gives the optimum startup schedule that can minimize the startup time and energy loss. Both can be reduced significantly compared with the conventional method. This means that the proposed method provides fast and economical plant startup under NO_x emission regulation and reliable machine life management. The simulation results along with the optimum startup schedule are shown in Figure 10.30. The load-up pattern of the gas turbine shown in Figure 10.30(a) is modified drastically compared with that of the conventional method, as shown in Figure 10.27(a).

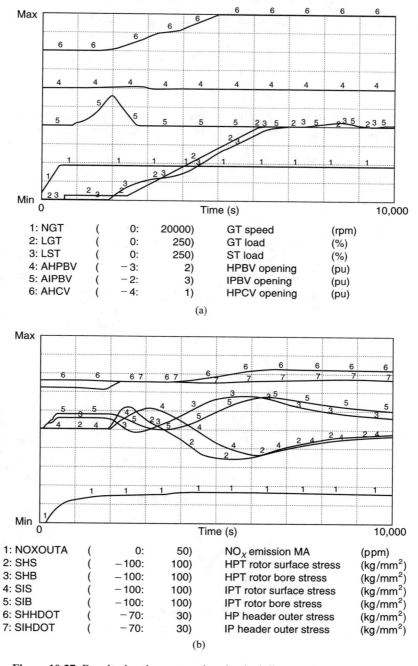

1: NGT	(	0:	20000)	GT speed	(rpm)
2: LGT	(	0:	250)	GT load	(%)
3: LST	(	0:	250)	ST load	(%)
4: AHPBV	(	−3:	2)	HPBV opening	(pu)
5: AIPBV	(	−2:	3)	IPBV opening	(pu)
6: AHCV	(	−4:	1)	HPCV opening	(pu)

(a)

1: NOXOUTA	(	0:	50)	NO$_x$ emission MA	(ppm)
2: SHS	(	−100:	100)	HPT rotor surface stress	(kg/mm^2)
3: SHB	(	−100:	100)	HPT rotor bore stress	(kg/mm^2)
4: SIS	(	−100:	100)	IPT rotor surface stress	(kg/mm^2)
5: SIB	(	−100:	100)	IPT rotor bore stress	(kg/mm^2)
6: SHHDOT	(	−70:	30)	HP header outer stress	(kg/mm^2)
7: SIHDOT	(	−70:	30)	IP header outer stress	(kg/mm^2)

(b)

Figure 10.27 Results by the conventional scheduling method: (a) startup schedule and (b) dynamics of stresses and NO$_x$.

TABLE 10.4 TUNING COEFFICIENT α

α	$\alpha(1)$	$\alpha(2)$	$\alpha(3)$	$\alpha(4)$	$\alpha(5)$	$\alpha(6)$	$\alpha(7)$	$\alpha(8)$	$\alpha(9)$	$\alpha(10)$	$\alpha(11)$	$\alpha(12)$
Initial	3	1	1	1.5	1.5	2	4	3	2	3	2	4
Step 1	3	1	1	1.5	1.5	2	2.5	3	2	3	2	4
Step 2	3	1	1	1.5	1.5	2	2.5	3	3	3	3	4

Computation Time. The computation time depends on the time range of the startup process and the number of iterations of schedule optimization. Using an engineering workstation (Hitachi Engineering Workstation 3050RX/230) that has a 105-MIPS computation speed takes about 2.5 min for the optimum schedule to be reached in step 3. This is fast enough for practical applications.

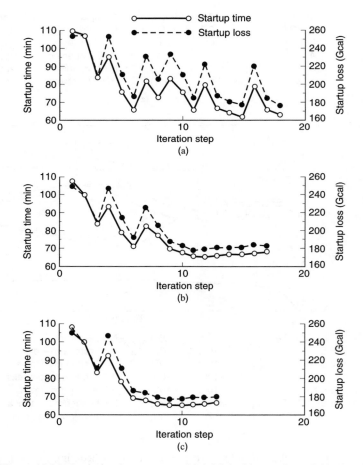

Figure 10.28 Optimization process: (a) initial; (b) step 1; and (c) step 2.

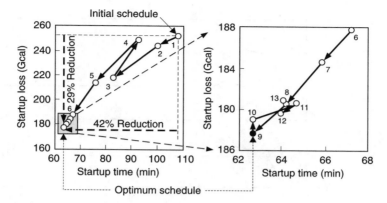

Figure 10.29 Convergence characteristics.

10.4.4. Conclusions

We have developed an expert system for optimizing the startup schedule of combined-cycle power plants. This system harmonizes machine operations to minimize startup time and energy loss to meet NO_x emission regulations and reduce machine stress. Plant dynamic models representing quantitative knowledge and fuzzy rules representing qualitative knowledge are used alternately to optimize schedule parameters that designate the startup pattern. Simulation results with this system show the following:

- Synergetic organization of fuzzy reasoning and plant dynamic models provide flexibility in minimizing the startup time automatically under various operating constraints.
- The schedule parameters converge more quickly to the optimum values compared with the conventional method.
- Startup energy loss was reduced due to the reduction in startup time.

The expert system can meet the requirements of medium-load operation with frequent startup and shutdown encountered in combined-cycle power plant operation.

10.5. DISCUSSION

10.5.1. Flexibilities of the Expert System

The expert systems deal with the stresses in the steam turbine and HRSG and NO_x emissions as the operational constraints. In a similar framework, other constraints such as differential expansion of the turbine rotor and casing, drum stress, and SO_x and CO emissions can be considered depending on plant types. Stresses can

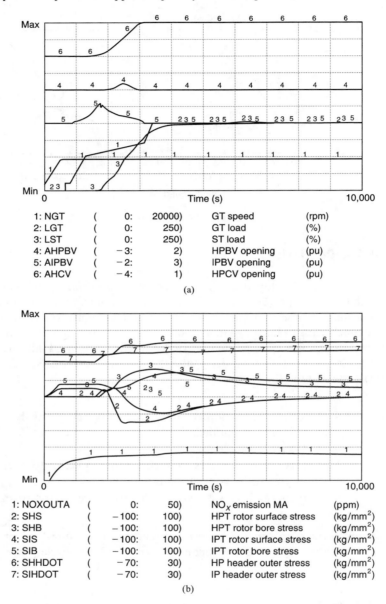

1: NGT	(	0:	20000)	GT speed	(rpm)
2: LGT	(	0:	250)	GT load	(%)
3: LST	(	0:	250)	ST load	(%)
4: AHPBV	(	− 3:	2)	HPBV opening	(pu)
5: AIPBV	(	− 2:	3)	IPBV opening	(pu)
6: AHCV	(	− 4:	1)	HPCV opening	(pu)

(a)

1: NOXOUTA	(	0:	50)	NO$_x$ emission MA	(ppm)
2: SHS	(	−100:	100)	HPT rotor surface stress	(kg/mm^2)
3: SHB	(	−100:	100)	HPT rotor bore stress	(kg/mm^2)
4: SIS	(	−100:	100)	IPT rotor surface stress	(kg/mm^2)
5: SIB	(	−100:	100)	IPT rotor bore stress	(kg/mm^2)
6: SHHDOT	(	−70:	30)	HP header outer stress	(kg/mm^2)
7: SIHDOT	(	−70:	30)	IP header outer stress	(kg/mm^2)

(b)

Figure 10.30 Results by the proposed schedule method: (a) optimum startup schedule and (b) dynamics of stresses and NO$_x$.

be managed indirectly using the temperature-varying speed of the steam and metal instead of the entire dynamic model directly.

The startup schedule is optimized on the basis of startup time in these expert systems as described earlier. Additional factors such as startup energy loss, life consumption rate of machines, or NO_x emission rate can be introduced to evaluate optimality using a weighted function. This allows flexible startup schedules to be obtained by varying the weights with environmental, seasonal, and daily requirements.

The optimum schedule of the combined-cycle power plant is intended to be able to keep the stresses and NO_x within their limits. Even then, an unscheduled speed hold or load hold might be caused accidentally after the gas turbine is ignited. The startup schedule can be reviewed to minimize the time deviation from the original schedule by the module for rescheduling in the same manner as for the optimization using the current states as the initial conditions for the dynamic model in the expert system for the conventional fossil power plant.

10.5.2. Potential Applications of the Optimization Algorithm

The optimization algorithm proposed in this chapter has the potential to be applied in various applications requiring parameter optimization in relation to plant dynamics, such as engineering tools for plant design and operation, training simulators for plant operators, and design tools for control systems.

In the engineering tools for plant design, the algorithm can function to optimize machine structure, strength, capacity, and materials considering cost benefit, reliability, environmental adaptability, controllability, and so on. In the basic planning of plant operation, this algorithm plays the part of experts to improve the reliability and economics of plant operation. The plant operators' work burden can be reduced in acquiring the knowledge and the skills needed for plant operation by using training simulators that apply to this algorithm. Control system designers can easily tune the logic parameters of the control system using this algorithm.

10.5.3. Future Improvements

The expert systems for conventional fossil power plants and combined-cycle power plants optimize the startup schedule with 7 and 12 objective parameters, respectively, as described earlier. The size of the rule tables for fuzzy reasoning is directly related to the number of the time period divisions to evaluate startup dynamics. Therefore, further study on the evaluation method, construction, and tuning method of fuzzy rules will help to reduce the table size without deteriorating convergence of the optimization algorithm.

10.6. CONCLUSIONS

This chapter has described expectations of fuzzy theory in fossil power plant operation and two types of expert systems applied to operation support systems for a conventional one-through-type supercritical pressure fossil power plant and a gas and steam turbine combined-cycle power plant.

Fuzzy models of expertise are introduced to modify the plant startup schedule.

The most remarkable feature of this approach is that the quantitative optimum schedule is obtained through iterative modification of the schedule from a combination of qualitative knowledge and plant dynamic simulations.

Simulation results with these two-operation support expert systems demonstrate the following:

- Plants are started quickly and accurately through the optimum startup schedule.
- Operators' work is reduced in monitoring and executing startup schedules with functions for on-line assessment and off-line learning of plant dynamics.
- Convergence of the schedule optimization is fast compared with conventional operations research techniques.

The expert systems are expected to contribute in reducing operators' work burden to harmonize machine operation not only for economical aspects but also for environmental advantages.

References

[1] T. Arakawa and Y. Hishinuma, "Thermal Power Generation Technology Giving Considerations for Global Environment," *Hitachi Hyoron*, Vol. 73, No. 11, 1992, pp. 4–8.

[2] T. Akiyama, T. Matsushima, N. Nagatuchi, A. Nakajima, H. Yoshizaki, and J. Matsumura, "Dynamic Simulation of an Advanced Combined Cycle Plant with Three Pressure and Reheat Cycle," paper presented at the JSME-ASME International Conference on Power Engineering-93 (ICOPE-93), Tokyo, Japan, Sept. 1993, pp. 221–226.

[3] H. Matsumoto, Y. Eki, A. Kagi, S. Nigawara, M. Tokuhira, and Y. Suzuki, "An Operation Support Expert System Based on On-line Dynamics Simulation and Fuzzy Reasoning for Startup Schedule Optimization in Fossil Power Plants," *IEEE Transactions on Energy Conversion*, Vol. 8, No. 4, 1993, pp. 674–680.

[4] H. Matsumoto. Y. Ohsawa, S. Takahashi, T. Akiyama, and O. Oshiguro, "An Expert System for Startup Optimization of Combined Cycle Power Plants under NO_x Emission Regulation and Machine Life Management," *IEEE Transactions on Energy Conversion*, Vol. 11, No. 2, 1996, pp. 414–422.

[5] F. J. Hanzalek and P. G. Ipsen, "Thermal Stress Influence Starting, Loading of Bigger Boilers," *Electrical World*, Vol. 165, No. 6, 1996, pp. 58–62.

[6] R. G. Livingston, "Computer Control of Turbine-Generators Startup Based on Rotor Stress," paper presented at the Joint Power Conference ASME and IEEE, New Orleans, LA, 1973, pp. 1–14.

[7] H. Matsumoto, Y. Sato, F. Kato, Y. Eki, K. Hisano, and K. Fukushima, "Turbine Control System Based on Prediction of Rotor Thermal Stress," *IEEE Transactions on Power Apparatus and Systems*, Vol. PAS-101, No. 8, 1982, pp. 2504–2512.

[8] A. Kaji, T. Marugama, and Y. Eki, "Development and Application of an Expert System (HITREX) for Plant Operational Support," paper presented at the Conference on Expert Systems Applications for the Electric Power Industry, Sponsored by EPRI, June 5–8, 1989.

S. M. Shahidehpour
R. W. Ferrero
*Department of Electrical
and Computer Engineering
Illinois Institute of Technology
Chicago, IL 60616*

Chapter 11

Fuzzy Systems Approach to Short-Term Power Purchases Considering Uncertain Prices

11.1. INTRODUCTION

The U.S. Energy Policy act (EPAct 92) requires electric utilities to provide transmission services to potential competitors. As a result, utilities may obtain significant savings in operation cost and improve service reliability by scheduling power purchases and local resources. The problem that utilities are facing is to model inherent characteristics of variables and to interpret and use the results from these models in the decision-making process.

In previous approaches to determining power purchases [1, 2], a fixed price was assumed and the uncertainty in input data was disregarded. In practice, system variables are associated with degrees of uncertainty that affect the decision-making process. For example, the estimate of market price would affect the optimal short-term generation scheduling in each utility, load forecasting accuracy would affect price decisions in a deregulated environment, and network reliability and component availability would affect buy/sell decisions as well as transmission access issues.

The uncertainty in power systems operation and control represents a certain type of qualitative uncertainty that may not follow the nature of a probabilistic function. This kind of uncertainty is best modeled by fuzzy logic for representing the relationship among different variables. Fuzzy logic describes properties of variables by associating them with semantic labels; the membership in a fuzzy set measures the conformance of a variable to a concept [3].

Besides the natural extension from semantic structures, several other advantages are obtained from the application of fuzzy numbers: In practical cases, limits imposed on specific power system variables (bus voltages, line flows, etc.) may be

321

considered as "soft," and small violations of these limits may be tolerable. An optimal solution may include small violations that a decision maker (DM) in a utility is willing to accept in order to improve the utility's operation or to decrease the utility's cost. For instance in the definition of purchase level, a DM may allow line flows that exceed the maximum capacity of the line (representing "soft" constraints) while the generator power is maintained at its maximum capacity ("hard" constraint) [4, 7].

For our application, a procedure is presented to evaluate power purchases considering variables uncertainty in the model. In the proposed method, offered prices for power, line flows, local generation, and load may be treated as fuzzy variables. Power flow constraints are considered using a DC load flow. The fuzzy nonlinear problem is converted to a crisp minimization model with a linear objective function and nonlinear constraints.

11.2. DESCRIPTION OF THE PROBLEM

11.2.1. Short-Term Power Purchases

Potential savings may be obtained by interchanging power with neighboring utilities because of dissimilarities in utilities' hourly loads and generation resources. The problem that we present in this chapter is illustrated in Figure 11.1.

Each utility is interconnected to other utilities and independent power producers at buses $1, 2, \ldots, j$. The index k stands for a purchase transaction; there may be more than one transaction (i.e., buy and sell) at each bus. The objective is to coordinate power purchases with local generation in order to supply the utility's load at minimum cost. We consider optimization at a particular time t; however, the proposed methodology may be used in longer optimization horizons for unit commitment.

Generation cost for each unit is represented with a second-order polynomial (notation is given in Appendix 1):

$$C(i) = a_i + b_i P(i) + c_i P(i)^2$$

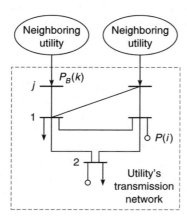

Figure 11.1 Power purchases in utility.

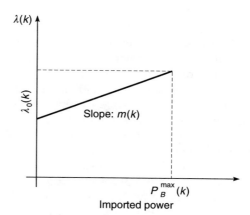

Figure 11.2 Price for purchased power.

Hence, the incremental cost of a generator is a linear function of generation. In general, the price per unit of purchased power in transaction k [i.e., $\lambda(k)$] is a function of the power level. We assume a linear function as in Figure 11.2.

In Figure 11.2, the price of import is at least $\lambda_0(k)$. The utility can buy up to $P_B^{\max}(k)$ at a price equal to $\lambda_0 + m(k)P_B^{\max}(k)$. The utility's expenditure for a power purchase $P_B(k)$ is

$$C_B(k) = [\lambda_0(k) + m(k)P_B(k)]P_B(k) = \lambda(k)P_B(k)$$

Hence, the optimal purchase problem in a utility is formulated as

$$\text{Minimize} \quad F = \sum_{i=1}^{N_G} C(i) + \sum_{k=1}^{N_B} C_B(k)$$

$$= \sum_{i=1}^{N_G} [a_i + b_i P(i) + c_i P(i)^2] + \sum_{k=1}^{N_B} [\lambda(k)P_B(k)] \qquad (11.1)$$

subject to the following:

- Power generation limits:

$$0 \le P(i) \le P^{\max}(i) \quad 0 \le P_B(k) \le P_B^{\max}(k)$$

$$i = 1, \ldots, N_G \quad k = 1, \ldots, N_B \qquad (11.2)$$

- Power balance at each bus:

$$\sum_{i \in \Phi_j} P(i) + \sum_{k \in \Omega_j} P_B(k) + \sum_{n \in \Psi_j} P_{n-j} = L_j \quad \text{for each } j \in \{1, \ldots, N_{\text{buses}}\} \qquad (11.3)$$

- Line flow constraints:

$$P_{n-j} = \frac{\theta_n - \theta_j}{X_{n-j}} \qquad P_{n-j}^{\min} \le P_{n-j} \le P_{n-j}^{\max} \quad \text{for all lines} \qquad (11.4)$$

Equations (11.3) and (11.4) correspond to Kirchhoff's laws in a DC load flow [8]. The problem formulated corresponds to a traditional economic dispatch in which the utility has local generation and defines the amount of power to be purchased from neighboring utilities. In this formulation, limits of the variables are rigid, and the solution will be reached as these variables remain within given ranges. This type of model may be incapable of representing practical situations. A more realistic model should simulate soft limits on specific variables to provide an optimal solution for an otherwise infeasible condition. A DM may, for instance, obtain an optimal solution in which line flows (as fuzzy variables) exceed limits in order to test the impact of network constraints on power purchases.

The minimization problem in Eqs. (11.1)–(11.4) is done for a particular price at which the utility knows a priori the price of purchased power. In practice, the price and amount of retail power offered in the near future may be forecasted by a particular utility for market analysis purposes. The price forecast has a degree of uncertainty usually expressed vaguely as "low prices" or "prices around a certain value." The solution of (11.1)–(11.4) gives no indication to a DM regarding variations in the price forecast and the impact of active constraints in the obtained dispatch.

11.2.2. Triangular Fuzzy Numbers

All uncertain variables are represented in our study with triangular fuzzy numbers (TFNs). A fuzzy number $\tilde{A}$ is said to be a TFN if

$$
\mu_{\tilde{A}}(x) = \begin{cases}
\dfrac{x - A + l_{\tilde{A}}}{l_{\tilde{A}}} & A - l_{\tilde{A}} \leq x \leq A \\[2mm]
1 - \dfrac{x - A}{r_{\tilde{A}}} & A \leq x \leq A + r_{\tilde{A}} \\[2mm]
0 & \text{otherwise}
\end{cases}
$$

where A is the mean value of $\tilde{A}$ and $l_{\tilde{A}} \geq 0, r_{\tilde{A}} \geq 0$ are left and right spreads, respectively (see Fig. 11.3). The TFN $\tilde{A}$ is symbolically written as $\tilde{A} = (A, l_{\tilde{A}}, r_{\tilde{A}})$. The maximum possible value of $\tilde{A}$ is equal to the mean plus the right spread and the minimum possible value is equal to the mean minus the left spread. Numbers above $\tilde{A} + r_{\tilde{A}}$ and below $\tilde{A} - l_{\tilde{A}}$ have zero membership function. When TFN $\tilde{A}$ is defined as a fuzzy number, there is a subjective valuation of this value as a crisp number A. The membership function $\mu_{\tilde{A}}(x)$ corresponds to the degree of possibility of occurrence of the corresponding crisp value x.

In the proposed method uncertainties are represented as TFN; for example, the statement "load would be around 20 MW, no more than 22 MW and no less than 17 MW" can be represented by the TFN $\tilde{L} = (20, 3, 2)[\text{MW}]$. We consider the price of purchased power, the local load, the local generation, and the power flow in lines as TFNs.

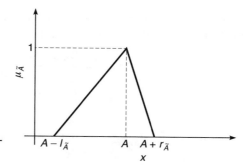

Figure 11.3 Triangular membership function.

11.2.3. Fuzzy Variables in the Problem

The decision-making process for power purchases in an interconnected system is quite complicated because of uncertainties associated with local demand and market price. Assume that a utility wants to optimize its generation and considers buying power using these data: "The minimum price per unit of power in the market is *about* λ_0 and we may buy $P_B^{\max}$ at a *price around* $\lambda_{\max}$" (see Fig. 11.4).

In our method, the market price for each level of power is modeled as a TFN with linear functions for left and right membership functions. The minimum price is $\tilde{\lambda}_0 = (\lambda_0, l_{\tilde{\lambda}_0}, r_{\tilde{\lambda}_0})$, and the offered price to import the maximum power is $\tilde{\lambda}_{\max} = (\lambda_{\max}, l_{\tilde{\lambda}_{\max}}, r_{\tilde{\lambda}_{\max}})$. For each power level, the mean value λ minus the left spread $l_{\tilde{\lambda}}$ can be interpreted as the minimum possible price; in the same way, the mean value plus the right spread $r_{\tilde{\lambda}}$ is the maximum possible price. Both values can be derived from the "best case" and the "worst-case" analyses.

As both maximum and minimum prices are fuzzy numbers, the slope joining the two values is also a fuzzy number $\tilde{m} = (m, l_{\tilde{m}}, r_{\tilde{m}})$. The price per unit of purchased power P_B is computed as $\tilde{\lambda} = (\lambda, l_{\tilde{\lambda}}, r_{\tilde{\lambda}}) = \tilde{\lambda}_0 + \tilde{m}P_B$. The expenditure for a purchase transaction is computed as $\tilde{C}_B(k) = \tilde{\lambda}(k)P_B(k)$, which is also a TFN. The expression of the fuzzy slope $\tilde{m}$ is computed as follows. In Figure 11.2 the mean value of the fuzzy slope is $m = (\lambda_{\max} - \lambda_0)/P_B^{\max}$. In Figure 11.4, $m_{\min}$ is the

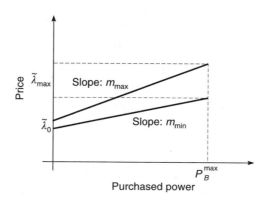

Figure 11.4 Uncertainty in the price for purchased power.

slope of the line joining the minimum possible values of $\tilde{\lambda}_0$ and $\tilde{\lambda}_{max}$; m_{max} corresponds to the slope of the line joining the maximum possible values of $\tilde{\lambda}_0$ and $\tilde{\lambda}_{max}$:

$$m_{min} = \frac{(\lambda_{max} - l_{\tilde{\lambda}_{max}}) - (\lambda_0 - l_{\tilde{\lambda}_0})}{P_B^{max}} \tag{11.5}$$

$$m_{max} = \frac{(\lambda_{max} + r_{\tilde{\lambda}_{max}}) - (\lambda_0 + r_{\tilde{\lambda}_0})}{P_B^{max}} \tag{11.6}$$

For a purchase level P_B, left and right spreads of the fuzzy slope $\tilde{m}$ are $l_{\tilde{m}} = m - m_{min}$ and $r_{\tilde{m}} = m_{max} - m$, respectively; replacing (11.5) and (11.6) in this equations, we obtain the expression for the fuzzy slope as

$$\tilde{m} = \left(\frac{\lambda_{max} - \lambda_0}{P_B^{max}}, \frac{l_{\tilde{\lambda}_{max}} - l_{\tilde{\lambda}_0}}{P_B^{max}}, \frac{r_{\tilde{\lambda}_{max}} - r_{\tilde{\lambda}_0}}{P_B^{max}} \right)$$

11.3. MATHEMATICAL FORMULATION

11.3.1. Mathematical Formulation Using Fuzzy Numbers

If we consider the price for purchased power and local generation as TFNs, the minimization (11.1)–(11.4) is formulated as

$$\text{Minimize} \quad \tilde{F} = \sum_{i=1}^{N_G} \tilde{C}(i) + \sum_{k=1}^{N_B} \tilde{C}_B(k)$$

$$= \sum_{i=1}^{N_G} (a_i + b_i \tilde{P}(i) + c_i \tilde{P}(i)^2) + \sum_{k=1}^{N_B} [\tilde{\lambda}(k) P_B(k)] \tag{11.7}$$

subject to the following:

• Power generation limits:

$$0 \leq \tilde{P}(i) \leq P^{max}(i) \qquad 0 \leq P_B(k) \leq P_B^{max}(k) \tag{11.8}$$

$$i = 1, \ldots, N_G \qquad k = 1, \ldots, N_B$$

• Power balance in each bus:

$$0 \tilde{\leq} \sum_{i \in \Phi_j} \tilde{P}(i) + \sum_{k \in \Omega_j} P_B(j) + \sum_{n \in \Psi_j} \tilde{P}_{n-j} - \tilde{L}_j \tilde{\leq} \tilde{\epsilon} \quad \text{for } j \in \{1, \ldots, N_{bus}\} \tag{11.9}$$

• Power flow constraints:

$$\tilde{P}_{n-j} = \frac{\tilde{\theta}_n - \tilde{\theta}_i}{X_{n-j}} \qquad \tilde{P}_{n-j}^{min} \tilde{\leq} \tilde{P}_{n-j} \tilde{\leq} \tilde{P}_{n-j}^{max} \quad \text{for all lines} \tag{11.10}$$

Equations (11.7)–(11.10) correspond to the fuzzified version of Eqs. (11.1)–(11.4). The formulation is a nonlinear optimization problem with fuzzy control variables and fuzzy coefficients in the objective function. It should be noted that constraints (11.8) are hard constraints; however, constraints (11.9) and (11.10) may be considered soft and will be satisfied up to a certain degree of acceptance defined by the DMs. In (11.9), $\tilde{\epsilon}$ is introduced as a tolerance for satisfying power balance equations.

11.3.2. Fuzzy Comparison

The proposed method is an extension of the procedure given elsewhere [12] for nonlinear objective functions and is based on the idea of "almost positive" ($\gtrsim$) fuzzy numbers. Considering the minimization criterion presented earlier, we encounter the problem of comparing fuzzy quantities. In order to compare two TFNs, say $\tilde{A}$ and $\tilde{B}$, we define a new TFN, $\tilde{Y} = \tilde{A} - \tilde{B}$, where $\tilde{A}$ is bigger than $\tilde{B}$ if $\tilde{Y} \gtrsim 0$.

Definition. TFN $\tilde{Y} = (Y, l_{\tilde{Y}}, r_{\tilde{Y}})$ is "almost positive," denoted by $\tilde{Y} \gtrsim 0$, if and only if $\mu_{\tilde{Y}}(0) \leq 1 - \delta$, $Y \geq 0$. In this definition, the larger the δ, the stronger the concept of "almost positive." This concept is illustrated in Figure 11.5, which will be used to compare solutions and compute the satisfaction of operational constraints. When $\delta = 1$, the TFN is strictly greater than or at least equal to zero; in other words, the minimum possible value of the TFN is greater than or at least equal to zero.

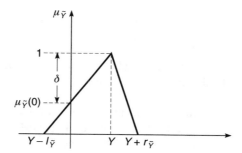

Figure 11.5 Exaplanation of $\tilde{Y}$ almost positive.

11.3.3. Goal Satisfaction

In our proposed procedure, fuzzy minimization (11.7) is replaced by a satisfaction or goal criterion. For instance, a utility may use the operation cost without transactions as an upper limit and try to define a set of purchase transactions that decrease the operation cost by a certain percentage. In our procedure, we define $\tilde{G}$ (satisfaction level or goal) as a TFN and include the fuzzy expression $\tilde{Y} = \tilde{G} - \tilde{F} \gtrsim 0$ in the set of constraints.

In the proposed method, fuzzy constraints are expressed as fuzzy comparisons and transformed into crisp functions using Definition 1 in the previous section. For instance, in (11.11) we express (11.7) as a fuzzy comparison: "Total operation cost should be lower than the goal with a degree of acceptance δ_1." Further development is shown in Appendix 2:

$$G - \sum_{i=1}^{N_G}(a_i + b_iP(i) + c_iP(i)^2) - \sum_{k=1}^{N_B}(\lambda_0^B(k)P_B(k) + m_kP_B(k)^2) - \delta_1$$

$$\times \left(l_{\tilde{G}} + \sum_{i=1}^{N_G}(b_ir_{\tilde{P}(i)} + 2c_iP(i)r_{\tilde{P}(i)}) + \sum_{k=1}^{N_B}(P_B(k)r_{\lambda_0^B(k)} + P_B(k)^2r_{\tilde{m}_k})\right) \geq 0 \quad (11.11)$$

$$P^{\max}(i) - P(i) - \delta_2r_{\tilde{P}(i)} \geq 0 \qquad P(i) - \delta_2l_{\tilde{P}(i)} \geq 0$$
$$P_B^{\max}(k) - P_B(k) \geq 0 \qquad P_B(k) \geq 0 \qquad (11.12)$$
$$i = 1, \ldots, N_G \qquad k = 1, \ldots, N_B$$

$$\epsilon - \left(\sum_{i\in\Phi_j}P(i) + \sum_{k\in\Omega_j}P_B(k) + \sum_{n\in\Psi_j}P_{p-j} - L_j\right)$$

$$-\delta_3\left(l_{\tilde{\epsilon}} + \sum_{i\in\Phi_k}r_{\tilde{P}(i)} + \sum_{n\in\Psi_j}r_{\tilde{P}_{n-j}} + r_{\tilde{L}_j}\right) \geq 0$$

$$\left(\sum_{i\in\Phi_j}P(i) + \sum_{k\in\Omega_j}P_B(k) + \sum_{n\in\Psi_j}P_{n-j} - L_j\right) - \delta_3\left(\sum_{i\in\Phi_j}l_{\tilde{P}(i)} + \sum_{n\in\Psi_j}l_{\tilde{P}_{n-j}} + l_{\tilde{L}_j}\right) \geq 0$$

$$\text{for each } k \in \{1, \ldots, N_{\text{buses}}\} \quad (11.13)$$

$$P_{n-j}^{\max} - P_{n-j} - \delta_4r_{\tilde{P}_{n-j}} \geq 0 \qquad P_{n-j} - P_{n-j}^{\min} - \delta_4l_{\tilde{P}_{n-j}} \geq 0 \quad \text{for each line} \quad (11.14)$$

Equations (11.11)–(11.14) correspond to the fuzzy comparison in (11.7)–(11.10), respectively. In these equations, left and right spreads are defined as nonnegative numbers.

In a fuzzy environment, control variables (local generation and purchased power) will be fuzzy in nature. The DM is interested in identifying the range of control variables that lead the system to a specific reduction in operation costs. The aim of our procedure is to identify the widest range in the mix of purchase levels and local generations that reduces the total operation cost up to a desired level and satisfies the operational constraints. The proposed method defines the crisp objective function as

$$\text{Maximize} \quad \sum_{i=1}^{N_G} w_{\tilde{P}(i)}^l l_{\tilde{P}(i)} + w_{\tilde{P}(i)}^r r_{\tilde{P}(i)} \quad (11.15)$$

where $w_{\tilde{P}(i)}^l$ and $w_{\tilde{P}(i)}^r$ are the weighting factors for the left and right spreads of the generation of unit i, respectively. The weighting factors enhance the role of certain variables in the solution. The problem is to solve (11.15) subject to (11.11)–(11.14).

In (11.11)–(11.14), values of $\delta_1, \ldots, \delta_4$, measure the degree of acceptance of the "almost positive" assumption in the fuzzy expressions; hence, the objective in the proposed procedure is to define limits in the operation schedule that meets the goal and constraints corresponding to the given degree of acceptance δ. For instance, (11.14) restricts the fuzzy line flow so that its mean value plus δ_4 times the right spread is lower than or at least equal to the maximum power flow in the line. For a line in which the maximum capacity limit is critical, a DM would define $\delta_4 = 1$ so even the maximum possible value of the fuzzy line flow (that is, mean value plus the right spread) is below the maximum capacity. On the other hand, for a line that may be overloaded during short periods of time, a DM may define a different value of δ_4. In (11.14), with $\delta_4 = 0$, the mean value of the fuzzy line flow will be lower than the line capacity; however, the maximum line flow may exceed the line capacity. By adjusting δ_4, DM decides up to what extent the constraint on the line should be enforced. The DM may set δ in (11.11)–(11.14) according to the degree of acceptance expected for the respective equations. The problem can be extended to consider fuzzy coefficients in the constraints using the transformations given in Appendix 2.

11.3.4. Minimum-Cost Scheduling Using Fuzzy Expressions

The goal is defined as the upper limit for the total operation cost. The solution of Eqs. (11.11)–(11.15) gives the range of scheduled transactions with operation costs lower than the goal. However, a DM may later be even interested in learning if there is a transaction schedule with even lower operation cost; hence, a reduced goal will be defined and Eqs. (11.11)–(11.15) will be solved again. In order to find the feasible schedule with the minimum cost, the goal is reduced iteratively and a new solution is computed in each iteration.

The index for iterations in Figure 11.6 is γ. In the first iteration ($\gamma = 1$) a feasible schedule is introduced as well as a desired reduction in the operation cost. The optimal local generation without transactions is used as the initial feasible schedule. Equations (11.11)–(11.15) are solved for the desired goal and constraints. The obtained schedule is saved in each iteration. When no further reductions in operation costs can be achieved, a DM may consider either accepting the latest feasible schedule or relaxing a few constraints. For instance, a DM may be willing to allow a few line flows to exceed their limits in order to obtain additional savings in the operation cost. The latest feasible schedule corresponds to the minimum operation cost that satisfies the set of constraints with a certain degree of acceptance. The optimal local generation and purchase level obtained from the proposed procedure are expressed in terms of fuzzy numbers; hence, the DM should defuzzify [1] these values in order to express control variables as crisp numbers. In Section 11.4, we offer additional discussion of the effect of relaxing constraints in the problem. The minimum-cost solution when right and left spreads in variables are null corresponds to the solution of (11.1)–(11.4); in other words, when uncertainty is neglected, the proposed procedure corresponds to the crisp minimization problem.

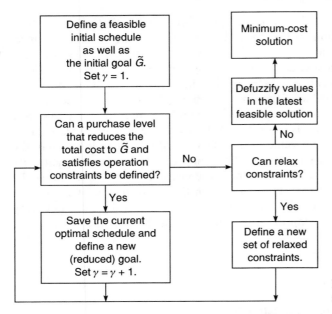

Figure 11.6 Decision process for the minimum total cost.

11.4. TEST RESULTS

11.4.1. System Characteristics

A five-bus utility with two local generators (buses 3 and 5) is used to test the proposed method. There are two offers from neighboring utilities. The topological connections, capacity, and reactance of lines are shown in Figure 11.7. Load at each bus is also shown.

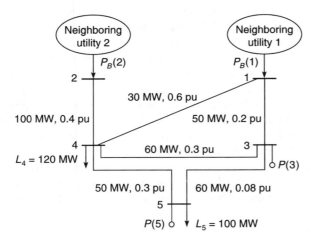

Figure 11.7 Topological connections.

We solve the problem for an initial goal, defined as a certain reduction in operation costs. We reduce this goal iteratively in order to find the operation schedule with minimum operation costs. The impact of the constraints on the solution is also shown in the example.

We consider the price of offered power (at buses 1 and 2), line flows, and local generation (at buses 3 and 5) as TFNs. The characteristics of local generators are shown in Table 11.1.

The price of offered power by the neighboring utilities is shown in Table 11.2.

In Table 11.2, the DM is assuming that the lowest price at bus 1 ranges between 1.7 and 2.3 \$/MW·h; the utility will be able to buy a maximum of 250 MW at a price ranging from 2.35 to 3.15 \$/MW·h.

TABLE 11.1   CHARACTERISTICS OF LOCAL GENERATORS

Bus Number	a (\$/h)	b (\$/MW·h)	c (\$/MW^2·h)	P^{max} (MW)
3	0	1	0.008	130
5	0	3	0.015	110

TABLE 11.2   OFFERED POWER

Offer at Bus Number	λ_0 (\$/MW·h)	$l_{\lambda 0}$ (\$/MW·h)	$r_{\lambda 0}$ (\$/MW·h)	λ_{max} (\$/MW·h)	$l_{\lambda, max}$ (\$/MW·h)	$r_{\lambda, max}$ (\$/MW·h)	P^B_{max} (MW)
1	2.0	0.3	0.3	2.75	0.4	0.4	250
2	2.0	0.2	0.2	2.50	0.3	0.3	100

11.4.2. Initial Solution, $\gamma = 1$ in Figure 11.6

Without importing transactions, the minimum operation cost is 656.7 \$/h. With uncertain information on offered power, the utility is interested to find the generation range that leads to a 10% reduction in operation costs with respect to that without transactions while all constraints are met; hence the goal is defined as $\tilde{G} = (591, 0, 0)$ \$/h. The degree of acceptance δ_1 for the goal is set to 0.8; we define $\delta_2 = 1$ in (11.12), δ_3 is set to 0.9 in (11.13), and $\delta_4 = 0.7$ in (11.14). Tolerance in (11.13) is defined as $\tilde{\epsilon} = (1, 0.2, 0.2)$ MW for power unbalance in each bus. The total operation cost is obtained by solving (11.11)–(11.15); $\tilde{F} = (550.58, 50.83, 50.53)$ \$/h and $\tilde{G}$ are shown in Figure 11.8.

From Figure 11.8 we learn that the operation cost can be higher than the goal [as we defined $\delta_1 = 0.8$ in (11.11)]. The operation cost is lower than the defined goal with a degree of acceptance at least equal to 0.8. Likewise, fuzzy constraints (11.11)–(11.14) are satisfied corresponding to $\delta_1, \ldots, \delta_4$.

The generation for each unit and line flows are shown in Table 11.3. Total operation cost is reduced by 10% (with a degree of acceptance of $\delta_1 = 0.8$) if generation at bus 3 is between 17.82 MW (mean minus left spread) and 20.25

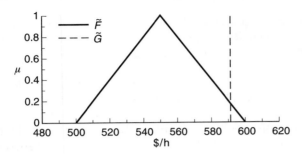

Figure 11.8 Goal and operation cost.

TABLE 11.3 GENERATION AND FLOW IN LINES FOR THE INITIAL SOLUTION

Parameter	Mean Value (MW)	Left Spread (MW)	Right Spread (MW)
$P_B(1)$	68.71		
$P_B(2)$	91.28		
$\tilde{P}(3)$	19.15	1.32	1.10
$\tilde{P}(5)$	42.05	0.00	0.00
$\tilde{P}_{4-5}$	3.54	0.00	0.00
$\tilde{P}_{3-5}$	55.23	0.91	0.00
$\tilde{P}_{3-4}$	11.18	0.24	0.00
$\tilde{P}_{1-3}$	47.09	0.00	0.36
$\tilde{P}_{1-4}$	21.29	0.00	0.00
$\tilde{P}_{2-4}$	91.28	0.00	0.67

MW (mean plus right spread) and generation at bus 5 is 42.05 MW, the remaining power imported from the neighboring utilities. Power unbalance at each bus will be below the specified tolerance and network constraints will be satisfied with their respective degrees of acceptance.

By adjusting the price of purchased power and local generation, the operation cost will be within the range of values given in Figure 11.8. For instance, when the price of purchased power is maximum and we use the generation in Table 11.3, the cost will correspond to the maximum value of $\tilde{F}$. Also, using the generation in Table 11.3, when the price of purchased power is minimum, the operation cost will be the minimum of $\tilde{F}$. Hence, the proposed procedure allows us to perform an analysis for the expected range of uncertain variables without repeating the computation for different combinations of price and generation.

11.4.3. Minimum-Cost Solution, Iterations with $\gamma > 1$ in Figure 11.6

In order to find the minimum-cost solution within the uncertain price information, we reduce the goal and keep the remaining parameters fixed. The mean value of the total operation cost for different desired cost reductions is shown in Figure 11.9. We repeat the optimization procedure (11.11)–(11.15) for different desired

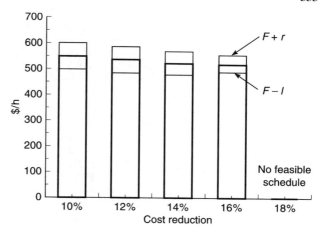

Figure 11.9 Mean value of the total operation cost.

reductions in the total operation costs until no further reduction is obtained. The results for cost reductions in the range 10–18% are computed. The maximum possible values of the total operation costs are also shown.

In Figure 11.9 there is no feasible operation state based on uncertain prices that would reduce the total operation cost by 18% or more while satisfying the set of constraints. The minimum operation cost is $\tilde{F} = (550.58, 50.83, 50.52)$ \$/h. A wider range of control variables will provide a better chance of satisfying the desired goal and operational constraints as we account for contingencies. However, there is a compromise between the requested goal and the range of control variables. In Figure 11.9, the higher the requested reduction in operation costs, the narrower $\tilde{F}$ and the narrower the range of control variables. For instance, for a 10% cost reduction l represents 9.2% of F, while for a 16% cost reduction l is 5.94% of F. We learn that for a higher requested reduction in cost a DM will find a narrower range of control variables that satisfy all constraints.

11.5. DISCUSSION

We examine the impact of parameters on the solution obtained in Section 11.4.2.

11.5.1. Effect of Changing δ_1 in the Goal

By adjusting δ_1 in (11.11), we define the extent of certainty in obtaining the desired cost reduction. For $\delta_1 = 1$, the solution is within the range of control variables in which the DM has full certainty that a 10% cost reduction will be obtained while satisfying all constraints with the defined degree of acceptance. The total operation cost is $\tilde{F} = (534.94, 49.87, 53.14)$ \$/h. The values of generation and line flows are shown in Table 11.4.

TABLE 11.4   GENERATION AND FLOW IN LINES FOR $\delta = 1$ IN EQ. (11.11)

Parameter	Mean Value (MW)	Left Spread (MW)	Right Spread (MW)
$P_B(1)$	70.00		
$P_B(2)$	92.11		
$\tilde{P}(3)$	21.73	0.00	2.43
$\tilde{P}(5)$	36.16	0.00	0.00
$\tilde{P}_{4-5}$	4.75	0.00	0.00
$\tilde{P}_{3-5}$	59.91	0.91	0.00
$\tilde{P}_{3-4}$	11.23	0.24	0.00
$\tilde{P}_{1-3}$	48.05	0.00	0.36
$\tilde{P}_{1-4}$	21.63	0.00	0.00
$\tilde{P}_{2-4}$	92.11	0.00	0.67

In Tables 11.3 and 11.4, we learn that by reducing the generation level in bus 5 and increasing the import and local generation in bus 3, even the maximum possible operation cost is below the goal.

11.5.2. Effect of Network Constraints

Suppose the connection between buses 2 and 4 is reduced to 50% of its capacity because of a contingency while the remaining parameters remain fixed, as those of the base case. In Table 11.5, the results obtained from the proposed procedure are shown when we decrease the capacity of line 2–4 to 50 MW. For the reduced network, the maximum cost reduction is 16%.

In Table 11.5, P_{2-4} is between 46.84 and 51.36 MW. The maximum value of the fuzzy line flow between buses 2 and 4 is higher than the line capacity (50 MW) because we define $\delta_4 = 0.7$ in (11.14). If we adjust $\delta_4 = 1$, even the maximum

TABLE 11.5   GENERATION AND FLOW IN LINES FOR CONTINGENCY IN LINE 2–4

Parameter	Mean Value (MW)	Left Spread (MW)	Right Spread (MW)
$P_B(1)$	38.27		
$P_B(2)$	47.45		
$\tilde{P}(3)$	77.01	0.00	0.02
$\tilde{P}(5)$	57.26	0.00	0.00
$\tilde{P}_{4-5}$	16.44	0.00	0.90
$\tilde{P}_{3-5}$	59.99	0.00	0.00
$\tilde{P}_{3-4}$	32.44	0.00	0.90
$\tilde{P}_{1-3}$	16.23	0.00	0.00
$\tilde{P}_{1-4}$	21.63	0.00	0.45
$\tilde{P}_{2-4}$	46.84	0.00	4.52

value of the fuzzy line flow between buses 2 and 4 will be smaller than the line capacity. However, there will be no feasible state that reduces the total cost by 16% and satisfy the constraints. In order to define a feasible state, we either relax the constraints (adjusting δ in each equation) or define a smaller reduction in operation costs. By adjusting δ in all equations, a DM decides which operational constraint is to be violated (and up to what extent) and obtains, from the proposed procedure, the operation state that satisfies all the constraints up to this degree. Otherwise, the DM may maintain δ in all equations and decrease the aspired cost reduction in order to obtain a feasible solution.

11.6. CONCLUSIONS

In this chapter, we have presented a methodology to evaluate power purchases in an uncertain environment. Power generation, line flows, and prices are considered as TFNs. Local generation and power purchases are control variables in the optimization procedure. The proposed methodology computes the range of control variables that satisfies the set of constraints as well as a certain reduction in the operation cost.

In the obtained results, the range of control variables is correlated with the desired cost reduction (goal). The lower the desired cost reduction, the narrower the range of control variables that satisfies the set of constraints. A DM in utility reduces the goal iteratively until no feasible solution is found, obtaining the lowest operation cost while satisfying the operational constraints. The degree of acceptance of variables in the problem can be measured with the left and right spreads. When more uncertain variables are introduced in the problem, the degree of uncertainty of the obtained solution grows.

A DM in the utility can define a degree of acceptance for individual constraints, in which hard (always satisfied) constraints and soft (violated up to a certain degree) constraints can be considered in the solution. With these procedures, the DM has the tools to analyze situations in which operational constraints could be violated during short intervals. The proposed procedure allows the DM to perform analyses within the range of uncertain variables without repeating the computation for different combinations of prices.

Network constraints also affect the obtained solution. By adjusting the values of the degree of acceptance in each constraint related to the power flow in lines, we can analyze the influence in the results of constraints in each line of the utility. Lower operation costs may be obtained by allowing the power flow in some lines to be above the maximum capacity.

The proposed method can be extended to include extra constraints related to operational practices in utilities. For instance, we can introduce an additional constraint, with its respective degree of acceptance, for the total generation in some particular area (reserve). Conflicting multiple objectives, such as cost minimization and emission minimization, may be included in the model as additional constraints.

APPENDIX 1: NOTATION

$P(i)$ Generation level at utility's unit i (MW)

a_i, b_i, c_i Cost coefficients at generator i

$C(i)$ Generation cost of unit i for utility ($/h)

$P_B(k)$ Power from purchase transaction k (MW)

$\lambda_0(k)$ Minimum price per unit of power in transaction k ($/h)

$\lambda(k)$ Per-unit price of purchased power $P_B(k)$ in transaction k ($/h)

$m(k)$ Slope of the price curve in purchase transaction k

$C_B(k)$ Cost of purchase transaction k ($/h)

F Total operation costs for utility ($/h)

N_G Number of generators in the utility

N_B Number of purchase transactions

$P^{\max}(i)$ Maximum generation capacity at unit i

$P_B^{\max}(k)$ Maximum purchased power from transaction k

Φ_j Set of utility's units at bus j

Ω_j Set of purchase transactions at bus j

Ψ_j Set of buses directly connected to bus j

P_{n-j} Power flow from bus n to bus j

L_j Load at bus j

N_{bus} Number of buses in the utility's transmission network

θ_j Voltage angle at bus j

X_{n-j} Reactance of line between bus n and bus j

$P_{n-j}^{\min}$ Maximum power flow from bus n to bus j

$P_{n-j}^{\min}$ Minimum power flow from bus n to bus j

All variables with a tilde correspond to a triangular fuzzy number (TFN).

APPENDIX 2: ALMOST POSITIVE FUZZY EXPRESSIONS

Based on ideas presented in ref. 12 and using the arithmetic operations defined in ref. 11, formulations are presented to define the "almost positive" character of linear and nonlinear expressions with fuzzy variables and fuzzy parameters.

A2.1. Fuzzy Goal

In Section 11.3.1 the objective function is presented for the optimal purchase level in uncertain environments. The fuzzy minimization is replaced by defining the decision variables that make $\tilde{F}$ lower than or at least equal to a defined fuzzy goal $\tilde{G}$, expressed as a TFN:

$$\tilde{F} \tilde{\leq} \tilde{G} \tag{A2.1}$$

In the most general case, $\tilde{F}$ will be a quadratic function with fuzzy variables and fuzzy parameters:

$$\tilde{F} = \sum_i (\tilde{a}_i + \tilde{b}_i \tilde{P}_i + \tilde{c}_i \tilde{P}_i^2)$$

where $\tilde{a}_i = (a_i, l_{\tilde{a}_i}, r_{\tilde{a}_i})$, $\tilde{b}_i = (b_i, l_{\tilde{b}_i}, r_{\tilde{b}_i})$ and $\tilde{c}_i = (c_i, l_{\tilde{c}_i}, r_{\tilde{c}_i})$ are the fuzzy coefficients of the cost function and $\tilde{P}_i = (P_i, l_{\tilde{P}_i}, r_{\tilde{P}_i})$ are either the generation from the local generators or the imported power, given as fuzzy numbers.

Using the arithmetic defined in ref. 9, we can define the fuzzy number $\tilde{Y}$ as

$$\tilde{Y} = \tilde{G} - \tilde{F} = (G, l_{\tilde{G}}, r_{\tilde{G}}) - (F, l_{\tilde{F}}, r_{\tilde{F}}) = (Y, l_{\tilde{Y}}, r_{\tilde{Y}})$$

where

$$Y = G - \sum_i (a_i + b_i P_i + c_i P_i^2)$$

$$l_{\tilde{Y}} = l_{\tilde{G}} + \sum_i (r_{\tilde{a}_i} + b_i r_{\tilde{P}_i} + r_{\tilde{b}_i} P_i + 2c_i r_{\tilde{P}_i} P_i + P_i^2 r_{\tilde{c}_i})$$

$$r_{\tilde{Y}} = r_{\tilde{G}} + \sum_i (l_{\tilde{a}_i} + b_i l_{\tilde{P}_i} + l_{\tilde{b}_i} P_i + 2c_i l_{\tilde{P}_i} P_i + P_i^2 l_{\tilde{c}_i})$$

Equation (A2.1) is equivalent to $\tilde{Y} \gtrsim 0$; from the definition of a TFN, $\mu_{\tilde{Y}}(0) = 1 - Y/l_{\tilde{Y}}$. Using the definition in Section 11.3.2, $\tilde{Y}$ will be almost positive with a degree of acceptance δ if and only if

$$Y - \delta l_Y \geq 0$$

Hence, Eq. (A2.1) with a degree of acceptance δ is replaced with the crisp expression

$$G - \sum_i (a_i + b_i P_i + c_i P_i^2) - \delta \left(l_{\tilde{G}} + \sum_i (r_{\tilde{a}_i} + b_i r_{\tilde{P}_i} + r_{\tilde{b}_i} P_i + 2c_i r_{\tilde{P}_i} P_i + P_i^2 r_{\tilde{c}_i}) \right) \geq 0$$

$$(A2.2)$$

A2.2. Fuzzy Inequalities

Using the same reasoning employed in the previous section, inequalities of the type $\sum_i \tilde{a}_i \tilde{P}_i \lesssim \tilde{b}$ are derived as crisp equations by defining $\tilde{Y} = \tilde{b} - \sum_i \tilde{a}_i \tilde{P}_i$. The value $\tilde{b} = (b, l_{\tilde{b}}, r_{\tilde{b}})$ represents, for instance, the maximum power flow in a line in the network.

The statement "$\tilde{Y}$ almost positive" with a degree of acceptance δ is given by the crisp equation

$$b - \sum_i a_i P_i - \delta \left(l_{\tilde{b}} + \sum_i (a_i r_{\tilde{P}_i} + P_i r_{\tilde{a}_i}) \right) \geq 0 \qquad (A2.3)$$

By defining $\tilde{Y} = \sum_i \tilde{a}_i \tilde{P}_i - \tilde{b}$, constraints of the type $\sum_i \tilde{a}_i \tilde{P}_i \geq \tilde{b}$ are transformed into crisp functions and considered in the proposed methodology

$$\sum_i a_i P_i - b - \delta \left(r_{\tilde{b}} + \sum_i (a_i l_{\tilde{P}_i} + P_i l_{\tilde{a}_i}) \right) \geq 0 \qquad (A2.4)$$

Using Eqs. (A2.2)–(A2.4) for defined values of δ, the constraints for the maximization stated in Section 11.3.3 are built. It should be noted that different values of δ can be defined for different constraints, representing the desired degree of acceptance for the constraints.

References

[1] L. Zhang, P. B. Luh, X. Guan, and G. Merchel, "Optimization-Based Inter-utility Power Purchase," *IEEE Transactions on Power Systems*, Vol. 9, No. 2, 1994, pp. 891–897.

[2] R. N. Wu, T. H. Lee, and E. F. Hill, "Effect of Interchange on Short-Term Hydro-thermal Scheduling," *IEEE Transactions on Power Systems*, Vol. 6, No. 3, 1991, pp. 1217–1223.

[3] E. Cox, *The Fuzzy Systems Handbook: A Practitioner's Guide to Building, Using, and Maintaining Fuzzy Systems*, AP Professional, Harcourt Brace & Company, Boston, 1994.

[4] K. H. Abdul-Rahman and S. M. Shahidehpour, "A Fuzzy-Based Optimal Reactive Power Control," *IEEE Transactions on Power Systems*, Vol. 8, No. 2, 1993, pp. 662–670.

[5] K. H. Abdul-Rahman and S. M. Shahidehpour, "Reactive Power Optimization Using Fuzzy Load Representation," *IEEE Transactions on Power Systems*, Vol. 9, No. 2, 1994, pp. 898–905.

[6] V. Miranda and J. T. Saravia, "Fuzzy Modeling of Power System Optimal Load Flow," *IEEE Transactions on Power Systems*, Vol. 7, No. 2, 1992, pp. 843–849.

[7] R. Yokoyama, T. Niimura, and Y. Nakanishi, "A Coordinated Control of Voltage and Reactive Power by Heuristic Modeling and Approximate Reasoning," *IEEE Transactions on Power Systems*, Vol. 8, No. 2, 1993, pp. 636–645.

[8] A. Wood and B. Wollenberg, *Power Generation, Operation, and Control*, Wiley, New York, 1984.

[9] L. A. Zadeh, "Fuzzy Sets," *Information and Control*, Vol. 8, 1965, pp. 338–353.

[10] M. Sakawa, *Fuzzy Sets and Interactive Multiobjective Optimization*, Plenum, New York, 1993.

[11] A. Kaufmann and M. Gupta, *Introduction to Fuzzy Arithmetic*, Van Nostrand Reinhold, New York, 1985.

[12] H. Tanaka and K. Asai, "Fuzzy Solution in Fuzzy Linear Programming Problems," *IEEE Transactions on Systems, Man and Cybernetics*, Vol. SMC-14, No. 2, 1984, pp. 325–328.

Index

About the Editor

M. E. El-Hawary was born in Egypt and received the bachelor of engineering degree in electrical engineering (First Class Honours) from the University of Alexandria in Egypt in 1965 and the Ph.D. in electrical engineering from the University of Alberta, Edmonton, Alberta, Canada, in 1972. His teaching experience includes service at the Universities of Alexandria and Alberta. In 1972–1973 he was an associate professor at the Graduate School for Engineering at the Federal University of Rio de Janeiro in Brazil. He joined Memorial University of Newfoundland in 1974 and served as chairman of Electrical Engineering during 1976–1981. He has been a professor of electrical engineering at the Technical University of Nova Scotia (now DalTech) since 1981. He is currently associate dean of Engineering at DalTech of Dalhousie University.

Dr. El-Hawary has written over 150 technical papers, mainly in power system engineering. He is the author of three textbooks on power systems analysis, principles of electric machines, and control system engineering, all published by Prentice-Hall. He is also coauthor of two research monographs on economic operation of power systems. In the IEEE Power Engineering Society, he is chair of the Operating Economics Subcommittee, Life Long Learning Subcommittee, and Awards and Recognitions of the System Operations and Dynamic Performance Committees. Within IEEE, he is the chair of the Prize Papers/Scholarship Committee and a member of the Awards Board. Dr. El-Hawary is also associate editor of *Electric Machines and Power Systems* and *International Journal of Electric Energy and Power Systems*. He is a member of the IEEE Press Board and serves as editor of the Understanding Science and Technology Series.

Dr. El-Hawary is a Fellow of both the IEEE and the Engineering Institute of Canada. He is married to Dr. Ferial El-Hawary (nee El-Bibany), and together they have a daughter and two sons.

343